Chapter Reviews and Working Papers
With Peachtree and Spreadsheet Guides

GLENCOE
Accounting
Real-World Applications & Connections

First-Year Course
Fifth Edition

Chapter Reviews and Working Papers
With Peachtree and Spreadsheet Guides

GLENCOE
Accounting
Real-World Applications & Connections

First-Year Course
Fifth Edition

Donald J. Guerrieri
Norwin High School
North Huntingdon, Pennsylvania

F. Barry Haber
Fayetteville State University
Fayetteville, North Carolina

William B. Hoyt
Wilton High School
Wilton, Connecticut

Robert E. Turner
University of Louisiana at Monroe
Monroe, Louisiana

 Glencoe

New York, New York   Columbus, Ohio   Chicago, Illinois   Peoria, Illinois   Woodland Hills, California

Glencoe

The *McGraw·Hill* Companies

Printed in the United States of America.

Send inquiries to:

Glencoe/McGraw-Hill
8787 Orion Place
Columbus, OH 43240

ISBN 0-07-846098-0 (Student Edition)
ISBN 0-07-846099-9 (Teacher Annotated Edition)

10 11 12 13 14 QDB 14 13 12 11

CONTENTS

Working Papers *for Section Problems*

Problem 1-1   Studying Yourself

List at least five personal interests or skills.

_____

_____

_____

_____

_____

Identify one or more careers that match the interests and skills you listed above.

_____

_____

_____

_____

hoose one of the above careers and write a description of how your skills nd interests fit into this career.

_____

_____

_____

_____

_____

_____

_____

_____

_____

_____

_____

_____

Problem 1-2   Gathering Career Resources

Personal Career Profile Form	
Name: **Date:**	**Career:**
Your Values	Career Values
Your Interests	Career Duties and Responsibilities
Your Personality	Personality Type Needed
Skills and Aptitudes	Skills and Aptitudes Required
Education/Training Acceptable	Education/Training Required

Problem 1-2   (continued)

Personal Career Profile Form	
Name: **Date:**	**Career:**
Your Values	Career Values
Your Interests	Career Duties and Responsibilities
Your Personality	Personality Type Needed
Skills and Aptitudes	Skills and Aptitudes Required
Education/Training Acceptable	Education/Training Required

Problem 1-2   (concluded)

Personal Career Profile Form	
Name: **Date:** Career:	
Your Values	Career Values
Your Interests	Career Duties and Responsibilities
Your Personality	Personality Type Needed
Skills and Aptitudes	Skills and Aptitudes Required
Education/Training Acceptable	Education/Training Required

Problem 1-3   Checking Out Accounting Careers

Career	Formal Training Needed	Work Experience Needed

Which of the careers listed above do you prefer? _____

Why does this career appeal to you?

_____

_____

_____

_____

_____

_____

Problem 1-4 Matching Interests and Careers

List the personal interests and skills of the accountants described in the text.

_____

_____

_____

_____

List three types of businesses (or actual companies) for which you might
like to work.

_____

_____

_____

How would you learn about accounting career opportunities in the above
companies?

_____

_____

_____

_____

_____

Aside from pursuing needed training and education, what else would you
do to prepare to work in that career?

_____

_____

_____

_____

_____

Problem 1-5   Researching Public Accounting Firms

Company's name:	Services provided:
Web site:	
Company's name:	Services provided:
Web site:	
Company's name:	Services provided:
Web site:	
Company's name:	Services provided:
Web site:	

Problem 1-6   Interviewing Accountants

_____

_____

_____

_____

_____

_____

_____

_____

_____

_____

Working Papers *for End-of-Chapter Problems*

Problem 1-7   Researching Careers in Your Library

_____

_____

_____

_____

Problem 1-8   Researching Careers in Your Local Newspaper

Job Title	Skills Required	Education Required

Problem 1-9   Assessing Your Skills and Interests

What are your aptitudes and abilities?

_____

_____

_____

What are your interests?

_____

_____

_____

What are your values?

_____

_____

_____

Do you like working with people?

_____

_____

_____

Do you like working with data?

_____

_____

_____

Do you like working with numbers?

_____

_____

_____

Using your answers above, list three careers that might match your skills
and interests.

_____

_____

_____

Problem 1-10   Working with Others

Situation	Skills Needed	How can you get those skills?
Training new hires in the accounting department		
Discuss project cost overruns with a department manager		
Present operating results to senior managers		

Problem 1-11   Summarizing Personal Traits

1. _____ _____ _____
2. _____ _____ _____
3. _____ _____ _____
4. _____ _____ _____
5. _____ _____ _____
6. _____ _____ _____
7. _____ _____ _____
8. _____ _____ _____
9. _____ _____ _____
10. _____ _____ _____

Which five characteristics were mentioned most often?

_____

_____

_____

Do these descriptions match your self-perception? Why or why not?

_____

_____

_____

Problem 1-12   Gathering Career Information

Problem 1-13   Exploring Careers in Accounting

_____

_____

_____

_____

_____

_____

_____

_____

_____

_____

_____

Problem 1-14   Exploring Global Careers

_____

_____

_____

_____

_____

_____

_____

_____

_____

_____

_____

_____

CHAPTER 1 You and the World of Accounting

Self-Test

Part A True or False

Directions: *Circle the letter* T *in the Answer column if the statement is true; circle the letter* F *if the statement is false.*

Answer

T F **1.** You are more likely to enjoy a career that uses your individual interests, skills, and traits.

T F **2.** Values are activities you do well.

T F **3.** Assessing your interests, skills, and personality traits will help you to determine a career path.

T F **4.** Your personality is what makes you the same as other people.

T F **5.** Your lifestyle is the way you use your time, energy, and resources.

T F **6.** Networking is not a good way to find out about a particular career you are interested in.

T F **7.** When planning a career, it is not important to consider the education and training needed for that career.

T F **8.** The Internet and professional organizations are good sources for career information.

T F **9.** A majority of businesses in the United States are for-profit businesses.

T F **10.** Not-for-profit businesses operate for reasons other than to earn a profit, such as to protect the environment.

Part B Matching

Directions: *Determine whether each business listed below is a for-profit or not-for-profit business. Write the letter corresponding to your choice in the blank next to the item.*

A. for-profit	**B.** not-for-profit

Answer

_____ **1.** General Motors

_____ **2.** United Way

_____ **3.** Virgin Atlantic Airlines

_____ **4.** Columbia Pictures

_____ **5.** World Wildlife Fund

_____ **6.** MCA Records

_____ **7.** Barnes & Noble

_____ **8.** Audubon Society

_____ **9.** Boeing

_____ **10.** Coca-Cola

_____ **11.** American Cancer Society

CHAPTER  The World of Business and Accounting

Self-Test

Part A   True or False

Directions: *Circle the letter* T *in the Answer column if the statement is true; circle the letter* F *if the statement is false.*

Answer

T   F   **1.** The United States' economy is referred to as a private enterprise system.

T   F   **2.** Financial accounting is often referred to as accounting for internal users.

T   F   **3.** The period of time covered by an accounting report is the accounting period.

T   F   **4.** Manufacturers make products and sell them while service businesses buy products and sell them.

T   F   **5.** Only through profits can businesses continue to operate.

T   F   **6.** Entrepreneurs are people willing to take risks.

T   F   **7.** Going into business for yourself has few drawbacks.

T   F   **8.** A sole proprietorship is the most difficult business to start.

T   F   **9.** GAAP stands for Generally Accepted Accounting Procedures.

T   F   **10.** Accountants assume that a business will operate forever.

Part B   Matching

Directions: *Using the business types listed below, write the letter of the term you have chosen in the space provided.*

| **A.** Sole Proprietorship | **B.** Partnership | **C.** Corporation |

Answer

_____   **1.** The business has one owner

_____   **2.** Easy to start

_____   **3.** Easy to transfer ownership

_____   **4.** Must share profits

_____   **5.** Owner has all the risks

_____   **6.** Complex to organize

_____   **7.** Owner has total control

_____   **8.** Pays higher taxes

_____   **9.** Hard to raise money

_____   **10.** Losses limited to investment

_____   **11.** Limited expertise

Computerized Accounting Using Spreadsheets

Learning to Use a Spreadsheet Program

Introduction

Accountants use several tools to help them evaluate and present various financial information. One tool that has proven to be invaluable to accountants is the **electronic spreadsheet**. An electronic spreadsheet is simply a computerized version of the paper work sheet with which you are probably already familiar. The advantage of using an electronic spreadsheet is that changes and corrections can be made to the spreadsheet very quickly.

Before you begin using spreadsheet software, there are some terms with which you should become familiar:

Column: a vertical area of varying width that is labeled with a letter.

Row: a horizontal area that is labeled with a number.

Cell: the intersection of a column and a row. The intersection is referenced by the column and row. For example, cell C12 is the point at which column C and row 12 intersect.

Cell pointer: a rectangular block that highlights the current cell. When you want to enter data in a cell, first use the mouse or arrow keys to move the cell pointer to that position.

Template: a spreadsheet that contains formulas, labels, and formatting codes; the template can be used simply by entering information in the appropriate cells.

Work sheet: a spreadsheet document. The cells where you enter labels, amounts, and formulas are collectively referred to as a work sheet.

Opening a Spreadsheet Problem Using Glencoe Accounting: Electronic Learning Center

Step 1   Turn on your computer.

Step 2   Open the Glencoe Accounting: Electronic Learning Center software.

Step 3   From the Program Menu, click the **Peachtree Complete® Accounting Software and Spreadsheet Applications** icon.

Step 4   Log onto the Accounting Management System by typing in your user name and password.

Step 5   Under the Problems & Tutorials tab, select the chapter and problem you want to work on. Click **Launch.**

Step 6   Your spreadsheet application will launch and the template will load.

Step 7   Complete the spreadsheet problem according to the instructions in the problem's *Spreadsheet Guide.*

Continuing a Session

- If you've been directed to save your work to a network, select the problem from the Problems & Tutorials list. The system will automatically retrieve your files from the previous session.

- If you've been directed to save your work to a floppy disk, and you want to continue to work on the problem you've saved, be sure the correct floppy is in the drive and select *Continue working on this problem.* If the Management System doesn't find a file on the floppy corresponding to the problem you've selected, it assumes you want to start the problem with a fresh template, in which case you may lose any work you previously completed.

Spreadsheet Guide

Entering Data into a Spreadsheet

- **Navigation.** As you work with a template, you will be required to enter text and numbers into various cells to complete a problem. When instructed to "enter" information, use the mouse or the arrow keys to move the cell pointer to the specified cell, key the data (text or numbers), and then press the **ENTER** key.

- **Types of Data.** There are two kinds of data that you will enter—numbers and labels. When you type *425* the software knows that you are entering a number. Numeric data begins with a digit (0–9) or a symbol (+ – (. @ $ # or any currency symbol). All other symbols signify a label.

- **Cell Protection.** Depending on the spreadsheet program you are using, many of the cells are "protected" to prevent you from accidentally erasing information. If you attempt to enter a number or label in a protected cell, the spreadsheet program displays a message indicating the cell is locked or protected.

- **Errors.** If you make a mistake while keying the information and you have not yet pressed **ENTER**, simply use the **BACKSPACE** key to erase the incorrect data and rekey the entry. When the entry is correct, press **ENTER** to accept the data. If you notice an error after you press **ENTER**, simply re-enter the data.

- **Overflows.** Sometimes, the data you enter may be too wide to fit into one cell. When you press **ENTER**, the data "overflows" from the current cell into the next cell. As long as data has not been entered in the adjoining cell, the entire cell contents appear on the work sheet.

- **Placeholders.** Placeholders appear in each template to identify where to enter your name and the date. Simply move to the cell with *(name)* and type your name. Then, move to the cell with *(date)* and type the date.

Saving a Spreadsheet

After you complete a problem, or if you need to save your work before you finish an activity, use the **Save** option from the *File* menu. When closing the problem, you will be asked to save your work to the network or to floppy disks. It is good practice to always move your cursor to cell A1 before saving your spreadsheet. This will ensure that your spreadsheet opens in the first cell when your reopen it. For many spreadsheet applications, a shortcut for moving to cell A1 is by holding down the **CTRL** key and pressing the **HOME** key. Check your software's Help file for its specific shortcut commands.

Printing a Spreadsheet

Use your software's Print command to print your completed spreadsheet. Access your software's Help file for detailed printing instructions.

Sometimes a spreadsheet will be too wide to fit vertically on an 8½" x 11" piece of paper. If your spreadsheet is too wide to fit on an 8½-inch wide piece of paper, you can change your print settings to print the worksheet *landscape*. Landscape means that the worksheet will be printed broadside on the page. Some spreadsheet applications also allow you to choose a "fit to page" option. This function will reduce the width and/or depth of the worksheet to fit on one page.

Ending a Session

After working with the software, you should exit the program. Remove the template disk from the drive (if necessary) and turn off the computer.

Problem 3-9   Determining the Effects of Business Transactions on the Accounting Equation

Completing the Spreadsheet

Step 1   Read the instructions for Problem 3-9 in your textbook.

Step 2   Open the Glencoe Accounting: Electronic Learning Center software.

Step 3   From the Program Menu, click on the **Peachtree Complete®
Accounting Software and Spreadsheet Applications** icon.

Step 4   Log onto the Management System by typing your user name
and password.

Step 5   Under the **Problems and Tutorials** tab, select template 3-9 from
the Chapter 3 drop-down menu. The template should look like the
one shown below.

```
PROBLEM 3-9
DETERMINING THE EFFECTS OF BUSINESS
TRANSACTIONS ON THE ACCOUNTING EQUATION

(name)
(date)
```

			ASSETS			LIABILITIES	OWNER'S EQUITY
Transaction	Cash in Bank	Accounts Receivable	Hiking Equipment	Rafting Equipment	Office Equipment =	Accounts Payable +	Juanita Ortega, Capital
1							
2							
3							
4							
5							
6							
7							
8							
9							
10							
BALANCE	$0	$0	$0	$0	$0	$0	$0

```
TOTAL ASSETS                          $0

TOTAL LIABILITIES                     $0
TOTAL OWNER'S EQUITY                  $0
TOTAL LIABILITIES + OWNER'S EQUITY    $0
```

Step 6   Key your name in the cell containing the *(name)* placeholder. After
you key your name and press **ENTER,** the *(name)* placeholder will be
replaced by the information you just keyed.

Step 7   Key today's date in the cell containing the *(date)* placeholder. After you
key the date and press **ENTER,** the *(date)* placeholder will be replaced
by the information you just keyed. When you work with any of the
other spreadsheet templates, your name and date should always be
keyed in the cells containing the *(name)* and *(date)* placeholders.

Spreadsheet Guide

TIP: Options you select in the Regional Settings of Control Panel determine the default format for the current date and time and the characters recognized as date and time separators—for example, the slash (/) and colon (:) on United States-based systems. Be careful not to enter an equal sign before entering a date with slashes, as your spreadsheet application may view the date as a formula and the slashes as division symbols.

Step 8 In the first transaction, Ms. Ortega opened a checking account for the business. Cash in Bank is increasing, and Juanita Ortega, Capital, is increasing. To record this transaction in the spreadsheet template, move the cell pointer to cell B12 and enter **60000**.

TIP: To enter data into the cell, you must first key the data and then press **ENTER**. Do *not* enter a dollar sign or a comma when you enter the data—the spreadsheet template will automatically format the data when it is entered.

TIP: Depending on the spreadsheet program you are using, the spreadsheet templates may be formatted to protect you from accidentally erasing information in selected cells. For example, the column headings in the spreadsheet for Problem 3-9 are protected. If you attempt to key a number or label in a protected cell, the spreadsheet program displays a message indicating that the cell is locked or protected. Simply move to the correct cell and re-enter the information.

Step 9 Next, move the cell pointer to cell J12. Enter **60000** in cell J12 to record the increase in Juanita Ortega, Capital. Again, do *not* include a dollar sign or a comma as part of the cell entry—the spreadsheet template will automatically format the data when it is entered. Move the cell pointer to cell J23. Notice that the spreadsheet automatically calculates the balance in each account as you enter the data.

Step 10 To check your work, look at rows 26 through 30 in column D. Total assets equal $60,000. Total liabilities plus owner's equity also equal $60,000. The accounting equation is in balance.

Step 11 Analyze the remaining transactions in Problem 3-9 and enter the appropriate data into the spreadsheet template.

TIP: To decrease an account balance, precede the amount entered by a minus sign. For example, to decrease Cash in Bank by $3,000, enter **–3000** in the Cash in Bank column.

Check the totals at the bottom of the spreadsheet after each transaction has been entered. Remember, total assets should always equal total liabilities plus owner's equity. If the accounting equation becomes out of balance, check your work to find the error.

Step 12 Save the spreadsheet using the **Save** option from the *File* menu. You should accept the default location for the save as this is handled by the management system.

TIP: It is good practice to always move your cursor to cell A1 before saving your spreadsheet. This will ensure that your spreadsheet opens in the first cell when you reopen it. For many spreadsheet applications, a shortcut for moving to cell A1 is by holding down the **CTRL** key and pressing the **HOME** key. Check your software's Help file for its specific shortcut commands.

Step 13 Print the completed spreadsheet.

TIP: If your spreadsheet is too wide to fit on an 8.5-inch wide piece of paper, you can change your print settings to print the worksheet *landscape*. Landscape means that the worksheet will be printed broadside on the page. Some spreadsheet applications also allow you to choose a "fit to page" option. This function will reduce the width and/or depth of the worksheet to fit on one page.

Step 14 Exit the spreadsheet program.
Step 15 In the Close Options box, select the location where you would like to save your work.
Step 16 Answer the Analyze question from your textbook for this problem.

What-If Analysis

TIP: Always save your work before performing What-If Analysis. It is not necessary to save your work after performing What-If Analysis unless your teacher instructs you to do so. If you are required to save your work after performing What-If Analysis, be sure to rename the spreadsheet to avoid saving over your original work.

If Ms. Ortega withdrew an additional $1,500 from the business for personal use, what would the balance in the Juanita Ortega, Capital account be?

_____

CHAPTER 3 Business Transactions and the Accounting Equation

Self-Test

Part A True or False

Directions: *Circle the letter* T *in the Answer column if the statement is true; circle the letter* F *if the statement is false.*

Answer

T	F	**1.** The accounting equation should remain in balance after each transaction.
T	F	**2.** A business transaction affects at least two accounts.
T	F	**3.** "Assets + Liabilities = Owner's Equity" is another way to express the accounting equation.
T	F	**4.** The increases and decreases caused by business transactions are recorded in specific accounts.
T	F	**5.** The private enterprise system is based on the right to own property.
T	F	**6.** The owner's personal financial transactions are part of the business's records.
T	F	**7.** The total financial claims to the assets of a business are referred to as equity.
T	F	**8.** The owner's claims to the assets of a business are liabilities.
T	F	**9.** When a business transaction occurs, the financial position of the business changes.
T	F	**10.** A creditor has a financial claim to the assets of a business.
T	F	**11.** An account is a record of only the increases in the balance of a specific item such as cash or equipment.
T	F	**12.** The total financial claims do not have to equal the total cost of the property.

Part B Multiple Choice

Directions: *Only one of the choices given with each of the following statements is correct. Write the letter of the correct answer in the Answer column.*

Answer

1. If the creditor's financial claim to property totals $1,000 and the owner's financial claim to property totals $11,000, the property value is
   (A) $10,000.   (C) $12,000.
   (B) $11,000.   (D) $1,000.

2. The account Accounts Receivable is an example of a(n)
   (A) asset.   (C) owner's equity.
   (B) liability.   (D) none of the above.

3. All of the following account names are asset names, except
   (A) Office Furniture. (C) Cash in Bank.
   (B) Accounts Payable. (D) Equipment.

4. If a business has assets of $5,600 and liabilities of $900, the owner's equity is
   (A) $6,500.   (C) $4,700.
   (B) $900.   (D) $5,600.

5. A business transaction that involves a purchase on account is considered to be a(n)
   (A) cash transaction.   (C) investment by the owner.
   (B) credit transaction.   (D) expense transaction.

6. If a business purchases a calculator on account, the accounts affected by this transaction are.
   (A) Cash in Bank and Accounts Payable.
   (B) Office Equipment and Accounts Receivable.
   (C) Office Equipment and Cash in Bank.
   (D) Office Equipment and Accounts Payable.

7. Each of the following is a business expense, except payment for
   (A) advertising.   (C) utility bills.
   (B) monthly rent.   (D) equipment.

8. The purchase of a desk on account will increase Office Furniture and will also increase
   (A) Cash in Bank.   (C) Accounts Receivable.
   (B) Accounts Payable.   (D) Jon McIvey, Capital.

Computerized Accounting Using Spreadsheets

Problem 4-6   Analyzing Transactions into Debit and Credit Parts

Completing the Spreadsheet

Step 1   Read the instructions for Problem 4-6 in your textbook.

Step 2   Open the Glencoe Accounting: Electronic Learning Center software.

Step 3   From the Program Menu, click on the **Peachtree Complete®
Accounting Software and Spreadsheet Applications** icon.

Step 4   Log onto the Management System by typing your user name
and password.

Step 5   Under the **Problems and Tutorials** tab, select template 4-6 from
the Chapter 4 drop-down menu. The template should look like the
one shown below.

```
PROBLEM 4-6
ANALYZING TRANSACTIONS INTO
DEBIT AND CREDIT PARTS

(name)
(date)


          Cash in Bank              Accounts Receivable -
                                       Mary Johnson              Office Equipment


               0                            0                           0


       Computer Equipment            Hiking Equipment            Rafting Equipment


               0                            0                           0


      Accounts Payable -            Accounts Payable -            Juanita Ortega,
        Peak Equipment              Premier Processors              Capital


               0                            0                           0


SUM OF DEBIT BALANCES                       $0
SUM OF CREDIT BALANCES                      $0
```

Step 6   Key your name and today's date in the cells containing the *(name)* and
(date) placeholders.

Step 7   In the first transaction, Juanita Ortega transferred an additional $53,250
from her personal account to the business. Two accounts are affected
by this transaction: Cash in Bank and Juanita Ortega, Capital. To record
this transaction, move to cell A10, the first cell on the debit side of
the Cash in Bank T account, and enter **53250**.

       Chapter 4 ■ 49

TIP: To enter data into a cell, you must first key the data and then press **ENTER.** Do *not* enter a comma when you enter the data.

Step 8   Next, move to cell H30, the first cell on the credit side of the Juanita Ortega, Capital T account. Enter **53250** in cell H30 to record the credit to Juanita Ortega, Capital. Move the cell pointer to cell H35. Notice that the spreadsheet automatically calculates the balance in each T account.

Step 9   To check your work, look at cells D39 and D40. The sum of debit balances equals $53,250. The sum of credit balances also equals $53,250.

Step 10   Analyze the remaining transactions in Problem 4-6 and enter the appropriate data into the spreadsheet template.

      Check the totals at the bottom of the spreadsheet after each transaction has been entered. Remember, the sum of debit balances should always equal the sum of credit balances. If the debit and credit balances become out of balance, check your work to find the errors.

Step 11   Save the spreadsheet using the **Save** option from the *File* menu. You should accept the default location for the save as this is handled by the management system.

Step 12   Print the completed spreadsheet.

Step 13   Exit the spreadsheet program.

Step 14   In the Close Options box, select the location where you would like to save your work.

Step 15   Answer the Analyze question from your textbook for this problem.

What-If Analysis

TIP: Always save your work before performing What-If Analysis. It is not necessary to save your work after performing What-If Analysis unless your teacher instructs you to do so. If you are required to save your work after performing What-If Analysis, be sure to rename the spreadsheet to avoid saving over your original work.

If Juanita Ortega purchased a computer for $1,500 cash, what would the balance in the Cash in Bank account be?

CHAPTER 4 — Transactions That Affect Assets, Liabilities, and Owner's Equity

Self-Test

Part A   True or False

Directions: *Circle the letter* T *in the Answer column if the statement is true; circle the letter* F *if the statement is false.*

Answer

T F **1.** The normal balance side for an asset account is the debit side.

T F **2.** "Debit" means the increase side of an account.

T F **3.** A credit to a liability account decreases the account balance.

T F **4.** Assets are increased on the debit side.

T F **5.** Capital is increased on the credit side.

T F **6.** Liabilities are decreased on the credit side.

T F **7.** The basic accounting equation may be expressed as $A - L = OE$

T F **8.** The right side of a T account is always the debit side.

T F **9.** For every debit there must be an equal credit.

T F **10.** A debit to one asset account and a credit to another asset account will result in the basic accounting equation being out of balance.

T F **11.** The left side of a T account is always the credit side.

T F **12.** Credit means to decrease a liability.

Part B   Identify the Normal Balance

Directions: *For each T account below, indicate with an (N) the normal balance side. The first account has been completed as an example.*

Computer Equipment		Accounts Payable		Cash in Bank	
Debit	Credit	Debit	Credit	Debit	Credit
(N)					

Accounts Receivable		Abe Dunn, Capital		Office Equipment	
Debit	Credit	Debit	Credit	Debit	Credit

Part C   Complete the T Account

Directions: *Analyze the transactions below and enter them in the T accounts provided.*

1. Ms. Adams invested $12,000 cash in the business.
2. Bought office equipment for cash, $1,000.
3. Bought a computer on account, $3,000.

Cash in Bank		Office Equipment		Computer Equipment	

Accounts Payable		J. Adams, Capital	

CHAPTER 5   Transactions That Affect Revenue, Expenses, and Withdrawals

Study Plan

Check Your Understanding

Section 1

Read Section 1 on pages 96–102 and complete the following exercises on page 103.
- ❏ Thinking Critically
- ❏ Communicating Accounting
- ❏ Problem 5-1   *Applying the Rules of Debit and Credit*

Section 2

Read Section 2 on pages 104–108 and complete the following exercises on page 109.
- ❏ Thinking Critically
- ❏ Analyzing Accounting
- ❏ Problem 5-2   *Identifying Accounts Affected by Transactions*

Summary

Review the Chapter 5 Summary on page 111 in your textbook.
- ❏ Key Concepts

Review and Activities

Complete the following questions and exercises on pages 112–113 in your textbook.
- ❏ Using Key Terms
- ❏ Understanding Accounting Concepts and Procedures
- ❏ Case Study
- ❏ Conducting an Audit with Alex
- ❏ Internet Connection
- ❏ Workplace Skills

Computerized Accounting

Read the Computerized Accounting information on page 114 in your textbook.
- ❏ *Making the Transition from a Manual to a Computerized System*
- ❏ *Setting Up General Ledger Accounts*

Problems

Complete the following end-of-chapter problems for Chapter 5 in your textbook.
- ❏ Problem 5-3   *Identifying Increases and Decreases in Accounts*
- ❏ Problem 5-4   *Using T Accounts to Analyze Transactions*
- ❏ Problem 5-5   *Analyzing Transactions into Debit and Credit Parts*
- ❏ Problem 5-6   *Analyzing Transactions into Debit and Credit Parts*
- ❏ Problem 5-7   *Analyzing Transactions*

Challenge Problem
- ❏ Problem 5-8   *Completing the Accounting Equation*

Chapter Reviews and Working Papers

Complete the following exercises for Chapter 5 in your Chapter Reviews and Working Papers.
- ❏ Chapter Review
- ❏ Self-Test

CHAPTER 5 REVIEW — Transactions That Affect Revenue, Expenses, and Withdrawals

Part 1 — Accounting Vocabulary (6 points)

Directions: *Using terms from the following list, complete the sentences below. Write the letter of the term you have chosen in the space provided.*

Total Points	62
Student's Score	

A. capital	**D.** revenue accounts	**F.** temporary capital accounts
B. expense accounts	**E.** revenue recognition principle	**G.** withdrawal
C. permanent accounts		

_____G_____   **0.** An amount of money taken out of the business by the owner is a _____.

_____   **1.** _____ record business income only.

_____   **2.** _____ are used to record information for only one accounting period.

_____   **3.** The _____ account shows the amount of the owner's investment, or equity, in a business.

_____   **4.** Accounts that are used to record information continuously from one accounting period to the next are called _____.

_____   **5.** _____ are used to record the costs and services used by a business.

_____   **6.** Recognizing and recording revenue on the date it is earned even if cash has not been received on that date is known as the _____.

Part 2 — Effects of a Transaction on an Account (24 points)

Directions: *For each of the business transactions below, indicate whether the left or right side of the account is affected and whether the account balance is increased or decreased.*

	Left	Right	Increase	Decrease
0. A credit of $850 to Accounts Payable		✓	✓	
1. A debit of $400 to B. Barns, Withdrawals				
2. A debit of $200 to Advertising Expense				
3. A credit of $300 to Cash in Bank				
4. A credit of $450 to Fees				
5. A debit of $650 to Rent Expense				
6. A credit to B. Barns, Capital of $1,500				
7. A credit to Accounts Receivable of $925				
8. A debit to Office Supplies of $40				
9. A debit of $3,000 to B. Barns, Capital				
10. A debit of $150 to Accounts Payable				
11. A debit to Accounts Receivable of $2,000				
12. A debit to Cash in Bank of $750				

Computerized Accounting Using Peachtree Complete® Accounting

Software Objectives

When you have completed this chapter, you will be able to use Peachtree to:

1. Change the company name.
2. Set the system date.
3. Print a Chart of Accounts report.
4. Enter a new general ledger account.
5. Record a beginning balance for a general ledger account.
6. Explain the purpose of the account type settings.

Problem 5-3   Identifying Increases and Decreases in Accounts

Follow the instructions provided below to print a Chart of Accounts report for Wilderness Rentals. Although the instructions in your textbook direct you to analyze the transactions, you will not use Peachtree to complete this analysis. In the next chapter, however, you will use the Wilderness Rentals general ledger accounts to enter transactions.

INSTRUCTIONS

Beginning a Session

Step 1   Open the Glencoe Accounting: Electronic Learning Center software and click on the **Peachtree Complete® Accounting Software and Spreadsheet Applications** icon.

Step 2   Log onto the system by entering your user name and password.

Step 3   From the scrolling list of chapter problems, select the problem set: Wilderness Rentals (Prob. 5-3).

Step 4   Rename the company by adding your initials, e.g., Wilderness (Prob. 5-3: XXX).

- Choose **Company Information** from the ***Maintain*** menu.
- Review the information in the Maintain Company Information window. (See Figure 5-3A.)
- Click in the *Company Name* field.
- Change the company name by adding your initials as shown in the figure.
- Click [OK] to record the new company name.

DO YOU HAVE A QUESTION

Q. *Are there any differences between a manual system and computerized system when you set up a chart of accounts?*

A. When you set up a chart of accounts using Peachtree, you must enter an account number and a title for each general ledger account just like you would in a manual system. Using a computerized system such as Peachtree, however, you must categorize each account by assigning an account type (e.g., asset, liability, equity, income, expense, etc.).

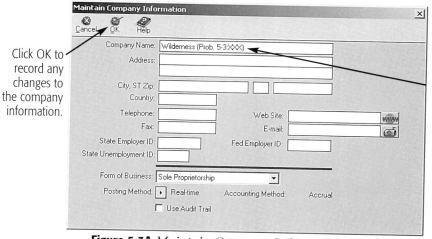

Click OK to record any changes to the company information.

Add your initials to the company name.

Figure 5-3A *Maintain Company Information Window*

Step 5 Set the system date to January 31, 2008.

- Click the *Options* menu and choose **Change System Date**.
- Type **1/31/08**.
- Click **OK** to record the new system date.

Preparing a Report

Step 6 Print a Chart of Accounts report.

To print a Chart of Accounts report:

- Choose **General Ledger** from the *Reports* menu to display the Select a Report window. (See Figure 5-3B.)

Notes

You must set the accounting date each time you begin working with Peachtree.

TIP: You can double-click a report title to go directly to that report, skipping the options window.

- Select Chart of Accounts in the report list.
- Click [Preview] and then click **OK** to display the Chart of Accounts report.
- Review the report as shown in Figure 5-3C and then click [Print] to print the report.
- Click [Close] to close the report window.

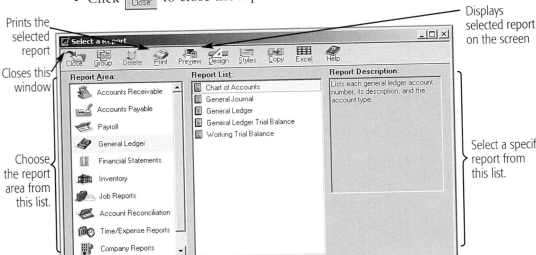

Prints the selected report

Closes this window

Choose the report area from this list.

Displays selected report on the screen

Select a specific report from this list.

Figure 5-3B *Select a Report Window with the General Ledger Reports*

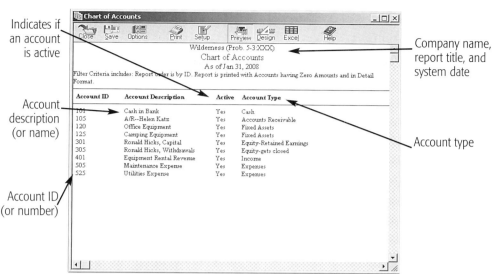

Indicates if an account is active

Company name, report title, and system date

Account description (or name)

Account type

Account ID (or number)

Figure 5-3C *Chart of Accounts Report*

Step 7   Review the information shown on the report.

As you can see, the Chart of Accounts report shows the general ledger account numbers and account titles for Wilderness Rentals. The report also includes two other columns—*Active* and *Account Type.*

The *Active* column includes information that identifies whether or not an account is active. You can post transactions to an active account, but not to an inactive account. The account type identifies the account category (e.g., Cash, Accounts Receivable, Fixed Assets, Equity—Retained Earnings, Equity—gets closed, Income, and Expenses). Peachtree uses this information to group accounts for financial reports and to determine which accounts are permanent and which are temporary when it closes an accounting period.

Ending the Session

When you finish the problem or if you must stop at the end of a class period, follow the instructions given here to end the session.

Step 8   Click the **Close Problem** button in the Glencoe Smart Guide window and select the appropriate save option as directed by your teacher.

Continuing a Problem from a Previous Session

- If you were previously directed to save your work on the network, select the problem from the scrolling menu and click **OK.** The system will retrieve your files from your last session.
- If you were previously directed to save your work on a floppy disk, insert the floppy, select the corresponding problem from the scrolling menu and click **OK.** The system will retrieve your files from the floppy disk.

Mastering Peachtree

What account types are used by the Peachtree software? On a separate sheet of paper, list these account types.

TIP: Search the Help Index to learn about the account types.

Peachtree Guide

Peachtree Guide

Problem 5-4   Using T Accounts to Analyze Transactions

Follow the instructions provided below to print a Chart of Accounts report for Hot Suds Car Wash. You will use the general ledger accounts to enter transactions in the next chapter.

INSTRUCTIONS

Beginning a Session

Step 1   Open the Glencoe Accounting: Electronic Learning Center software and click on the **Peachtree Complete® Accounting Software and Spreadsheet Applications** icon.

Step 2   Log onto the system by entering your user name and password.

Step 3   From the scrolling list of chapter problems, select the problem set: Hot Suds Car Wash (Prob. 5-4).

Step 4   Rename the company by adding your initials, e.g., Hot Suds (Prob. 5-4: XXX).

Step 5   Set the system date to January 31, 2008.

Preparing a Report

Step 6   Print a Chart of Accounts report.

TIP: You can use the General Ledger Navigation Aid at the bottom of the Peachtree main window to access the General Ledger Reports including the Chart of Accounts.

Step 7   Review the information shown on the report.

Ending the Session

Step 8   Click the **Close Problem** button in the Glencoe Smart Guide window and select the appropriate save option as directed by your teacher.

DO YOU HAVE A QUESTION

Q. *What is the purpose of the "Filter Criteria includes: …" section at the top of the Chart of Accounts report?*

A. Almost every report you print with Peachtree includes the report criteria at the top of the report. This information identifies which filters (or options) are set for the report. For example, you could choose to print a Chart of Accounts with only the general ledger accounts that have a balance. Or, you could choose to print only the asset accounts. The information at the top of the report reflects any special options you may have set. On some reports, the filter criteria may run past the edge of the page. The content of the report is not affected when this occurs.

Notes

   Refer to the instructions in Problem 5-3 if you need help installing a problem set, changing the company name, or setting the system date.

Problem 5-5 Analyzing Transactions into Debit and Credit Parts

Follow the instructions provided below to add new general ledger accounts for Kits & Pups Grooming. You will also learn how to enter the beginning balance for an account.

INSTRUCTIONS

Beginning a Session

Step 1 Open the Glencoe Accounting: Electronic Learning Center software and click on the **Peachtree Complete® Accounting Software and Spreadsheet Applications** icon.

Step 2 Log onto the system by entering your user name and password.

Step 3 From the scrolling list of chapter problems, select the problem set: Kits & Pups Grooming (Prob. 5-5).

Step 4 Rename the company and set the system date to January 31, 2008.

Completing the Accounting Problem

Step 5 Add the **Advertising Expense** general ledger account.

To add a new account:

- Choose **Chart of Accounts** from the *Maintain* menu.
- Click 🔍 in the Maintain Chart of Accounts window to view those accounts already recorded for the company.
- Type **501** in the *Account ID* field and press **TAB** to move to the next field.

    As you type an account number, Peachtree displays a list of the general ledger accounts and highlights the first account that matches what you have entered. This feature is helpful if you are entering an account number for an account you want to change. **Note:** This feature may be disabled if your teacher changed the Peachtree preferences.

TIP: Press **TAB** to move to the next field in a data entry window and press **SHIFT+TAB** to move to the previous field.

- Type **Advertising Expense** in the *Description* field.
- Click the down arrow *Account Type* in the field and select **Expenses** since this account is an expense.
- Review the information you just entered. (See Figure 5-5A.)

    You do not have to complete any of the other fields shown in the Maintain Chart of Accounts window. The account activity appears in this window as a company uses the software to record its transactions. You can also enter budget amounts, as you will learn in a later chapter.

- Click 💾 or press **ALT+S** to record the new account.
- Click 📄 to clear the data entry fields in preparation to enter a new account.

DO YOU HAVE A QUESTION

Q. *When you add a new account or enter a transaction using the Peachtree software, is it necessary to manually save your work?*

A. Some applications, such as a word processor or a spreadsheet program, require that you choose to save your work by choosing the **Save** command or by clicking the **Save** button on a toolbar. Peachtree does not require you to manually choose to save your work. It automatically updates the company files for you. However, you must always be sure to properly exit Peachtree to avoid losing any data.

Notes

Refer to the instructions in Problem 5-3 if you need help installing a problem set, changing the company name, or setting the system date.

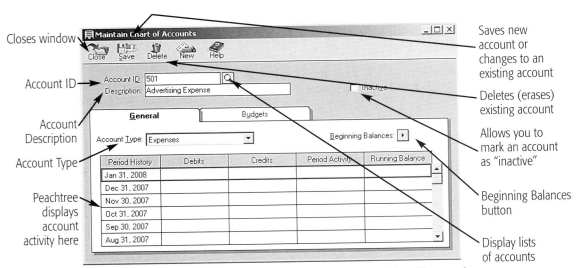

Figure 5-5A *Maintain Chart of Accounts Window (Advertising Expense)*

Step 6   Enter the remaining expense accounts shown below for Kits & Pups Grooming.

Account No.	Description	Account Type
505	Equipment Repair Expense	Expenses
510	Maintenance Expense	Expenses
520	Rent Expense	Expenses
530	Utilities Expense	Expenses

Note: To edit an account description or account type, enter the account number and then change the account information.

Click to save the changes. If you enter the wrong account number, enter or select the account. Then, click

[Delete] to remove it from the chart of accounts.

Step 7   Enter the beginning balances for **Cash in Bank** ($15,000) and **Abe Shultz** ($15,000).

To enter beginning balances:

- Click the **Beginning Balances** button in the Maintain Chart of Accounts window.
- Choose **From 1/1/08 through 1/31/08** from the Select Period window and click **OK**.
- Type **15000.00** in the *Cash in Bank* field. Be sure to enter the decimal point.
- Tab to the *Abe Shultz, Capital* field and enter **15000.00**.
- Review the entry. (See Figure 5-5B.)
- Click [OK] to record the beginning balances.

Step 8   Click [Close] to close the Maintain Chart of Accounts window.

Notes

If you do not click the New button, Peachtree does not clear the description field for the next new account. However, you can type over the account number and account description to enter a new account.

Records beginning balances

Accounting period for beginning balances

Describes the concept of the basic accounting equation

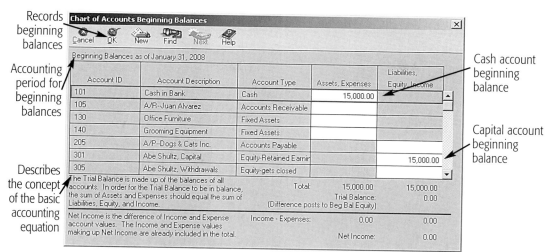

Cash account beginning balance

Capital account beginning balance

Figure 5-5B *Chart of Accounts Beginning Balances Window*

Preparing a Report

Step 9   Print a Chart of Accounts report.

Step 10   Review the information shown on the report.

Step 11   If you notice an error, use the **Chart of Accounts** option in the **Maintain** menu to edit/delete a general ledger account.

Ending the Session

Step 12   Click the **Close Problem** button in the Glencoe Smart Guide window and select the appropriate save option.

Problem 5-6   Analyzing Transactions into Debit and Credit Parts

Follow the instructions provided below to add new general ledger accounts for Outback Guide Service. You will enter transactions in the next chapter.

INSTRUCTIONS

Beginning a Session

Step 1   Open the Glencoe Accounting: Electronic Learning Center software and click on the **Peachtree Complete® Accounting Software and Spreadsheet Applications** icon.

Step 2   Log onto the system by entering your user name and password.

Step 3   From the scrolling list of chapter problems, select the problem set: Outback Guide Service (Prob. 5-6).

Step 4   Rename the company and set the system date to January 31, 2008.

Completing the Accounting Problem

Step 5   Enter the capital, income, and expense accounts.

Account No.	Description	Account Type
301	Juanita Ortega, Capital	Equity—Retained Earnings
302	Juanita Ortega, Withdrawals	Equity—gets closed
401	Guide Service Revenue	Income
505	Maintenance Expense	Expenses
515	Rent Expense	Expenses
525	Utilities Expense	Expenses

Notes

Refer to the instructions in Problem 5-3 if you need help installing a problem set, changing the company name, or setting the system date.

Notes

The withdrawal account is a temporary capital account and gets closed at the end of each accounting period.

TIP: Review the instructions for the previous problem if you need help entering a new account.

Step 6   Click [Close] to close the Maintain Chart of Accounts window.

Preparing a Report

Step 7   Print a Chart of Accounts report.

Step 8   Proof your work and make any corrections, as needed.

Ending the Session

Step 9   Click the **Close Problem** button in the Glencoe Smart Guide window and select a save option.

Problem 5-7   Analyzing Transactions

Follow the instructions provided below to add new general ledger accounts for Showbiz Video. You will enter transactions into the accounts in the next chapter.

INSTRUCTIONS

Beginning a Session

Step 1   Open the Glencoe Accounting: Electronic Learning Center software and click on the **Peachtree Complete® Accounting Software and Spreadsheet Applications** icon.

Step 2   Log onto the system by entering your user name and password.

Step 3   From the scrolling list of chapter problems, select the problem set: Showbiz Video (Prob. 5-7).

Step 4   Rename the company and set the system date to January 31, 2008.

Completing the Accounting Problem

Step 5   Review all of the accounts listed in your textbook for Showbiz Video.

Step 6   Determine the account type for each account.

Step 7   Enter all of the new accounts.

TIP: You must abbreviate some account names to fit in the *Description* field. For example, use *A/R* instead of *Accounts Receivable*.

Preparing a Report

Step 8   Print a Chart of Accounts report.

Step 9   Proof your work and make any corrections, as needed.

Ending the Session

Step 10   Click the **Close Problem** button in the Glencoe Smart Guide window and select a save option.

Mastering Peachtree

Peachtree allows you to enter a budget amount for each general ledger account. On a separate sheet of paper, describe why a company would record budget figures.

Computerized Accounting Using Spreadsheets

Problem 5-8   Completing the Accounting Equation

Completing the Spreadsheet

Step 1   Read the instructions for Problem 5-8 in your textbook. This problem involves determining the missing amounts for the accounting equations given.

Step 2   Open the Glencoe Accounting: Electronic Learning Center software.

Step 3   From the Program Menu, click on the **Peachtree Complete® Accounting Software and Spreadsheet Applications** icon.

Step 4   Log onto the Management System by typing your user name and password.

Step 5   Under the **Problems and Tutorials** tab, select template 5-8 from the Chapter 5 drop-down menu. The template should look like the one shown below.

```
PROBLEM 5-8
COMPLETING THE ACCOUNTING EQUATION

(name)
(date)


                                     Owner's
          Assets    =   Liabilities  +  Equity   -  Withdrawals  +  Revenue   -   Expenses
    1     $64,400           $8,200      $56,300            $500      $10,000         $9,600
    2
    3
    4
    5
    6
    7
    8
    9

Total Assets                                                                        $64,400

Total Liabilities + Owner's Equity - Withdrawals + Revenue - Expenses               $64,400
```

Step 6   Key your name and today's date in the cells containing the *(name)* and *(date)* placeholders.

Step 7   The first equation is completed for you. Notice that Assets ($64,400) equal Liabilities ($8,200) + Owner's Equity ($56,300) – Withdrawals ($500) + Revenue ($10,000) – Expenses ($9,600).

Step 8   To complete the second equation, enter the amounts given in your textbook for Assets, Liabilities, Owner's Equity, Withdrawals, and Revenue in the appropriate cells. To calculate the amount for Expenses, add Liabilities ($525) + Owner's Equity ($18,800) – Withdrawals ($1,200) + Revenue ($12,100) to get a total of $30,225. Subtract Assets ($22,150) from this amount to get $8,075, the missing amount for Expenses. Enter **8075** in cell L12.

 TIP: Remember, do *not* enter a dollar sign or a comma when you enter the data—the spreadsheet template will automatically format the data when it is entered.

Step 9   To check your work, look at cells L23 and L25. Total Assets should equal Total Liabilities + Owner's Equity – Withdrawals + Revenue – Expenses.

Step 10   Complete the remaining equations in Problem 5-8 by entering the appropriate data from your text into the spreadsheet template and calculating the missing amounts.

        Check the totals at the bottom of the spreadsheet after the amounts have been entered to make sure they are in balance. If the totals do not balance, check your work to find the error.

Step 11   Save the spreadsheet using the **Save** option from the *File* menu. You should accept the default location for the save as this is handled by the management system.

Step 12   Print the completed spreadsheet.

Step 13   Exit the spreadsheet program.

Step 14   In the Close Options box, select the location where you would like to save your work.

Step 15   Answer the Analyze question from your textbook for this problem.

What-If Analysis

TIP: Always save your work before performing What-If Analysis. It is not necessary to save your work after performing What-If Analysis unless your teacher instructs you to do so. If you are required to save your work after performing What-If Analysis, be sure to rename the spreadsheet to avoid saving over your original work.

If Liabilities are $50,000, Owner's Equity is $39,250, Withdrawals are $1,176, Revenue is $15,802, and Expenses are $11,660, what are Assets?

TIP: Use row 11 of the spreadsheet template to answer this question. Enter the amounts for Liabilities, Owner's Equity, Withdrawals, Revenue, and Expenses. Note that the amount for Assets is automatically computed for you! This is because cell B11 contains a *formula* that automatically calculates the missing amount. Formulas are very useful in spreadsheets, saving time and improving accuracy.

CHAPTER 5 — Transactions That Affect Revenue, Expenses, and Withdrawals

Self-Test

Part A   True or False

Directions: *Circle the letter* T *in the Answer column if the statement is true; circle the letter* F *if the statement is false.*

Answer

T   F   **1.** The normal balance side for a revenue account is the debit side.

T   F   **2.** "Credit" means the increase side of an account.

T   F   **3.** A credit to an expense account decreases the account balance.

T   F   **4.** Withdrawals are increased on the debit side.

T   F   **5.** Revenue is increased on the credit side.

T   F   **6.** Expenses are decreased on the credit side.

T   F   **7.** The basic accounting equation may be expressed as A = L + OE.

T   F   **8.** The left side of a T account is always the debit side.

T   F   **9.** You may have two debits and one credit as long as the amounts are equal.

T   F   **10.** A debit to an expense account and a credit to a capital account will result in the basic accounting equation being out of balance.

T   F   **11.** Capital is always increased by credits.

T   F   **12.** Debits decrease the withdrawals account.

Part B Identify the Normal Balance

Directions: *For each T account below, indicate with an (N) the normal balance side. The first account has been completed as an example.*

Cash in Bank		Accounts Payable		Jones, Capital	
Debit	Credit	Debit	Credit	Debit	Credit
(N)					

Rent Expense		Fees Revenue		Jones, Withdrawals	
Debit	Credit	Debit	Credit	Debit	Credit

Part C Complete the T Account

Directions: *Analyze the transactions below and enter them in the T accounts provided.*

1. Ms. Adams invested $12,000 cash in the business.
2. Bought office equipment for cash, $1,000.
3. Bought a computer on account, $3,000.

Cash in Bank		Office Equipment		Computer Equipment	

Accounts Payable		Adams, Capital	

Computerized Accounting Using Peachtree

Software Objectives

When you have completed this chapter, you will be able to use Peachtree to:

1. Record and post a general journal entry.
2. Print a General Journal report.
3. Edit a general journal entry.
4. Print a General Ledger report.
5. Continue a problem from a previous session.

Problem 6-4   Recording General Journal Transactions

Review the accounts and the transactions listed in your textbook for Wilderness Rentals. You will record these transactions using the **General Journal Entry** option in the *Tasks* menu.

INSTRUCTIONS

Beginning a Session

Step 1 Open the Glencoe Accounting: Electronic Learning Center software and click on the **Peachtree Complete® Accounting Software and Spreadsheet Applications** icon.

Step 2 Log onto the system by entering your user name and password.

Step 3 From the scrolling list of chapter problems, select the problem set: Wilderness Rentals (Prob. 6-4).

Step 4 If you haven't already done so, follow these steps to rename the company by adding your initials, e.g., Wilderness (Prob. 6-4: XXX).

- Choose **Company Information** from the *Maintain* menu.
- Review the information in the Maintain Company Information window. (See Figure 6-4A.)
- Click in the *Company Name* field.
- Add your initials to the company name.
- Click [OK] to record the new company name.

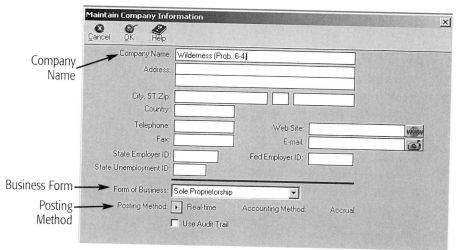

Figure 6-4A *Maintain Company Information Window*

Notes

*Using the **Company Information** option, you can also enter/change a company's address, telephone numbers, tax ID numbers, business form (sole proprietorship, corporation, partnership, S corporation, or limited liabiltiy company), and posting method (real-time or batch).*

Step 5 Set the system date to January 31, 2008.

- Click the *Options* menu and choose **Change System Date**.
- Type **1/31/08**.
- Click **OK** to record the new system date.

Completing the Accounting Problem

The instructions in this section explain how to enter and post general journal transactions using the Peachtree software.

Step 6 Review the transactions for Wilderness Rentals shown in your textbook.

Step 7 Enter the transaction for January 1.

January 1, Wrote Check 310 for the part-time secretary's salary, $270.

To enter the general journal transaction:

- Choose **General Journal Entry** from the *Tasks* menu to display the General Journal Entry window.
- Type **1/1/08** and press the **TAB** key to record the transaction date in the *Date* field.

  Make sure that Peachtree shows 2008 for the year. If not, close the window without recording the transaction and change the system date as outlined in the *Beginning a Session* instructions.

TIP: You can save time by entering just the day in a date field. Peachtree will automatically show the full date (e.g., Jan. 1, 2008).

- Type **Check 310** in the *Reference* field and then press **TAB** to move to the next field.
- Enter **520** in the first *Account No.* field and press **TAB** to record the account number for **Salaries Expense**.

  As you type an account number, Peachtree displays a list of the general ledger accounts and highlights the first account that matches what you have entered. This feature is helpful if you do not know an account number. For example, what if you did not know the **Salaries Expense** account number? You do know, however, that this account is an expense account and that expense accounts begin with the digit **5**. When you type a **5**, Peachtree highlights the first account that begins with this digit. From this point, you can scroll through the list to locate the account you need. **Note:** This feature may be disabled if your teacher changed the Peachtree preferences.

- Type **Part-time secretary's salary** for the description.

  Every time you type or enter information in a field, you must press **TAB** or **ENTER** to record the information. You can also use these keys to move from field to field. Press **SHIFT+TAB** to move backwards.

> **Notes**
>
> *Changing the system date in Peachtree does not affect the clock settings for your computer.*

- Type **270.00** in the   *it* field to record the debit to **Salaries Expense**.

TIP: Unless your teacher changed the   ult Peachtree settings, you must include the decimal point whe   ou enter an amount. If you type 270 without the period, the p.   m will automatically format the amount to $2.70.

- Press **TAB** to move to the *Account No.* field on the   ' line.
- Click Q to display a Lookup list that shows the gene   dger account numbers.
- Use the arrow keys or the mouse to highlight **101 Cash in**   **\k**. Click [OK] or press **ENTER** to select this account.
- Type **Part-time secretary's salary** for the description for this part of the transaction, too.
- Move to the *Credit* field and enter **270.00**.
- Verify that the *Out of Balance* amount shows 0.00. If it does not, check the transaction amounts.
- Proof the information you just recorded. Check the account numbers, descriptions, and amounts. If you notice a mistake, move to that field and make the correction. Compare the information on your screen to the completed transaction shown in Figure 6-4B.

Notes

You do not have to enter information in the Job *field. It is used to link a transaction to a specific job or project.*

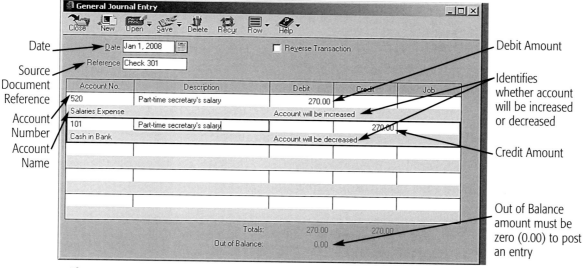

Figure 6-4B *Completed General Journal Entry (January 1, Check 301)*

- Click [Save] to post the transaction.

When you post a general journal transaction, the Peachtree software automatically updates the general ledger accounts.

Step 8   Enter the remaining transactions for the month.

> **IMPORTANT:** If you realize that you made a mistake after you have posted an entry, make a note to correct the entry after you record all of the transactions. Instructions to edit a general journal entry are provided in the next section.
>
> For each transaction:
>
> —Enter the date. Use the year 2008.
> —Enter the source document reference.
> —Enter the number of the account debited.
> —Enter a description.
> —Enter the amount of the debit.
> —Enter the number of the account credited.
> —Enter a description.
> —Enter the amount of the credit.
> —Verify that the _Out of Balance_ total is 0.00.
> —Proof your work.
> —Save the entry.

Step 9   Click to close the General Journal Entry window.

Preparing a Report and Proofing Your Work

After you enter the transactions, the next step is to print a report and proof your work. This section explains how to print a General Journal report. You will also learn how to edit a general journal entry.

Step 10   Print a General Journal report.

> _To print a General Journal report:_
>
> • Choose **General Ledger** from the **Reports** menu to display the Select a Report window. (See Figure 6-4C.)

TIP: As an alternative, you can click the ⟨General Ledger⟩ navigation aid and then choose the General Journal report to go directly to the report.

> • Select General Journal in the report list.
> • Click ⟨Preview⟩ and then click **OK** to display the General Journal report.
> • Review the report as shown in Figure 6-4D and then click ⟨Print⟩ to print the report.
> • Click ⟨Close⟩ to close the report window.

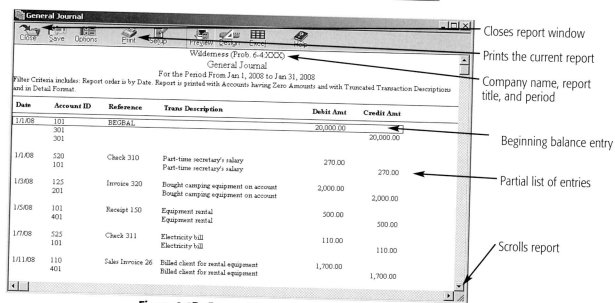

Prints the selected report

Displays selected report on the screen

Report Area:

General Ledger reports

Report area

Report description

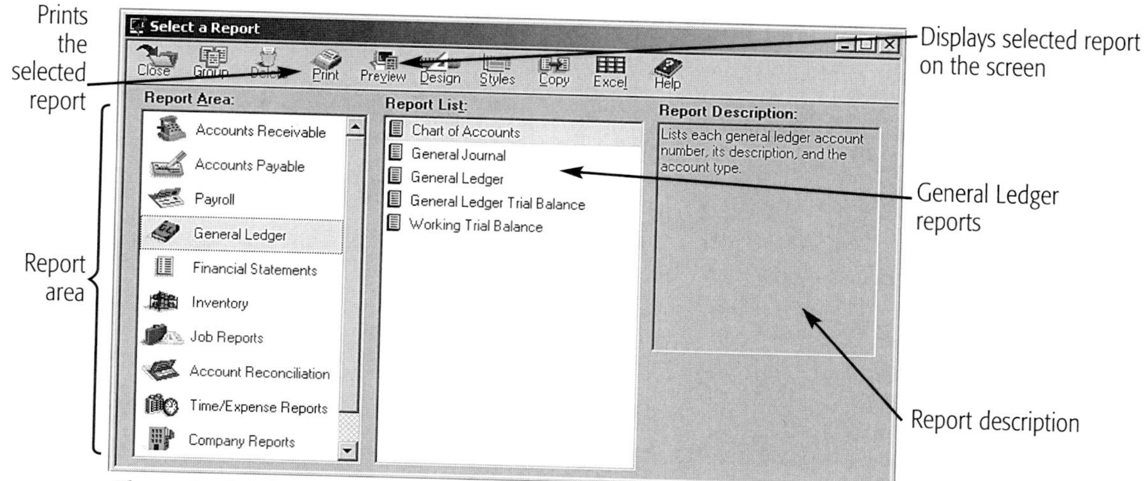

Figure 6-4C *Select a Report Window with the General Ledger Reports*

Closes report window

Prints the current report

Company name, report title, and period

Beginning balance entry

Partial list of entries

Scrolls report

Figure 6-4D *General Journal Report*

Step 11 Proof the information shown on the report.

Find the beginning balance (BEGBAL) entry for $20,000 on the General Journal report. This entry was included as part of the problem set file for Wilderness Rentals to record the owner's initial investment. As you complete other problems using the Peachtree software, you may notice other entries with a BEGBAL reference. These entries are used to establish the beginning balances for various general ledger accounts, but will not affect your work.

Step 12 If there are any corrections needed, follow these instructions to edit/delete a transaction.

- Choose **General Journal Entry** from the **Tasks** menu to display the General Journal Entry window.
- Click [Open] to display a list of the general journal entries. (See Figure 6-4E.)

Peachtree Guide

- Select the transaction that you want to edit/delete and click the **OK** button.
- To edit a transaction, simply make the necessary changes to the date, reference, account numbers, descriptions, and/or amounts. Post the transaction to record the change.
- To delete a transaction, click [Delete] and then confirm that you want to remove the transaction.
- Select another entry to edit or close the General Journal window if you are finished making changes.

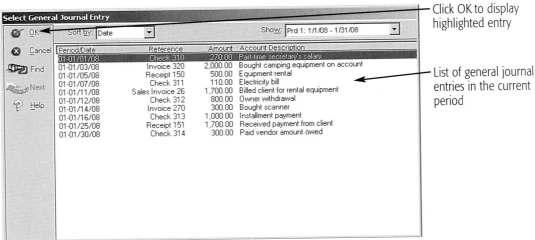

Figure **6-4E** *Select General Journal Entry Window*

Step 13 Print a revised General Journal report if you changed or deleted any of the transactions.

Analyzing Your Work

To answer the **Analyze** question, print a General Ledger report.

Step 14 Choose **General Ledger** from the *Reports* menu to display the Select a Report window.

Step 15 Double-click the General Ledger report title to display the report.

Step 16 Locate the **Cash in Bank** account on the General Ledger report. What is the sum of the credits to this account during January?

Ending the Session

Step 17 Click the **Close Problem** button in the Glencoe Smart Guide window and select a save option as directed by your teacher.

Continuing a Problem from a Previous Session

- If you were previously directed to save your work on the network, select the problem from the scrolling menu and click **OK.** The system will retrieve your files from your last session.
- If you were previously directed to save your work on a floppy disk, insert the floppy, select the corresponding problem from the scrolling menu and click **OK.** The system will retrieve your files from the floppy disk.

Mastering Peachtree

On a separate sheet of paper, answer the following questions:
How do you enter a beginning balance for a general ledger account? Why would a company need to enter beginning balances?

 TIP: Search the help information to learn how to enter beginning balances.

Problem 6-5   Recording General Journal Transactions

Review the accounts and the transactions listed in your textbook for Hot Suds Car Wash. Use the **General Journal Entry** option to record the transactions for the month of January.

INSTRUCTIONS

Beginning a Session

Step 1   Open the Glencoe Accounting: Electronic Learning Center software and click on the **Peachtree Complete® Accounting Software and Spreadsheet Applications** icon.

Step 2   Log onto the system by entering your user name and password.

Step 3   From the scrolling list of chapter problems, select the problem set: Hot Suds Car Wash (Prob. 6-5).

Step 4   Rename the company by adding your initials, e.g., Hot Suds (Prob. 6-5: XXX).

Step 5   Set the system date to January 31, 2008.

 Notes

Refer to Problem 6-4 if you need instructions to begin a session.

Completing the Accounting Problem

Step 6   Review the transactions shown in your textbook for Hot Suds Car Wash.

Step 7   Record all of the transactions using the **General Journal Entry** option.

 TIP: Remember to proof each general journal entry before you post it. Check the account numbers, descriptions, and amounts.

Step 8   Close the General Journal Entry window after you finish recording the transactions.

Preparing a Report and Proofing Your Work

Step 9   Print a General Journal report.

Step 10   Proof your work. Make any corrections as needed and print a revised report, if necessary.

 TIP: While viewing a General Journal report, you can double-click on an entry to display it in the General Journal Entry window. You can edit the transaction and then close the window to see an updated report.

Peachtree Guide

Analyzing Your Work

Step 11 Print a General Ledger report.

Step 12 Sum the asset account balances to answer the Analyze question.

Checking Your Work and Ending the Session

Step 13 Click the **Close Problem** button in the Glencoe Smart Guide window.

Step 14 If your teacher has asked you to check your solution, select *Check my answer to this problem.* Review, print, and close the report.

Step 15 Click the **Close Problem** button. In the Close Options box, select the save option as directed by your teacher. Click **OK.**

Mastering Peachtree

On a separate sheet of paper, answer the following questions:
Why does the General Journal Entry form include a *Job* field? How might a company use the information recorded in this field?

Problem 6-6   Recording General Journal Transactions

Review the accounts and the transactions listed in your textbook for Kits & Pups Grooming. Follow the instructions provided to record the transactions for the month of January.

INSTRUCTIONS

Beginning a Session

Step 1 Begin the session and select the problem set for Kits & Pups Grooming (Prob. 6-6).

Step 2 Rename the company and set the system date to January 31, 2008.

Completing the Accounting Problem

Step 3 Record all of the transactions.

TIP: Check the *Out of Balance* and *Totals* amounts for each entry before you save it.

Step 4 Print a General Journal report.

Step 5 Proof your work. Make any corrections as needed and print a revised report.

Step 6 Answer the Analyze question shown in your textbook.

Ending the Session

Step 7 Click the **Close Problem** button in the Glencoe Smart Guide window and select a save option.

Notes

Refer to Problem 6-4 if you need instructions to begin a session.

Problem 6-7   Recording General Journal Transactions

Review the accounts and the transactions listed in your textbook for Outback Guide Service. Follow the instructions provided to record the transactions for the month of January.

INSTRUCTIONS

Beginning a Session

Step 1   Begin the session and select the problem set for Outback Guide Service (Prob. 6-7).

Step 2   Rename the company and set the system date to January 31, 2008.

Completing the Accounting Problem

Step 3   Record all of the transactions.

The General Journal Entry window can accommodate multi-part transactions such as the January 1 transaction, which includes five parts—four debits and one credit. (See Figure 6-7A.)

What should you do if you accidentally omit a part of a multi-part entry? First, select the line just below where you want to insert the omitted part of the entry. Next, click [Row ▾], then [Add ▶] to insert a blank line in the entry and then enter the missing part.

If you type a line in an entry twice, highlight the occurrence you want to delete and click [Row ▾], then [Remove]. Peachtree removes the line from the entry.

TIP: As a shortcut, press **ALT+S** to post a transaction.

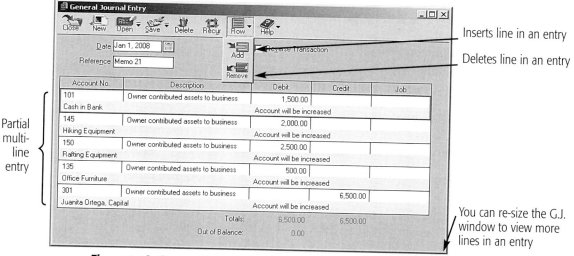

Figure 6-7A *General Journal Entry with Multi-Part Entry*

Step 4   Print a General Journal report.

Step 5   Proof your work. Make any corrections as needed and print a revised report.

If you need to end the session before you finish this problem, refer to the instructions given in Problem 6-4. Steps are also provided to continue a problem at the beginning of the next session.

Step 6 Answer the Analyze question.

Checking Your Work and Ending the Session

Step 7 Click the **Close Problem** button in the Glencoe Smart Guide window.

Step 8 If your teacher has asked you to check your solution, select *Check my answer to this problem*. Review, print, and close the report.

Step 9 Click the **Close Problem** button. From the Close Options box, select the Save Option as directed by your teacher. Click **OK.**

Problem 6-8   Recording General Journal Transactions

Review the accounts and the transactions listed in your textbook for Showbiz Video. Follow the instructions provided to record the transactions for the month of January.

INSTRUCTIONS

Beginning a Session

Step 1 Begin the session and select the problem set for Showbiz Video (Prob. 6-8).

Step 2 Rename the company and set the system date to January 31, 2008.

Completing the Accounting Problem

Step 3 Record all of the transactions.

Step 4 Print a General Journal report.

Step 5 Proof your work. Make any corrections as needed and print a revised report.

Step 6 Answer the Analyze question.

Ending the Session

Step 7 Click the **Close Problem** button and select a save option.

FAQs

How do you correct a general journal entry?

You can edit/delete a general journal entry at any time unless you have closed the current period. To make a change, choose the **General Journal Entry** option. Click the **Open** button and select the entry you want to update. Make the changes and post the corrected transaction. Peachtree automatically applies the corrections. To delete an entry, click the **Delete** button when the transaction is displayed in the General Journal Entry window.

Is there a way to turn off the automatic drop-down lists and the automatic field completion features?

As you type in an account field, Peachtree will show the accounts in a drop-down list and will automatically attempt to complete the field for you. Choose **Global** from the *Options* menu and then change Smart Data Entry options to enable/disable these features.

Problem 6-6   Recording General Journal Transactions

GENERAL JOURNAL

PAGE _____

DATE	DESCRIPTION	POST. REF.	DEBIT	CREDIT	
					1
					2
					3
					4
					5
					6
					7
					8
					9
					10
					11
					12
					13
					14
					15
					16
					17
					18
					19
					20
					21
					22
					23
					24
					25
					26
					27
					28
					29
					30
					31
					32
					33
					34
					35
					36
					37

Analyze: _____

Problem 6-7 Recording General Journal Transactions

GENERAL JOURNAL   PAGE _____

	DATE	DESCRIPTION	POST. REF.	DEBIT	CREDIT	
1						1
2						2
3						3
4						4
5						5
6						6
7						7
8						8
9						9
10						10
11						11
12						12
13						13
14						14
15						15
16						16
17						17
18						18
19						19
20						20
21						21
22						22
23						23
24						24
25						25
26						26
27						27
28						28
29						29
30						30
31						31
32						32
33						33
34						34
35						35
36						36
37						37
38						38
39						39

Problem 6-7   (concluded)

Analyze: _____

Problem 6-8   Source Documents

Instructions: *Use the following source documents to record the transactions for this business.*

SHOWBIZ VIDEO
7575 Ingram Blvd.
Spokane, WA 99204

RECEIPT
No. 435

January 1 20--

RECEIVED FROM *Cash Sales* $ *3,400.00*

Three thousand four hundred and ⁰⁰/₁₀₀ ————— DOLLARS

FOR *VCR Rentals $1,900.00 + Video Tape Rentals $1,500.00*

RECEIVED BY *Greg Failla*

NEW MEDIA SUPPLIERS
14308 San Mateo Blvd.
Los Angeles, CA 90016

INVOICE NO. NM101
DATE: *Jan. 7, 20--*
ORDER NO.:
SHIPPED BY: *UPS*
TERMS: *Balance payable in 30 days*

TO *Showbiz Video*
7575 Ingram Blvd.
Spokane, WA 99204

QTY.	ITEM	UNIT PRICE	TOTAL
8	VCRs -- model ALG45	$ 80.00	$ 640.00
4	13" TV/VCR -- model LX44	100.00	400.00
6	Camcorders -- model GR77	260.00	1,560.00
			$2,600.00
	Less down payment		- 600.00
			$2,000.00

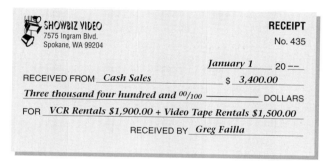

$ *325.00*		**No. 1250**
Date *January 3*		20--
To *Washington Repairs & Service*		
For *Equipment repairs*		
	Dollars	Cents
Balance brought forward	11,310	00
Add deposits *6/1*	3,400	00
Total	14,710	00
Less this check	325	00
Balance carried forward	14,385	00

SHOWBIZ VIDEO
7575 Ingram Blvd.
Spokane, WA 99204

INVOICE NO. 1650
DATE: *Jan. 10, 20--*
ORDER NO.:
SHIPPED BY: *Katie's Kouriers*
TERMS: *Payable in 30 days*

TO *Spring Branch School District*
2023 Sampson Drive
Spokane, WA 99204

QTY.	ITEM	UNIT PRICE	TOTAL
18	Video rental -- History & Government series	$100.00	$1,800.00

Palace Films
606 Lei Min Street
San Francisco, CA 94133

INVOICE NO. PF32
DATE: *Jan. 5, 20--*
ORDER NO.:
SHIPPED BY: *Freight Systems*
TERMS: *Payable in 30 days*

TO *Showbiz Video*
7575 Ingram Blvd.
Spokane, WA 99204

QTY.	ITEM	UNIT PRICE	TOTAL
40	Videos X117--X205	$ 8.50	$340.00
4	Videos VV27--VW29	15.00	60.00
			$400.00

$ *750.00*		**No. 1252**
Date *January 12*		20--
To *Computer Horizons*		
For *On account*		
	Dollars	Cents
Balance brought forward	13,785	00
Add deposits		
Total	13,785	00
Less this check	750	00
Balance carried forward	13,035	00

$ *600.00*		**No. 1251**
Date *January 7*		20--
To *New Media Suppliers*		
For *Down payment on account*		
	Dollars	Cents
Balance brought forward	14,385	00
Add deposits		
Total	14,385	00
Less this check	600	00
Balance carried forward	13,785	00

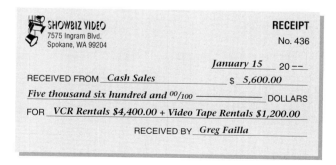

SHOWBIZ VIDEO
7575 Ingram Blvd.
Spokane, WA 99204

RECEIPT
No. 436

January 15 20--

RECEIVED FROM *Cash Sales* $ *5,600.00*

Five thousand six hundred and ⁰⁰/₁₀₀ ————— DOLLARS

FOR *VCR Rentals $4,400.00 + Video Tape Rentals $1,200.00*

RECEIVED BY *Greg Failla*

Problem 6-8   (continued)

$ 100.00		No. 1253
Date *January 18*		20 --
To *Clear Vue Window Cleaners*		
For *Maintenance*		

	Dollars	Cents
Balance brought forward	13,035	00
Add deposits 1/15	5,600	00
Total	18,635	00
Less this check	100	00
Balance carried forward	18,535	00

$ 1,000.00		No. 1254
Date *January 25*		20 --
To *New Media Suppliers*		
For *On account*		

	Dollars	Cents
Balance brought forward	18,535	00
Add deposits		
Total	18,535	00
Less this check	1,000	00
Balance carried forward	17,535	00

Problem 6-8   Recording General Journal Transactions

GENERAL JOURNAL   PAGE _____

	DATE	DESCRIPTION	POST. REF.	DEBIT	CREDIT	
1						1
2						2
3						3
4						4
5						5
6						6
7						7
8						8
9						9
10						10
11						11
12						12
13						13
14						14
15						15
16						16
17						17
18						18
19						19
20						20
21						21
22						22
23						23
24						24
25						25
26						26
27						27
28						28
29						29
30						30
31						31
32						32
33						33
34						34

Analyze: _____

_____

CHAPTER 6   Recording Transactions in a General Journal

Self-Test

Part A   True or False

Directions: *Circle the letter* T *in the Answer column if the statement is true; circle the letter* F *if the statement is false.*

Answer

T   F   **1.** A fiscal year is an accounting period that begins on January 1 and ends on December 31.

T   F   **2.** When recording a business transaction, the amount is always entered first because it is the most important part of the journal entry.

T   F   **3.** A source document is a business paper that proves a transaction occurred.

T   F   **4.** A journal is sometimes called a "book" of original entry because it is the only place in which the details of the transaction are recorded.

T   F   **5.** A check was written on April 15. It is being recorded today, April 17. The date used for recording the transaction is April 17.

T   F   **6.** The analysis of a business transaction into its debit and credit parts is required for both manual and computerized accounting systems.

T   F   **7.** The different kinds of source documents a business uses depends on the nature of the transaction.

T   F   **8.** If an error is discovered before posting occurs, simply draw a line through the incorrect item and write the correct item above it.

T   F   **9.** The debit part of a business transaction should be indented one-half inch from the left edge of the paper.

T   F   **10.** The accounting cycle is a series of activities a business completes over a period of time.

Part B Multiple Choice

Directions: *Choose the letter of the correct answer and write it in the space provided.*

Answer

1. Most businesses use an accounting period of:
   (A) one month
   (B) three months
   (C) six months
   (D) twelve months

2. A form listing specific information about a business transaction that involves the buying and selling of goods is a:
   (A) check
   (B) invoice
   (C) memorandum
   (D) receipt

3. Which of the following is a calendar year accounting period?
   (A) Jan. 1–Dec. 31
   (B) Feb. 1–Jan. 31
   (C) July 1–June 30
   (D) April 15–March 31

4. Which source document is used within the business?
   (A) check
   (B) invoice
   (C) memorandum
   (D) receipt

5. Which of the following is entered with each transaction?
   (A) date
   (B) debit account title and amount
   (C) credit account title and amount
   (D) source document or explanation
   (E) all of the above

Problem 7-4   Recording and Posting a Correcting Entry

GENERAL JOURNAL   PAGE ___5___

	DATE		DESCRIPTION	POST. REF.	DEBIT	CREDIT	
1	20--						1
2	July	3	Rent Expense	530	30000		2
3			Cash in Bank	101		30000	3
4			Check 1903				4
5							5
6							6
7							7
8							8
9							9

GENERAL LEDGER (PARTIAL)

ACCOUNT __Advertising Expense__   ACCOUNT NO. ___502___

DATE		DESCRIPTION	POST. REF.	DEBIT	CREDIT	BALANCE DEBIT	BALANCE CREDIT
20--							
July	1	Balance	✓			260000	

ACCOUNT __Rent Expense__   ACCOUNT NO. ___530___

DATE		DESCRIPTION	POST. REF.	DEBIT	CREDIT	BALANCE DEBIT	BALANCE CREDIT
20--							
July	1	Balance	✓			1500000	
	3		G5	30000		1530000	

Conducting an Audit with Alex

Palmer's Flying School

Trial Balance

Cash in Bank	4 7 0 0 00	
Accounts Receivable	1 6 0 0 00	
Airplanes	90 0 0 0 00	
Accounts Payable		2 6 8 0 00
Frank Palmer, Capital		68 6 5 0 00
Frank Palmer, Withdrawals	3 0 0 0 00	
Flying Fees		40 0 0 0 00
Advertising Expense	1 5 0 00	
Fuel and Oil Expense	5 6 0 0 00	
Repairs Expense	6 2 0 0 00	
Totals		

Computerized Accounting Using Peachtree

Software Objectives

When you have completed this chapter, you will be able to use Peachtree to:

1. Print a General Ledger report.
2. Print a General Ledger Trial Balance report.
3. Make corrections to General Journal entries.

Problem 7-5 Posting General Journal Transactions

INSTRUCTIONS

Beginning a Session

Step 1 Select the problem set: Wilderness Rentals (Prob. 7-5).
Step 2 Rename the company and set the system date to March 31, 2008.

Preparing a Report

Step 3 Print a General Ledger report.
Step 4 Review the information shown on the report.

A partial General Ledger report is shown in Figure 7-5A. As you can see, the General Ledger report is very similar to a General Ledger report you have learned how to prepare manually. One difference, however, is that Peachtree does not show a running balance after each posting. For each line item, Peachtree shows the transaction reference (e.g., Memo 21) and the journal (e.g., GENJ) from which the entry was posted. Unlike a manual General Ledger report, the Peachtree report includes the current period change (debits/credits) for each account.

> ### DO YOU HAVE A QUESTION
>
> **Q.** *Do you have to post general journal transactions to the general ledger?*
>
> **A.** No. Peachtree automatically updates the general ledger accounts every time you record (save) a general journal entry. A separate step to post the transactions is not required.

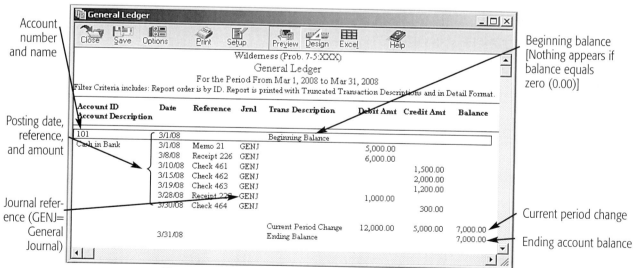

Figure 7-5A *General Ledger Report*

Step 5 Use the General Ledger report to answer the Analyze question shown in your textbook.

> ### Notes
>
> *Credit balances appear as negative amounts on a General Ledger report.*

Peachtree Guide

Ending the Session

When you finish the problem or if you must stop at the end of a class period, follow the instructions given here to end the session.

Step 6   Click the **Close Problem** button in the Glencoe Smart Guide window.

Continuing a Problem from a Previous Session

If you want to continue working on a problem that you did not complete in a previous session, follow step 1 on the previous page. If you saved your work to the network, the management system will retrieve your files from that session. If you saved your work to a floppy, insert the floppy. Your files will be retrieved from the floppy disk.

Mastering Peachtree

How do you print a General Ledger report with summary information only? Record your answer on a separate sheet of paper.

TIP: Explore the Help Index. Search for information on how to change the options for a General Ledger report.

Problem 7-6   Preparing a Trial Balance

INSTRUCTIONS

Beginning a Session

Step 1   Select the problem set: Hot Suds Car Wash (Prob. 7-6).
Step 2   Rename the company and set the system date to March 31, 2008.

Preparing a Report

Step 3   Print a General Ledger Trial Balance report.
Step 4   Review the information shown on the report.
Step 5   Answer the Analyze question.

Ending the Session

Step 6   Click the **Close Problem** button in the Glencoe Smart Guide window.

Problem 7-7   Journalizing and Posting Business Transactions

INSTRUCTIONS

Beginning a Session

Step 1   Select the problem set: Kits & Pups Grooming (Prob. 7-7).
Step 2   Rename the company and set the system date to March 31, 2008.

Completing the Accounting Problem

Step 3   Record all of the March transactions using the **General Journal Entry** task option.
Step 4   Print a General Journal report and proof your work.

Step 5   Print a General Ledger report summarized by transaction.

Sometimes, you may not need all of the information on a standard (default) report. Summary reports show the critical information, but leave out some of the detailed parts.

To print a summary General Ledger report:

- Choose **General Ledger** from the **Reports** menu.
- Select General Ledger report.
- Click .
- Choose **Summary by Transaction** from the Report Format drop-down list. Click **OK** to save changes.
- Click **OK** to print the summary report.

TIP: While viewing a report, you can click to change the report options.

✳Notes

Remember that Peachtree may sometimes truncate certain fields on a report depending on the space available.

Step 6   Print a General Ledger Trial Balance report.
Step 7   Answer the Analyze question.

Checking Your Work and Ending the Session

Step 8   Click the **Close Problem** button in the Glencoe Smart Guide window.
Step 9   If your teacher has asked you to check your solution, select *Check my answer to this problem.* Click **OK** and review, print, and close the report on your screen.
Step 10   Click the **Close Problem** button. In the Close Options box, select the save option as directed by your teacher. Click **OK.**

Mastering Peachtree

Answer the following question on a separate sheet of paper:
How do you print a General Ledger report for only one account?

Problem 7-8   Journalizing and Posting Business Transactions

INSTRUCTIONS

Beginning a Session

Step 1   Select the problem set: Outback Guide Service (Prob. 7-8).
Step 2   Rename the company and set the system date to March 31, 2008.

Completing the Accounting Problem

Step 3   Record the March transactions using the **General Journal Entry** task option.
Step 4   Print a General Journal report and proof your work.

TIP: While viewing a General Journal report, you can double-click on a transaction to display the General Journal Entry window where you can edit that transaction. Saving the correction automatically updates the General Journal report.

Step 5   Print a General Ledger report summarized by transaction.
Step 6   Print a General Ledger Trial Balance.
Step 7   Answer the Analyze question.

Ending the Session

Step 8   Click the **Close Problem** button in the Glencoe Smart Guide window and select a save option.

Mastering Peachtree

Can you customize a report, such as the General Journal report, to make more space for selected fields? Explain your answer on a separate sheet of paper. Also, print a General Journal report so that the information in the *Reference* column is not truncated.

Problem 7-9   Recording and Posting Correcting Entries

INSTRUCTIONS

Beginning a Session

Step 1   Select the problem set: Showbiz Video (Prob. 7-9).
Step 2   Rename the company and set the system date to March 31, 2008.

Completing the Accounting Problem

Step 3   Review the auditor's memo shown in your textbook.
Step 4   Correct the general journal transactions based on the auditor's comments.

> Use the **General Journal Entry** option in the *Tasks* menu to display the General Journal Entry window. Click [Open] to display a list of the general journal entries. Select an entry to edit, make your changes, and then post the corrected entry.

> **IMPORTANT:** The posting errors (March 7 and March 19) do not apply when you complete this problem with Peachtree. These errors noted in the auditor's memo deal with errors in the posting process to the general ledger. Since Peachtree automatically posts entries, these errors are not applicable.

Step 5   Print a General Journal report and proof your work.
Step 6   Print a General Ledger report.
Step 7   Answer the Analyze question.

Checking Your Work and Ending the Session

Step 8   Click the **Close Problem** button in the Glencoe Smart Guide window.
Step 9   If your teacher has asked you to check your solution, select *Check my answer to this problem*. Review, print, and close the report on your screen.
Step 10   Click the **Close Problem** button. Select the save option as directed by your teacher. Click **OK.**

Mastering Peachtree

Does Peachtree include any features that let you keep track of daily tasks—pay a vendor, place an ad in the local newspaper, or contact a client? On a separate sheet of paper, explain your answer.

DO YOU HAVE A QUESTION

Q. *What is the best method to make a correction to a general journal entry?*

A. In a manual system, you were taught how to make a correcting entry to correct a general journal entry error. Using Peachtree, you can make a correcting entry or you can edit an entry to correct it. Although changing an entry is often easier and faster than making a new entry, you will not have an audit trail that shows any corrections you may have made. Unless your teacher instructs you to make correcting entries, follow the instructions in this workbook to edit an entry if you identify a mistake.

Problem 7-7 Journalizing and Posting Business Transactions
(1), (3)

GENERAL LEDGER

ACCOUNT _____ ACCOUNT NO. _____

DATE	DESCRIPTION	POST. REF.	DEBIT	CREDIT	BALANCE	
					DEBIT	CREDIT

ACCOUNT _____ ACCOUNT NO. _____

DATE	DESCRIPTION	POST. REF.	DEBIT	CREDIT	BALANCE	
					DEBIT	CREDIT

ACCOUNT _____ ACCOUNT NO. _____

DATE	DESCRIPTION	POST. REF.	DEBIT	CREDIT	BALANCE	
					DEBIT	CREDIT

ACCOUNT _____ ACCOUNT NO. _____

DATE	DESCRIPTION	POST. REF.	DEBIT	CREDIT	BALANCE	
					DEBIT	CREDIT

ACCOUNT _____ ACCOUNT NO. _____

DATE	DESCRIPTION	POST. REF.	DEBIT	CREDIT	BALANCE	
					DEBIT	CREDIT

Problem 7-7 (continued)

ACCOUNT _____ ACCOUNT NO. _____

DATE	DESCRIPTION	POST. REF.	DEBIT	CREDIT	BALANCE	
					DEBIT	CREDIT

ACCOUNT _____ ACCOUNT NO. _____

DATE	DESCRIPTION	POST. REF.	DEBIT	CREDIT	BALANCE	
					DEBIT	CREDIT

ACCOUNT _____ ACCOUNT NO. _____

DATE	DESCRIPTION	POST. REF.	DEBIT	CREDIT	BALANCE	
					DEBIT	CREDIT

ACCOUNT _____ ACCOUNT NO. _____

DATE	DESCRIPTION	POST. REF.	DEBIT	CREDIT	BALANCE	
					DEBIT	CREDIT

ACCOUNT _____ ACCOUNT NO. _____

DATE	DESCRIPTION	POST. REF.	DEBIT	CREDIT	BALANCE	
					DEBIT	CREDIT

ACCOUNT _____ ACCOUNT NO. _____

DATE	DESCRIPTION	POST. REF.	DEBIT	CREDIT	BALANCE	
					DEBIT	CREDIT

Problem 7-7   (continued)

(2)

GENERAL JOURNAL   PAGE _____

	DATE	DESCRIPTION	POST. REF.	DEBIT	CREDIT	
1						1
2						2
3						3
4						4
5						5
6						6
7						7
8						8
9						9
10						10
11						11
12						12
13						13
14						14
15						15
16						16
17						17
18						18
19						19
20						20
21						21
22						22
23						23
24						24
25						25
26						26
27						27
28						28
29						29
30						30
31						31
32						32
33						33
34						34
35						35
36						36

Problem 7-7 (concluded)

(4)

Analyze: _____

_____

Problem 7-8 Source Documents

Instructions: *Use the following source documents to record the transactions for this problem.*

MEMORANDUM 35

Outback Guide Service
705 Fernhill Road
Encinitas, CA 92024

TO: *Accounting Clerk*
FROM: *Juanita Ortega*
DATE: *March 1, 20--*
SUBJECT: *Investment in business*

I have invested $20,000 in cash and rafting equipment valued at $5,000 in the business. Please record the journal entry.

PEAK EQUIPMENT INC.
402 Industry Blvd.
San Diego, CA 92122

INVOICE NO. 101

TO *Outback Guide Service*
    705 Fernhill Road
    Encinitas, CA 92024

DATE: *March 3, 20--*
ORDER NO.:
SHIPPED BY: *Speedy Delivery*
TERMS: *Payable in 30 days*

QTY.	ITEM	UNIT PRICE	TOTAL
3	Daypacks -- DP41714-0	$100.00	$300.00
2	Backpacks -- BP43714-1	150.00	300.00
			$600.00

Premier Processors
5775 Lemon Grove Drive
San Diego, CA 92107

INVOICE NO. 616

TO *Outback Guide Service*
    705 Fernhill Road
    Encinitas, CA 92024

DATE: *March 5, 20--*
ORDER NO.:
SHIPPED BY: *Pick up*
TERMS: *Payable in 60 days*

QTY.	ITEM	UNIT PRICE	TOTAL
1	Computer system -- IEF407	$2,800.00	$2,800.00

RECEIPT
No. 310

Outback Guide Service
705 Fernhill Road
Encinitas, CA 92024

March 2   20 --

RECEIVED FROM *Cathy & Jonathon Smith* $ *135.00*
One hundred thirty-five and 00/100 ——— DOLLARS
FOR *Guide Service*

RECEIVED BY *Juanita Ortega*

RECEIPT
No. 311

Outback Guide Service
705 Fernhill Road
Encinitas, CA 92024

March 2   20 --

RECEIVED FROM *Chad Schmidt* $ *80.00*
Eighty and 00/100 ——— DOLLARS
FOR *Guide Service*

RECEIVED BY *Juanita Ortega*

RECEIPT
No. 312

Outback Guide Service
705 Fernhill Road
Encinitas, CA 92024

March 3   20 --

RECEIVED FROM *Jason & Brittany Kelley* $ *135.00*
One hundred thirty-five and 00/100 ——— DOLLARS
FOR *Guide Service*

RECEIVED BY *Juanita Ortega*

RECEIPT
No. 313

Outback Guide Service
705 Fernhill Road
Encinitas, CA 92024

March 5   20 --

RECEIVED FROM *Clancey McMichael* $ *80.00*
Eighty and 00/100 ——— DOLLARS
FOR *Guide Service*

RECEIVED BY *Juanita Ortega*

RECEIPT
No. 314

Outback Guide Service
705 Fernhill Road
Encinitas, CA 92024

March 5   20 --

RECEIVED FROM *Louise Wicker & Dudley Hartel* $ *135.00*
One hundred thirty-five and 00/100 ——— DOLLARS
FOR *Guide Service*

RECEIVED BY *Juanita Ortega*

Problem 7-8 (continued)

RECEIPT

Outback Guide Service
705 Fernhill Road
Encinitas, CA 92024

No. 315

March 7 _____ 20 ––

RECEIVED FROM _Greg & Ann Ingram_ $ 135.00

One hundred thirty-five and 00/100 ——————— DOLLARS

FOR _Guide Service_

RECEIVED BY _Juanita Ortega_

$ 1,400.00   No. 654
Date March 18 _____ 20 ––
To Premier Processors
For On account

	Dollars	Cents
Balance brought forward	19,500	00
Add deposits		
Total	19,500	00
Less this check	1,400	00
Balance carried forward	18,100	00

$ 400.00   No. 652
Date March 9 _____ 20 ––
To Daily Courier
For Ad

	Dollars	Cents
Balance brought forward	0	00
Add deposits 2/1	20,000	00
2/7	700	00
Total	20,700	00
Less this check	400	00
Balance carried forward	20,300	00

RECEIPT

Outback Guide Service
705 Fernhill Road
Encinitas, CA 92024

No. 316

March 22 _____ 20 ––

RECEIVED FROM _Podaski Systems Inc._ $ 900.00

Nine hundred and 00/100 ——————— DOLLARS

FOR _Payment on account_

RECEIVED BY _Juanita Ortega_

INVOICE NO. 352

Outback Guide Service
705 Fernhill Road, Encinitas, CA 92024

DATE: March 12, 20––
ORDER NO.:
SHIPPED BY:

TO   Podaski Systems Inc.
     115 Beach Blvd.
     San Diego, CA 92103

TERMS: Payment due upon receipt

DATE	SERVICE	AMOUNT
3/10/––	Group rafting trip	$900.00

$ 500.00   No. 655
Date March 27 _____ 20 ––
To Live TV
For Ad

	Dollars	Cents
Balance brought forward	18,100	00
Add deposits 2/22	900	00
Total	19,000	00
Less this check	500	00
Balance carried forward	18,500	00

$ 600.00   No. 656
Date March 28 _____ 20 ––
To Peak Equipment Inc.
For On account

	Dollars	Cents
Balance brought forward	18,500	00
Add deposits		
Total	18,500	00
Less this check	600	00
Balance carried forward	17,900	00

$ 800.00   No. 653
Date March 15 _____ 20 ––
To Cash
For Personal use

	Dollars	Cents
Balance brought forward	20,300	00
Add deposits		
Total		
Less this check	800	00
Balance carried forward	19,500	00

Problem 7-8 Journalizing and Posting Business Transactions
(1), (3)

GENERAL LEDGER

ACCOUNT _____ ACCOUNT NO. _____

DATE	DESCRIPTION	POST. REF.	DEBIT	CREDIT	BALANCE DEBIT	BALANCE CREDIT

ACCOUNT _____ ACCOUNT NO. _____

DATE	DESCRIPTION	POST. REF.	DEBIT	CREDIT	BALANCE DEBIT	BALANCE CREDIT

ACCOUNT _____ ACCOUNT NO. _____

DATE	DESCRIPTION	POST. REF.	DEBIT	CREDIT	BALANCE DEBIT	BALANCE CREDIT

ACCOUNT _____ ACCOUNT NO. _____

DATE	DESCRIPTION	POST. REF.	DEBIT	CREDIT	BALANCE DEBIT	BALANCE CREDIT

ACCOUNT _____ ACCOUNT NO. _____

DATE	DESCRIPTION	POST. REF.	DEBIT	CREDIT	BALANCE DEBIT	BALANCE CREDIT

Problem 7-8   (continued)

ACCOUNT _____ ACCOUNT NO. _____

DATE	DESCRIPTION	POST. REF.	DEBIT	CREDIT	BALANCE	
					DEBIT	CREDIT

ACCOUNT _____ ACCOUNT NO. _____

DATE	DESCRIPTION	POST. REF.	DEBIT	CREDIT	BALANCE	
					DEBIT	CREDIT

ACCOUNT _____ ACCOUNT NO. _____

DATE	DESCRIPTION	POST. REF.	DEBIT	CREDIT	BALANCE	
					DEBIT	CREDIT

ACCOUNT _____ ACCOUNT NO. _____

DATE	DESCRIPTION	POST. REF.	DEBIT	CREDIT	BALANCE	
					DEBIT	CREDIT

ACCOUNT _____ ACCOUNT NO. _____

DATE	DESCRIPTION	POST. REF.	DEBIT	CREDIT	BALANCE	
					DEBIT	CREDIT

ACCOUNT _____ ACCOUNT NO. _____

DATE	DESCRIPTION	POST. REF.	DEBIT	CREDIT	BALANCE	
					DEBIT	CREDIT

Problem 7-8 (continued)
(2)

GENERAL JOURNAL   PAGE _____

	DATE		DESCRIPTION	POST. REF.	DEBIT	CREDIT	
1							1
2							2
3							3
4							4
5							5
6							6
7							7
8							8
9							9
10							10
11							11
12							12
13							13
14							14
15							15
16							16
17							17
18							18
19							19
20							20
21							21
22							22
23							23
24							24
25							25
26							26
27							27
28							28
29							29
30							30
31							31
32							32
33							33
34							34
35							35
36							36

Problem 7-8   (concluded)

(4)

Analyze: _____

_____

Problem 7-9 Recording and Posting Correcting Entries

GENERAL JOURNAL

	DATE		DESCRIPTION	POST. REF.	DEBIT	CREDIT	
1	20--						1
2	Mar.	3	Office Furniture	135	125 00		2
3			Cash in Bank	101		125 00	3
4			Check 1401				4
5		5	Cash in Bank	101	400 00		5
6			Accts. Rec.—James Coletti	110		400 00	6
7			Receipt 602				7
8		7	Accts. Pay.—Broad Street Office Supply	201	200 00		8
9			Cash in Bank	101		200 00	9
10			Check 1402				10
11		9	Office Furniture	135	500 00		11
12			Cash in Bank	101		500 00	12
13			Check 1403				13
14		13	Greg Failla, Capital	301	1200 00		14
15			Cash in Bank	101		1200 00	15
16			Check 1404				16
17		17	Cash in Bank	101	2000 00		17
18			Greg Failla, Capital	301		2000 00	18
19			Receipt 603				19
20		19	Cash in Bank	101	75 00		20
21			Accts. Rec.—Shannon Flannery	113		75 00	21
22			Receipt 604				22
23		20	Cash in Bank	101	100 00		23
24			Accts. Rec.—James Coletti	110		100 00	24
25			Receipt 605				25
26		24	Utilities Expense	530	75 00		26
27			Cash in Bank	101		75 00	27
28			Check 1405				28
29		27	Cash in Bank	101	3000 00		29
30			Greg Failla, Withdrawals	305		3000 00	30
31			Memorandum 40				31
32		29	Cash in Bank	101	1000 00		32
33			VCR Rental Revenue	405		1000 00	33
34			Receipt 606				34
35							35
36							36

Problem 7-9 (continued)

GENERAL LEDGER

ACCOUNT *Cash in Bank* ACCOUNT NO. 101

DATE		DESCRIPTION	POST. REF.	DEBIT	CREDIT	BALANCE DEBIT	BALANCE CREDIT
20--							
Mar.	1	Balance	✓			9 855 00	
	3		G21		1 250 0	9 730 00	
	5		G21	4 000 0		10 130 00	
	7		G21		2 000 0	9 930 00	
	9		G21		5 000 0	9 430 00	
	13		G21		1 200 0	8 230 00	
	17		G21	20 000 0		10 230 00	
	19		G21	750 0		10 305 00	
	20		G21	1 000 0		10 405 00	
	24		G21		750 0	10 330 00	
	27		G21	30 000 0		13 330 00	
	29		G21	10 000 0		14 330 00	

ACCOUNT *Accounts Receivable—Shannon Flannery* ACCOUNT NO. 113

DATE		DESCRIPTION	POST. REF.	DEBIT	CREDIT	BALANCE DEBIT	BALANCE CREDIT
20--							
Mar.	1	Balance	✓			300 00	
	19		G21		57 00	243 00	

ACCOUNT *Office Supplies* ACCOUNT NO. 120

DATE		DESCRIPTION	POST. REF.	DEBIT	CREDIT	BALANCE DEBIT	BALANCE CREDIT
20--							
Mar.	1	Balance	✓			120 00	

ACCOUNT *Office Furniture* ACCOUNT NO. 135

DATE		DESCRIPTION	POST. REF.	DEBIT	CREDIT	BALANCE DEBIT	BALANCE CREDIT
20--							
Mar.	1	Balance	✓			1 500 00	
	3		G21	125 00		1 625 00	
	9		G21	500 00		2 125 00	

Problem 7-9 (continued)

ACCOUNT __Accounts Payable—Broad Street Office Supply__ ACCOUNT NO. __201__

DATE		DESCRIPTION	POST. REF.	DEBIT	CREDIT	BALANCE DEBIT	BALANCE CREDIT
20--							
Mar.	1	Balance	✓				2 2 0 0 00

ACCOUNT __Greg Failla, Capital__ ACCOUNT NO. __301__

DATE		DESCRIPTION	POST. REF.	DEBIT	CREDIT	BALANCE DEBIT	BALANCE CREDIT
20--							
Mar.	1	Balance	✓				13 0 0 0 00
	13		G21	1 2 0 0 00			11 8 0 0 00
	17		G21		2 0 0 0 00		13 8 0 0 00

ACCOUNT __Greg Failla, Withdrawals__ ACCOUNT NO. __305__

DATE		DESCRIPTION	POST. REF.	DEBIT	CREDIT	BALANCE DEBIT	BALANCE CREDIT
20--							
Mar.	27		G21		3 0 0 0 00	3 0 0 0 00	

ACCOUNT __Video Rental Revenue__ ACCOUNT NO. __401__

DATE	DESCRIPTION	POST. REF.	DEBIT	CREDIT	BALANCE DEBIT	BALANCE CREDIT

ACCOUNT __VCR Rental Revenue__ ACCOUNT NO. __405__

DATE		DESCRIPTION	POST. REF.	DEBIT	CREDIT	BALANCE DEBIT	BALANCE CREDIT
20--							
Mar.	29		G21		1 0 0 0 00		1 0 0 0 00

Problem 7-9   (concluded)

GENERAL JOURNAL   PAGE _____

	DATE	DESCRIPTION	POST. REF.	DEBIT	CREDIT	
1						1
2						2
3						3
4						4
5						5
6						6
7						7
8						8
9						9
10						10
11						11
12						12
13						13
14						14
15						15
16						16
17						17
18						18
19						19
20						20
21						21
22						22
23						23
24						24
25						25
26						26
27						27
28						28
29						29
30						30
31						31
32						32
33						33
34						34

Analyze: _____

_____

CHAPTER 7   Posting Journal Entries to General Ledger Accounts

Self-Test

Part A   True or False

Directions: *Read each of the following statements to determine whether the statement is true or false. Write your answer in the space provided.*

Answer

_____   **1.** The accounts in a business are kept in a book called a ledger.

_____   **2.** Posting is the process of transferring information from the journal to the ledger accounts.

_____   **3.** Every posting requires the year, month, and day to be entered in the Date column of the ledger account for every transaction.

_____   **4.** Every journal entry requires a posting to at least two accounts.

_____   **5.** If a transaction is journalized on the 6th, but not posted until the 8th, the date of the posting should be the 8th.

_____   **6.** If an account has a zero balance, it is not necessary to list it on the trial balance.

_____   **7.** A ledger is sometimes called a book of "final" entry.

_____   **8.** Ideally, all businesses should post on a daily basis; however, businesses having few transactions may post only once a week.

_____   **9.** An example of a transposition error is writing the number 45 when you should have written 54.

_____   **10.** If you discover an error before posting, a correcting entry is required.

Part B   Multiple Choice

Directions: *Choose the letter of the correct answer and write it in the space provided.*

Answer

_____ **1.** The first step in the posting process is to:
(A) post the amount.
(B) enter the page number in the Posting Reference column of the ledger account.
(C) enter the date in the Date column of the ledger account.
(D) compute the new balance.

_____ **2.** If the total debits and total credits of the trial balance do not agree, the first step in locating the error is to:
(A) re-add the debit and credit columns.
(B) find the amount you are out of balance.
(C) make certain you copied the amount correctly from the ledger account to the trial balance.
(D) divide the amount you are out of balance by 2.

_____ **3.** All of the following about a trial balance are true *except:*
(A) it includes all general ledger accounts.
(B) it includes only the permanent accounts.
(C) it is completed after posting.
(D) it proves the ledger.

_____ **4.** Transposition errors are evenly divisible by the number:
(A) 2
(B) 3
(C) 4
(D) 9

_____ **5.** Which of the following steps in the accounting cycle is in the correct order?
(A) Post to the ledger, prepare a trial balance, journalize, analyze each transaction, and collect source documents.
(B) Analyze each transaction, journalize, collect source documents, prepare a trial balance, and post.
(C) Collect and verify source documents, analyze each transaction, journalize each transaction, post to the ledger, and prepare a trial balance.
(D) Collect and verify source documents, analyze each transaction, prepare a trial balance, journalize each transaction, and post to the ledger.

MINI PRACTICE SET

Canyon.com Web Sites

CHART OF ACCOUNTS

ASSETS
101   Cash in Bank
105   Accounts Receivable—Andrew Hospital
110   Accounts Receivable—Indiana Trucking
115   Accounts Receivable—Sunshine Products
130   Office Supplies
135   Office Equipment
140   Office Furniture
145   Web Server

LIABILITIES
205   Accounts Payable—Computer Specialists Inc.
210   Accounts Payable—Office Systems
215   Accounts Payable—Service Plus Software Inc.

OWNER'S EQUITY
301   Jack Hines, Capital
305   Jack Hines, Withdrawals

REVENUE
401   Web Service Fees

EXPENSES
505   Membership Expense
506   Telecommunications Expense
507   Rent Expense
508   Utilities Expense

Mini Practice Set 1   Source Documents

Instructions: *Use the following source documents to record the transactions for this practice set.*

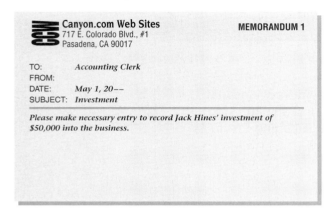

Canyon.com Web Sites
717 E. Colorado Blvd., #1
Pasadena, CA 90017

MEMORANDUM 1

TO: *Accounting Clerk*
FROM:
DATE: *May 1, 20--*
SUBJECT: *Investment*

Please make necessary entry to record Jack Hines' investment of $50,000 into the business.

COMPUTER SPECIALISTS INC.
1231 Reseda Blvd., #2A
Reseda, CA 91124

INVOICE NO. WS4658421

DATE: *May 7, 20--*
ORDER NO.:
SHIPPED BY:
TERMS:

TO   *Canyon.com Web Sites*
     717 E. Colorado Blvd., #1
     Pasadena, CA 90017

QTY.	ITEM	UNIT PRICE	TOTAL
1	Web server	$35,000.00	$35,000.00

Canyon.com Web Sites
717 E. Colorado Blvd., #1
Pasadena, CA 90017

MEMORANDUM 2

TO: *Accounting Clerk*
FROM:
DATE: *May 2, 20--*
SUBJECT: *Additional investment*

Record additional investment of a desktop computer and printer (office equipment) worth $3,500 made by Jack Hines.

Canyon.com Web Sites
717 E. Colorado Blvd., #1
Pasadena, CA 90017

RECEIPT
No. 101

May 9   20--

RECEIVED FROM *James Market* $ _1,000.00_

One thousand and ⁰⁰/₁₀₀ ———————— DOLLARS

FOR _Web site services_

RECEIVED BY *Jack Hines*

Canyon.com Web Sites
717 E. Colorado Blvd., #1
Pasadena, CA 90017

101

90-7177
3222

DATE *May 2*   20--

PAY TO THE
ORDER OF *Office Mart* $ *125.00*

One hundred twenty-five and ⁰⁰/₁₀₀ ———— DOLLARS

1st *First Bank*

MEMO *Office supplies*   *Jack Hines*

⑈322271779⑈ 0710613 ⑈101

Canyon.com Web Sites
717 E. Colorado Blvd., #1
Pasadena, CA 90017

INVOICE NO. 101

DATE: *May 11, 20--*
ORDER NO.:
SHIPPED BY:
TERMS:

TO   *Andrew Hospital*
     1314 Sherman Way
     Van Nuys, CA 91331

QTY.	ITEM	UNIT PRICE	TOTAL
N/A	Web site design services	$3,000.00	$3,000.00

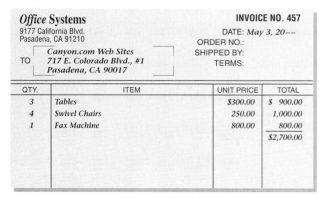

Office Systems
9177 California Blvd.
Pasadena, CA 91210

INVOICE NO. 457

DATE: *May 3, 20--*
ORDER NO.:
SHIPPED BY:
TERMS:

TO   *Canyon.com Web Sites*
     717 E. Colorado Blvd., #1
     Pasadena, CA 90017

QTY.	ITEM	UNIT PRICE	TOTAL
3	Tables	$300.00	$ 900.00
4	Swivel Chairs	250.00	1,000.00
1	Fax Machine	800.00	800.00
			$2,700.00

Service Plus Software Inc.
616 Cordova Street, #7
Glendale, CA 90121

INVOICE NO. 876

DATE: *May 12, 20--*
ORDER NO.:
SHIPPED BY:
TERMS:

TO   *Canyon.com Web Sites*
     717 E. Colorado Blvd., #1
     Pasadena, CA 90017

QTY.	ITEM	UNIT PRICE	TOTAL
N/A	Software for Web server	$10,000.00	$10,000.00

Mini Practice Set 1   (continued)

Canyon.com Web Sites
717 E. Colorado Blvd., #1
Pasadena, CA 90017

102
90-7177
3222

DATE _May 14_ 20--

PAY TO THE ORDER OF _DWP_ $ _118.00_

One hundred eighteen and 00/100 ——————— DOLLARS

1st First Bank

MEMO _Electric bill_   _Jack Hines_

⑆322271779⑆ 0710613 ⑈102

Canyon.com Web Sites
717 E. Colorado Blvd., #1
Pasadena, CA 90017

RECEIPT
No. 102

May 19 20 --

RECEIVED FROM _Intercom Inc._ $ _4,000.00_

Four thousand and 00/100 ——————— DOLLARS

FOR _Web site maintenance for one year_

RECEIVED BY _Jack Hines_

Canyon.com Web Sites
717 E. Colorado Blvd., #1
Pasadena, CA 90017

103
90-7177
3222

DATE _May 15_ 20--

PAY TO THE ORDER OF _Jack Hines_ $ _2,500.00_

Two thousand five hundred and 00/100 ——————— DOLLARS

1st First Bank

MEMO _Personal withdrawal_   _Jack Hines_

⑆322271779⑆ 0710613 ⑈103

Canyon.com Web Sites
717 E. Colorado Blvd., #1
Pasadena, CA 90017

INVOICE NO. 103

DATE: _May 20, 20--_
ORDER NO.:
SHIPPED BY:
TERMS:

TO _Indiana Trucking_
28111 Soledad Canyon
Newhall, CA 90011

QTY.	ITEM	UNIT PRICE	TOTAL
N/A	Design services	$2,000.00	$2,000.00

Canyon.com Web Sites
717 E. Colorado Blvd., #1
Pasadena, CA 90017

INVOICE NO. 102

DATE: _May 17, 20--_
ORDER NO.:
SHIPPED BY:
TERMS:

TO _Sunshine Products_
1213 Oceanview Street
Santa Monica, CA 90171

QTY.	ITEM	UNIT PRICE	TOTAL
N/A	Web site design	$5,000.00	$5,000.00

Canyon.com Web Sites
717 E. Colorado Blvd., #1
Pasadena, CA 90017

RECEIPT
No. 103

May 21 20 --

RECEIVED FROM _Sunshine Products_ $ _2,500.00_

Two thousand five hundred and 00/100 ——————— DOLLARS

FOR _Payment on account_

RECEIVED BY _Jack Hines_

Canyon.com Web Sites
717 E. Colorado Blvd., #1
Pasadena, CA 90017

104
90-7177
3222

DATE _May 18_ 20--

PAY TO THE ORDER OF _Office Mart_ $ _275.00_

Two hundred seventy-five and 00/100 ——————— DOLLARS

1st First Bank

MEMO _Filing cabinet_   _Jack Hines_

⑆322271779⑆ 0710613 ⑈104

Canyon.com Web Sites
717 E. Colorado Blvd., #1
Pasadena, CA 90017

105
90-7177
3222

DATE _May 22_ 20--

PAY TO THE ORDER OF _Telecom_ $ _4,900.00_

Four thousand nine hundred and 00/100 ——————— DOLLARS

1st First Bank

MEMO _Telephone services_   _Jack Hines_

⑆322271779⑆ 0710613 ⑈105

Mini Practice Set 1   (continued)

Canyon.com Web Sites	**106**
717 E. Colorado Blvd., #1	90-7177
Pasadena, CA 90017	3222

DATE _May 22_ 20 --

PAY TO THE ORDER OF _Service Plus Software Inc._ $ _3,333.00_

Three thousand three hundred thirty-three and 00/100 — DOLLARS

1st _First Bank_

MEMO _On account_   _Jack Hines_

⑆322271779⑆ 0710613 ⑈106

Canyon.com Web Sites	**109**
717 E. Colorado Blvd., #1	90-7177
Pasadena, CA 90017	3222

DATE _May 30_ 20 --

PAY TO THE ORDER OF _Property Management_ $ _750.00_

Seven hundred fifty and 00/100 — DOLLARS

1st _First Bank_

MEMO _Rent_   _Jack Hines_

⑆322271779⑆ 0710613 ⑈109

Canyon.com Web Sites	**107**
717 E. Colorado Blvd., #1	90-7177
Pasadena, CA 90017	3222

DATE _May 25_ 20 --

PAY TO THE ORDER OF _Office Systems_ $ _2,000.00_

Two thousand and 00/100 — DOLLARS

1st _First Bank_

MEMO _On account_   _Jack Hines_

⑆322271779⑆ 0710613 ⑈107

Canyon.com Web Sites	**110**
717 E. Colorado Blvd., #1	90-7177
Pasadena, CA 90017	3222

DATE _May 30_ 20 --

PAY TO THE ORDER OF _Jack Hines_ $ _2,500.00_

Two thousand five hundred and 00/100 — DOLLARS

1st _First Bank_

MEMO _Personal withdrawal_   _Jack Hines_

⑆322271779⑆ 0710613 ⑈110

Canyon.com Web Sites	**RECEIPT**
717 E. Colorado Blvd., #1	No. 104
Pasadena, CA 90017	

May 26 20 --

RECEIVED FROM _Job Info. Services_ $ _1,000.00_

One thousand and 00/100 — DOLLARS

FOR _2 months of Web site services_

RECEIVED BY _Jack Hines_

Canyon.com Web Sites	**111**
717 E. Colorado Blvd., #1	90-7177
Pasadena, CA 90017	3222

DATE _May 30_ 20 --

PAY TO THE ORDER OF _Computer Specialists Inc._ $ _25,000.00_

Twenty-five thousand and 00/100 — DOLLARS

1st _First Bank_

MEMO _On account_   _Jack Hines_

⑆322271779⑆ 0710613 ⑈111

Canyon.com Web Sites	**108**
717 E. Colorado Blvd., #1	90-7177
Pasadena, CA 90017	3222

DATE _May 27_ 20 --

PAY TO THE ORDER OF _All Inclusive Group_ $ _7,000.00_

Seven thousand and 00/100 — DOLLARS

1st _First Bank_

MEMO _Membership dues_   _Jack Hines_

⑆322271779⑆ 0710613 ⑈108

Computerized Accounting Using Peachtree

Mini Practice Set 1

INSTRUCTIONS

Beginning a Session

Step 1   Open the Glencoe Accounting: Electronic Learning Center software.

Step 2   From the Program Menu, click on the **Peachtree Complete®
Accounting Software and Spreadsheet Applications** icon.

Step 3   Log onto the Management System by typing your user name and
password.

Step 4   Under the **Problems & Tutorials** tab, select problem set
Canyon.com Web Sites (MP-1).

Step 5   Rename the company by adding your initials, e.g., Canyon
(MP-1: XXX).

Step 6   Set the system date to May 31, 2008.

Completing the Accounting Problem

Step 7   Review the transactions shown in your textbook for Canyon.com
Web Sites.

Step 8   Record all of the transactions using the **General Journal
Entry** option.

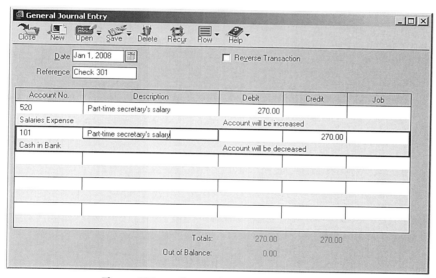

Figure MP1-1 *General Journal Entry*

TIP: Proof each general journal entry before you post it. Check the
account numbers, descriptions, and amounts.

Preparing Reports and Proofing Your Work

Step 9   Print a General Journal report.

Step 10   Proof your work. Make any corrections as needed and print a revised
report, if necessary.

TIP: While viewing a General Journal report, you can double-click on an entry to display it in the General Journal Entry window. You can edit the transaction and then close the window to see an updated report.

Step 11 Print a General Ledger report.
Step 12 Print a General Ledger Trial Balance.

Analyzing Your Work

Step 13 Answer the Analyze question.
Step 14 Complete the Audit Test.

Checking Your Work and Ending the Session

Step 15 Click the **Close Problem** button in the Glencoe Smart Guide window.
Step 16 If your teacher has asked you to check your solution, select *Check my answer to this problem.* Review, print, and close the report.
Step 17 Click the **Close Problem** button. Select the close option as directed by your teacher. Click **OK.**

Notes

Use the reports you prepared to answer the Analyze question and to complete the Audit Test.

Continuing from a Previous Session

If you want to continue from a previous session, follow steps 1–4 on the previous page. If you saved your work to the network, the management system will retrieve your files. If you saved your work to a floppy disk, insert the disk. The system will then retrieve your files from the floppy disk.

FAQs

Does Peachtree produce an audit trail if you edit a general journal entry?

Yes. Peachtree can create an audit trail when you edit a general journal entry. Select **Company Information** from the ***Maintain*** menu, select the Use Audit Trail check box and click **OK**.

Why doesn't Peachtree allow you to access the General Journal Entry window directly from a customized General Journal report?

When you display the standard (default) General Journal report, you can double-click an entry to access the General Journal Entry window where you can edit the selected transaction. If you change the standard report format (e.g., print a summary report), Peachtree will not allow you to edit transactions in this manner.

Peachtree Guide

Mini Practice Set 1   (continued)

GENERAL JOURNAL   PAGE _____

	DATE		DESCRIPTION	POST. REF.	DEBIT	CREDIT	
1							1
2							2
3							3
4							4
5							5
6							6
7							7
8							8
9							9
10							10
11							11
12							12
13							13
14							14
15							15
16							16
17							17
18							18
19							19
20							20
21							21
22							22
23							23
24							24
25							25
26							26
27							27
28							28
29							29
30							30
31							31
32							32
33							33
34							34
35							35
36							36
37							37
38							38
39							39

Mini Practice Set 1 (continued)

GENERAL JOURNAL   PAGE _____

	DATE		DESCRIPTION	POST. REF.	DEBIT	CREDIT	
1							1
2							2
3							3
4							4
5							5
6							6
7							7
8							8
9							9
10							10
11							11
12							12
13							13
14							14
15							15
16							16
17							17
18							18
19							19
20							20
21							21
22							22
23							23
24							24
25							25
26							26
27							27
28							28
29							29
30							30
31							31
32							32
33							33
34							34
35							35
36							36
37							37
38							38
39							39

GENERAL LEDGER

ACCOUNT _____ ACCOUNT NO. _____

DATE	DESCRIPTION	POST. REF.	DEBIT	CREDIT	BALANCE	
					DEBIT	CREDIT

ACCOUNT _____ ACCOUNT NO. _____

DATE	DESCRIPTION	POST. REF.	DEBIT	CREDIT	BALANCE	
					DEBIT	CREDIT

ACCOUNT _____ ACCOUNT NO. _____

DATE	DESCRIPTION	POST. REF.	DEBIT	CREDIT	BALANCE	
					DEBIT	CREDIT

ACCOUNT _____ ACCOUNT NO. _____

DATE	DESCRIPTION	POST. REF.	DEBIT	CREDIT	BALANCE	
					DEBIT	CREDIT

Mini Practice Set 1 (continued)

ACCOUNT _____ ACCOUNT NO. _____

DATE	DESCRIPTION	POST. REF.	DEBIT	CREDIT	BALANCE DEBIT	BALANCE CREDIT

ACCOUNT _____ ACCOUNT NO. _____

DATE	DESCRIPTION	POST. REF.	DEBIT	CREDIT	BALANCE DEBIT	BALANCE CREDIT

ACCOUNT _____ ACCOUNT NO. _____

DATE	DESCRIPTION	POST. REF.	DEBIT	CREDIT	BALANCE DEBIT	BALANCE CREDIT

ACCOUNT _____ ACCOUNT NO. _____

DATE	DESCRIPTION	POST. REF.	DEBIT	CREDIT	BALANCE DEBIT	BALANCE CREDIT

ACCOUNT _____ ACCOUNT NO. _____

DATE	DESCRIPTION	POST. REF.	DEBIT	CREDIT	BALANCE DEBIT	BALANCE CREDIT

ACCOUNT _____ ACCOUNT NO. _____

DATE	DESCRIPTION	POST. REF.	DEBIT	CREDIT	BALANCE DEBIT	BALANCE CREDIT

Mini Practice Set 1   (continued)

ACCOUNT _____ ACCOUNT NO. _____

DATE	DESCRIPTION	POST. REF.	DEBIT	CREDIT	BALANCE	
					DEBIT	CREDIT

ACCOUNT _____ ACCOUNT NO. _____

DATE	DESCRIPTION	POST. REF.	DEBIT	CREDIT	BALANCE	
					DEBIT	CREDIT

ACCOUNT _____ ACCOUNT NO. _____

DATE	DESCRIPTION	POST. REF.	DEBIT	CREDIT	BALANCE	
					DEBIT	CREDIT

ACCOUNT _____ ACCOUNT NO. _____

DATE	DESCRIPTION	POST. REF.	DEBIT	CREDIT	BALANCE	
					DEBIT	CREDIT

Mini Practice Set 1   (continued)

ACCOUNT _____ ACCOUNT NO. _____

DATE	DESCRIPTION	POST. REF.	DEBIT	CREDIT	BALANCE	
					DEBIT	CREDIT

ACCOUNT _____ ACCOUNT NO. _____

DATE	DESCRIPTION	POST. REF.	DEBIT	CREDIT	BALANCE	
					DEBIT	CREDIT

ACCOUNT _____ ACCOUNT NO. _____

DATE	DESCRIPTION	POST. REF.	DEBIT	CREDIT	BALANCE	
					DEBIT	CREDIT

ACCOUNT _____ ACCOUNT NO. _____

DATE	DESCRIPTION	POST. REF.	DEBIT	CREDIT	BALANCE	
					DEBIT	CREDIT

Mini Practice Set 1 (continued)

Mini Practice Set 1   (concluded)

Analyze: _____

_____

_____

_____

_____

_____

_____

_____

_____

_____

_____

_____

MINI PRACTICE SET 1

Canyon.com Web Sites

Audit Test

Directions: *Use your completed solutions to answer the following questions. Write the answer in the space to the left of each question.*

_____ **1.** In the entry to record the May 1 transaction, which account was debited?

_____ **2.** Were assets increased, decreased, or unaffected by the May 3 transaction?

_____ **3.** What type of account is Web Server?

_____ **4.** Which account was credited in the May 9, 17, and 26 transactions?

_____ **5.** What account was credited for the purchase of the Web server on May 7?

_____ **6.** What was the source document for the May 17 transaction?

_____ **7.** How does the May 20 transaction affect the owner's capital account?

_____ **8.** Was Accounts Receivable—Sunshine Products increased or decreased by the transaction on May 21?

_____ **9.** What was the balance of Cash in Bank on May 27?

_____ **10.** What were the account numbers entered in the posting Reference column of the general journal for the May 25 transaction?

_____ **11.** Which account was debited to record the issue of Check 110?

_____ **12.** What is the ending balance of the Utilities Expense account?

_____ **13.** Has the amount owed to Office Systems been paid off?

_____ **14.** How many transactions recorded during May affected the Cash in Bank account?

_____ **15.** What was the total cost of the office equipment purchased during the month?

_____ **16.** How many checks were written by the business during May?

_____ **17.** What was the total amount credited to Web Services Fees during May?

_____ **18.** What was the total amount debited to Cash in Bank for May?

_____ **19.** At the end of the month, did the Jack Hines, Withdrawals account have a debit or credit balance?

_____ **20.** On May 30, what was the total amount owed to Canyon.com Web Sites for services performed for clients?

_____ **21.** What was the date of the trial balance?

_____ **22.** What was the amount of the debit and credit totals on the trial balance?

_____ **23.** How many accounts are listed in the trial balance for Canyon.com Web Sites?

_____ **24.** How many accounts on the trial balance have debit balances?

_____ **25.** Which account on the trial balance has the largest balance?

Computerized Accounting Using Peachtree

Software Objectives

When you have completed this chapter, you will be able to use Peachtree to:

1. Print a General Ledger Trial Balance report.
2. Verify the information on a General Ledger Trial Balance.

Problem 8-5   Preparing a Six-Column Work Sheet

INSTRUCTIONS

Beginning a Session

Step 1   Select the problem set: Hot Suds Car Wash (Prob. 8-5).
Step 2   Rename the company and set the system date to May 31, 2008.

Preparing a Report

Step 3   Print a General Ledger Trial Balance report.

TIP: You can access the General Ledger Trial Balance report using the General Ledger navigation aid.

Step 4   Review the General Ledger Trial Balance report. Verify that the debit/credit column totals are the same. The total debits/credits should be $49,862.

Ending the Session

Step 5   Click the **Close Problem** button in the Glencoe Smart Guide window.

Mastering Peachtree

Use the report design button to widen the *Account Description* column on the General Ledger Trial Balance report. Print a revised report.

Problem 8-6   Preparing a Six-Column Work Sheet

INSTRUCTIONS

Beginning a Session

Step 1   Select the problem set: Kits & Pups Grooming (Prob. 8-6).
Step 2   Rename the company and set the system date to May 31, 2008.

Preparing a Report

Step 3   Print a General Ledger Trial Balance report.

TIP: Choose **General Ledger** from the *Reports* menu to print a General Ledger Trial Balance.

Ending the Session

Step 4   Click the **Close Problem** button in the Glencoe Smart Guide window.

DO YOU HAVE A QUESTION

Q. *Why doesn't Peachtree include a six-column work sheet report?*

A. The six-column work sheet is a convenient way to manually prepare the income statement and balance sheet. When you use accounting software, such as Peachtree, the program automatically generates these financial statements. A six-column work sheet is not needed to prepare these reports.

Notes

Some of the longer account names may be truncated (or cut off) on the General Ledger Trial Balance to fit in the space available.

Problem 8-7   Preparing a Six-Column Work Sheet

INSTRUCTIONS

Beginning a Session

Step 1   Select the problem set: Outback Guide Service (Prob. 8-7).

Step 2   Rename the company and set the system date to May 31, 2008.

Preparing a Report

Step 3   Print a General Ledger Trial Balance report.

Ending the Session

Step 4   Click the **Close Problem** button in the Glencoe Smart Guide window.

═══ FAQs ═══

Does Peachtree include an option to print a six-column work sheet?

No. Peachtree does not provide an option to print a work sheet. The program allows you to print an Income Statement and a Balance Sheet. To print an Income Statement, Statement of Changes in Owner's Equity, or a Balance Sheet, choose the **Financial Statements** option from the ***Reports*** menu.

Computerized Accounting Using Spreadsheets

Problem 8-4   Preparing a Six-Column Work Sheet

Completing the Spreadsheet

Step 1   Read the instructions for Problem 8-4 in your textbook. This problem involves preparing a six-column work sheet for Wilderness Rentals.

Step 2   Open the Glencoe Accounting: Electronic Learning Center software.

Step 3   From the Program Menu, click on the **Peachtree Complete® Accounting Software and Spreadsheet Applications** icon.

Step 4   Log onto the Management System by typing your user name and password.

Step 5   Under the **Problems & Tutorials** tab, select template 8-4 from the Chapter 8 drop-down menu. The template should look like the one shown below.

```
PROBLEM 8-4
PREPARING A SIX-COLUMN WORK SHEET

(name)
(date)


WILDERNESS RENTALS
WORK SHEET
FOR THE MONTH ENDED MAY 31, 20—
```

ACCOUNT NUMBER	ACCOUNT NAME	TRIAL BALANCE DEBIT	TRIAL BALANCE CREDIT	INCOME STATEMENT DEBIT	INCOME STATEMENT CREDIT	BALANCE SHEET DEBIT	BALANCE SHEET CREDIT
101	Cash in Bank					0.00	
105	Accounts Receivable – Helen Katz					0.00	
110	Accounts Receivable – Polk and Co.					0.00	
115	Office Supplies					0.00	
120	Office Equipment					0.00	
125	Camping Equipment					0.00	
201	Accounts Payable – Adventure Equip. Inc.						0.00
203	Accounts Payable – Digital Tech Computers						0.00
205	Accounts Payable – Greg Mollaro						0.00
301	Ronald Hicks, Capital						0.00
305	Ronald Hicks, Withdrawals					0.00	
310	Income Summary	----	----	----	----		
401	Equipment Rental Revenue				0.00		
501	Advertising Expense			0.00			
505	Maintenance Expense			0.00			
515	Rent Expense			0.00			
525	Utilities Expense			0.00			
		0.00	0.00	0.00	0.00	0.00	0.00
	Net Income			0.00			0.00
				0.00	0.00	0.00	0.00

Step 6   Key your name and today's date in the cells containing the *(name)* and *(date)* placeholders.

Step 7   The first account, Cash in Bank, has a month-end debit balance of $5,814. Move the cell pointer to cell C14 and enter the account balance into the Trial Balance section of the spreadsheet template: **5814**.

TIP: It is not necessary to include a comma or the decimal point and two zeroes as part of the amount. The spreadsheet will automatically format the data when it is entered.

Step 8 Enter the remaining balances into the Trial Balance section of the spreadsheet template. When you have entered all of the balances, move the cell pointer into the Income Statement and Balance Sheet sections of the spreadsheet template. Notice that the amounts for the Income Statement and Balance Sheet are automatically entered. As you enter the data into the Trial Balance section of the spreadsheet template, the program automatically calculates the remaining sections of the work sheet for you. The program also calculates the column totals and the net income for Wilderness Rentals.

Step 9 Save the spreadsheet using the **Save** option from the *File* menu. You should accept the default location of the save as this is handled by the management system.

Step 10 Print the completed spreadsheet.

TIP: If your spreadsheet is too wide to fit on an 8.5-inch wide piece of paper, you can change your print settings to print the worksheet *landscape*. Landscape means that the worksheet will be printed broadside on the page. Some spreadsheet applications also allow you to choose a "fit to page" option. This function will reduce the width and/or depth of the worksheet to fit on one page.

Step 11 Exit the spreadsheet program.

Step 12 In the Close Options box, select the location where you would like to save your work.

Step 13 Answer the Analyze question from your textbook for this problem.

What-If Analysis

TIP: Always save your work before performing What-If Analysis. It is not necessary to save your work after performing What-If Analysis unless your teacher instructs you to do so. If you are required to save your work after performing What-If Analysis, be sure to rename the spreadsheet to avoid saving over your original work.

If Cash in Bank were $4,314 and Rent Expense were $5,000, what would Wilderness Rentals' net income (or net loss) be?

Problem 8-8   Completing the Work Sheet

Completing the Spreadsheet

Step 1 Read the instructions for Problem 8-8 in your textbook. This problem involves completing a six-column work sheet for Job Connect.

Step 2 Open the Glencoe Accounting: Electronic Learning Center software.

Step 3 From the Program Menu, click on the **Peachtree Complete® Accounting Software and Spreadsheet Applications** icon.

Step 4 Log onto the Management System by typing your user name and password.

Step 5 Under the **Problems & Tutorials** tab, select template 8-8 from the Chapter 8 drop-down menu. The template should look like the one shown below.

```
PROBLEM 8-8
COMPLETING THE WORK SHEET

(name)
(date)


JOB CONNECT
WORK SHEET
FOR THE MONTH ENDED MAY 31, 20—
```

ACCOUNT NUMBER	ACCOUNT NAME	TRIAL BALANCE DEBIT	CREDIT	INCOME STATEMENT DEBIT	CREDIT	BALANCE SHEET DEBIT	CREDIT
101	Cash in Bank	18,972.00				18,972.00	
105	Accounts Receivable – CompuRite Systems	765.00				765.00	
110	Accounts Receivable – Marquez Manufacturing	AMOUNT				908.00	
113	Accounts Receivable – Roaring Rivers Water Park	1,268.00				AMOUNT	
115	Accounts Receivable – Melanie Spencer	AMOUNT				86.00	
120	Training Class Supplies	AMOUNT				413.00	
125	Office Supplies	3,061.00				AMOUNT	
130	Office Equipment	4,719.00				AMOUNT	
135	Office Furniture	AMOUNT				19,960.00	
140	Computer Equipment	9,382.00				AMOUNT	
201	Accounts Payable – Micro Solutions, Inc.		AMOUNT				3,019.00
205	Accounts Payable – Vega Internet Services		AMOUNT				8,397.00
207	Accounts Payable – Wildwood Furniture		AMOUNT				5,284.00
301	Richard Tang, Capital		AMOUNT				41,500.00
302	Richard Tang, Withdrawals	1,500.00				AMOUNT	
303	Income Summary	----	----	----	----		
401	Placement Fees Revenue		3,385.00		AMOUNT		
405	Technology Classes Revenue		7,600.00		7,600.00		
501	Advertising Expense	2,174.00		2,174.00			
505	Maintenance Expense	AMOUNT		1,385.00			
510	Miscellaneous Expense	AMOUNT		206.00			
520	Rent Expense	4,100.00		AMOUNT			
530	Utilities Expense	286.00		286.00			
		46,227.00	10,985.00	4,051.00	7,600.00	41,104.00	58,200.00
	Net Income			3,549.00			3,549.00
				7,600.00	7,600.00	41,104.00	61,749.00

Step 6 Key your name and today's date in the cells containing the *(name)* and *(date)* placeholders.

Step 7 The work sheet for Job Connect is given in the spreadsheet template. However, several amounts are missing from various columns. Calculate the missing amounts and enter them in the cells containing the AMOUNT placeholders. For example, the first amount missing is the Trial Balance debit amount for Accounts Receivable—Marquez Manufacturing. By looking in the Balance Sheet debit column, you can see this amount is $908.00. Enter **908** in cell C16. Remember, it is not necessary to enter a dollar sign or the decimal point and ending zeroes.

Step 8 Enter the remaining missing amounts into the work sheet. Notice that the template recalculates the column totals and the net income for Job Connect as you enter the missing amounts.

Step 9 Save the spreadsheet using the **Save** option from the *File* menu. You should accept the default location of the save as this is handled by the management system.

Step 10 Print the completed spreadsheet.

Step 11 Exit the spreadsheet program.

Step 12 In the Close Options box, select the location where you would like to save your work.

Step 13 Answer the Analyze question from your textbook for this problem.

Problem 8-6   Preparing a Six-Column Work Sheet

ACCT. NO.	ACCOUNT NAME	TRIAL BALANCE DEBIT	TRIAL BALANCE CREDIT	INCOME STATEMENT DEBIT	INCOME STATEMENT CREDIT	BALANCE SHEET DEBIT	BALANCE SHEET CREDIT
1							
2							
3							
4							
5							
6							
7							
8							
9							
10							
11							
12							
13							
14							
15							
16							
17							
18							
19							
20							
21							
22							
23							
24							
25							
26							
27							

Analyze:

Problem 8-7   Preparing a Six-Column Work Sheet

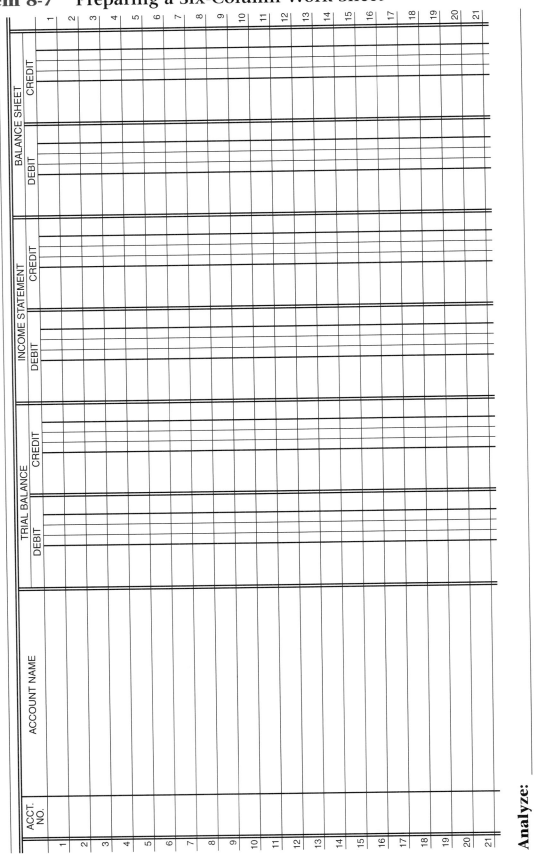

Analyze:

Problem 8-8   Completing the Work Sheet

Job Connect
Work Sheet
For the Month Ended May 31, 20--

ACCT. NO.	ACCOUNT NAME	TRIAL BALANCE DEBIT	CREDIT	INCOME STATEMENT DEBIT	CREDIT	BALANCE SHEET DEBIT	CREDIT	
101	Cash in Bank	1897200				1897200		1
105	Accts. Rec.—CompuRite Systems	76500				76500		2
110	Accts. Rec.—Marquez Manuf.	90800				90800		3
113	Accts. Rec.—Roaring Rivers	126800						4
115	Accts. Rec.—Melanie Spencer	8600				8600		5
120	Training Class Supplies	41300				41300		6
125	Office Supplies	306100						7
130	Office Equipment	471900						8
135	Office Furniture	1996000						9
140	Computer Equipment	938200						10
201	Accts. Pay.—Micro Solutions Inc.		301900				301900	11
205	Accts. Pay.—Vega Internet Svcs.		839700					12
207	Accts. Pay.—Wildwood Furniture		528400				528400	13
301	Richard Tang, Capital		4150000				4150000	14
302	Richard Tang, Withdrawals	150000						15
303	Income Summary							16
401	Placement Fees Revenue		338500					17
405	Technology Classes Revenue		760000					18
501	Advertising Expense	217400		217400				19
505	Maintenance Expense	138500		138500				20
510	Miscellaneous Expense	20600		20600				21
520	Rent Expense	410000						22
530	Utilities Expense	28600		28600				23
		6918500	6918500					24
	Net Income							25
								26
								27

Analyze:

CHAPTER 8 The Six-Column Work Sheet

Self-Test

Part A   True or False

Directions: *Circle the letter* T *in the Answer column if the statement is true; circle the letter* F *if the statement is false.*

Answer

T   F   **1.** A work sheet always covers a period of one month.

T   F   **2.** A net loss decreases the balance in the owner's equity account.

T   F   **3.** A net income for the period is the amount left after the expenses for the period have been subtracted from revenue.

T   F   **4.** Account names are listed on the work sheet in alphabetical order.

T   F   **5.** A net loss for the period is entered in the debit column of the Balance Sheet section.

T   F   **6.** Amounts from the Trial Balance section are extended first to the Income Statement section.

T   F   **7.** Total expenses for the period are reflected in the totals of the credit column of the Income Statement section and the credit column of the Balance Sheet section.

T   F   **8.** All liability accounts are listed in the credit column of the Income Statement section.

T   F   **9.** The Trial Balance section will have entries for all accounts in the general ledger including those with zero balances.

T   F   **10.** All asset accounts are extended to the Balance Sheet section.

Part B Fill in the Missing Term

Directions: *In the Answer column, write the letter of the word or phrase that best completes the sentence. Some answers may be used more than once.*

A. Balance Sheet section	**E.** extending	**I.** net income
B. capital	**F.** heading	**J.** net loss
C. credit	**G.** Income Statement section	**K.** trial balance
D. debit	**H.** matching principle	**L.** work sheet

Answer

_____ 1. The first two columns of the work sheet are used to enter the _____.

_____ 2. The permanent general ledger accounts are extended to the _____ of the work sheet.

_____ 3. A(n) _____ results when revenue is larger than expenses.

_____ 4. A net loss is entered in the _____ column of the Income Statement section.

_____ 5. If the total of the credit column of the Income Statement section is less than the debit column, there is a _____ for the period.

_____ 6. The amount of net income for the period is added to the total of the credit column of the Balance Sheet section because it increases the balance in the _____ account.

_____ 7. The _____ of the work sheet answers the questions "who," "what," and "when."

_____ 8. A paper used to collect information from the general ledger accounts is a(n) _____.

_____ 9. _____ is transferring balances from the Trial Balance section of the work sheet to either the Balance Sheet section or the Income Statement section.

_____ 10. The _____ allows a business to match revenue against expenses as a means of measuring profit for the period.

CHAPTER  Financial Statements for a Sole Proprietorship

Study Plan

Check Your Understanding

Section 1	*Read Section 1 on pages 204–208 and complete the questions and problems on page 209.* ❑ Thinking Critically ❑ Communicating Accounting ❑ Problem 9-1 *Analyzing a Source Document*
Section 2	*Read Section 2 on pages 210–213 and complete the questions and problems on page 214.* ❑ Thinking Critically ❑ Analyzing Accounting ❑ Problem 9-2 *Determining Ending Capital Balances*
Section 3	*Read Section 3 on pages 216–221 and complete the questions and problems on page 222.* ❑ Thinking Critically ❑ Communicating Accounting ❑ Problem 9-3 *Calculating Return on Sales*
Summary	*Review the Chapter 9 Summary on page 223 in your textbook.* ❑ Key Concepts
Review and Activities	*Complete the following questions and exercises on pages 224–225 in your textbook.* ❑ Using Key Terms ❑ Understanding Accounting Concepts and Procedures ❑ Case Study ❑ Conducting an Audit with Alex ❑ Internet Connection ❑ Workplace Skills
Computerized Accounting	*Read the Computerized Accounting information on page 226 in your textbook.* ❑ *Making the Transition from a Manual to a Computerized System* ❑ *Preparing Financial Statements in Peachtree*
Problems	*Complete the following end-of-chapter problems for Chapter 9 in your textbook.* ❑ Problem 9-4 *Preparing an Income Statement* ❑ Problem 9-5 *Preparing a Statement of Changes in Owner's Equity* ❑ Problem 9-6 *Preparing Financial Statements* ❑ Problem 9-7 *Preparing Financial Statements*
Challenge Problem	❑ Problem 9-8 *Preparing a Statement of Changes in Owner's Equity*
Chapter Reviews and Working Papers	*Complete the following exercises for Chapter 9 in your Chapter Reviews and Working Papers.* ❑ Chapter Review ❑ Self-Test

CHAPTER 9 REVIEW — Financial Statements for a Sole Proprietorship

Part 1   Accounting Vocabulary (24 points)

Total Points 66

Student's Score

Directions: *Using terms from the following list, complete the sentences below. Write the letter of the term you have chosen in the space provided.*

A. balance sheet	**F.** income statement	**H.** liquidity ratio	**L.** return on sales
B. current assets	**G.** Income Statement	**I.** net income	**M.** statement of changes
C. current liabilities	section of the	**J.** profitability ratios	in owner's equity
D. current ratio	work sheet	**K.** report form	**N.** work sheet
E. financial statements			

_____G_____   **0.** The source of information for completing the income statement is the _____.

_____   **1.** The financial statement that reports the net income or net loss for the fiscal period it covers is the _____.

_____   **2.** The classifications of balance sheet accounts are shown one under the other in the _____.

_____   **3.** Reports prepared to summarize the changes resulting from business transactions that have occurred during a fiscal period are called _____.

_____   **4.** A financial statement that is prepared to summarize the effects on the capital account of the various business transactions that occurred during the fiscal period is called a(n) _____.

_____   **5.** _____ occurs when total revenue is greater than total expenses.

_____   **6.** The net income or net loss amount shown on the income statement must agree with the amount shown on the _____.

_____   **7.** A financial statement that is a report of the final balances in all asset, liability, and owner's equity accounts at the end of the fiscal period is the _____.

_____   **8.** The _____ is the relationship between current assets and current liabilities.

_____   **9.** The _____ ratio is used to examine the portion of each sales dollar that represents profit.

_____   **10.** A _____ is a measure of a business's ability to pay its current debts as they become due and to provide for unexpected needs of cash.

_____   **11.** _____ are the debts of the business that must be paid within the next accounting period.

_____   **12.** _____ are assets used up or converted to cash during the normal operating cycle of the business.

_____   **13.** _____ are used to evaluate the earnings performance of the business during the accounting period.

Part 2   Determining Account Balances (17 points)

Directions: *The beginning capital balance for several different businesses appears in the first column below. The other columns list the withdrawals, investments, total revenue, total expenses, and ending capital balances for each business. Fill in the blank spaces by adding or subtracting across each line.*

	Capital Oct. 1	Withdrawals	Owner's Investment	Total Revenue	Total Expenses	Capital Oct. 31
0	$28,394	$500	$ 2,000	$7,394	*$4,203*	$33,085
1	$36,495	$300	$ 0	$4,395	$3,127	
2	$84,393	$600	$ 1,000	$5,584		$86,410
3	$52,815	$700		$8,721	$5,906	$55,780
4		$400	$ 0	$6,849	$6,127	$13,943
5	$19,302	$600	$ 0		$3,833	$18,229
6	$31,304		$ 1,000	$9,494	$8,048	$33,250

Computerized Accounting Using Peachtree

Software Objectives

When you have completed this chapter, you will be able to use Peachtree to:
1. Print a Trial Balance report.
2. Print an Income Statement.
3. Print a Balance Sheet.
4. Print a Statement of Changes in Owner's Equity.

Problem 9-4   **Preparing an Income Statement**
Problem 9-5   **Preparing a Statement of Changes in Owner's Equity**

INSTRUCTIONS

Beginning a Session

Step 1   Select the problem set: Wilderness Rentals (Prob. 9-4).
Step 2   Rename the company and set the system date to September 30, 2008.

Preparing Reports

Step 3   Print a General Ledger Trial Balance report.
Step 4   Print an Income Statement.

> *To print an Income Statement:*
>
> - Choose **Financial Statements** from the **Reports** Menu.
> - Select **Income Statement (Monthly)** shown on the report list. (See Figure 9-5A.)
> - Click [Preview].
> - Click **OK** to accept the report options.
> - Review the report.
> - Click [Print] to print the report.

DO YOU HAVE A QUESTION

Q. *Does the Peachtree software include an option to print a Statement of Changes in Owner's Equity?*

A. Yes. Peachtree does provide the capability to print the Statement of Changes in Owner's Equity report.

Notes

Peachtree lets you create customized financial statement reports. A customized report appears in the Report List with a special icon. (See Figure 9-5A.)

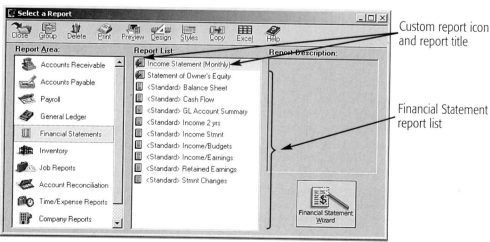

Custom report icon and report title

Financial Statement report list

Figure 9-5A   *Select a Report Window (Financial Statements)*

Step 5   Print the Balance Sheet. Select it from the **Financial Statements** option of the **Reports** menu.

Step 6   Answer the Analyze questions shown in your textbook.

TIP: Use the calculator accessory available on your computer to calculate the return on sales and the current ratio.

Ending the Session

Step 7   Click the **Close Problem** button in the Glencoe Smart Guide window.

Problem 9-6   Preparing Financial Statements

INSTRUCTIONS

Beginning a Session

Step 1   Select the problem set: Hot Suds Car Wash (Prob. 9-6).

Step 2   Rename the company and set the system date to September 30, 2008.

Preparing Reports

Step 3   Print a General Ledger Trial Balance report.

Step 4   Print the quarterly Income Statement report, not the <standard> report.

Step 5   Print the <Standard> Balance Sheet.

Step 6   Answer the Analyze question shown in your textbook.

Ending the Session

Step 7   Click the **Close Problem** button in the Glencoe Smart Guide window.

Mastering Peachtree

Customize the <standard> balance sheet to summarize the financial information. Change the layout of the report to combine all assets into one section. Also, combine the liabilities into one section. Print your results.

Problem 9-7   Preparing Financial Statements

INSTRUCTIONS

Beginning a Session

Step 1   Select the problem set: Kits & Pups Grooming (Prob. 9-7).

Step 2   Rename the company and set the system date to September 30, 2008.

Preparing Reports

Step 3   Print a General Ledger Trial Balance report.

Step 4   Print an Income Statement.

Step 5   Print a Balance Sheet.

Step 6   Print a Statement of Changes in Owner's Equity.

You can use the <Standard> Retained Earnings report and change the report title (customize) to Statement of Changes in Owner's Equity for all sole proprietorships and partnerships.

DO YOU HAVE A QUESTION

Q. *Can you use the <standard> income statement instead of the customized report?*

A. Yes. You can use the <standard> income statement. The only difference, however, is that the <standard> report includes year-to-date figures. For the problems in this chapter, the current month and year-to-date figures are the same.

Notes

A statement of changes in owner's equity can be customized from Peachtree's <Standard> Retained Earnings report.

How to change a financial statement report title:

- Choose **Financial Statements** from the **Reports** Menu.
- Select **<Standard> Retained Earnings** from the report list.
- Click **Design** button.
- Double-click **Text-Header** (for report title field).

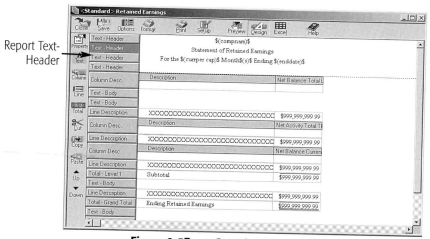

Figure 9-5B   *<Standard> Retained Earnings*

- In **Text to Print** field change report title to Statement of Changes in Owner's Equity.

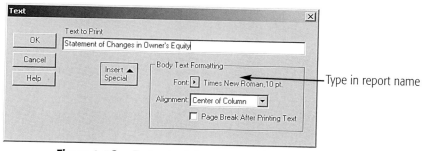

Figure 9-5C   *Enter text for report title*

- Click **OK**.
- Click **Preview**.
- Select period and print options and click **OK**.
- Click **Print** to print the report.
- Click **Close**, then **No**. You do not need to save changes to the modified report.

You can prepare the Statement of Changes in Owner's Equity manually by using capital balances from the Balance Sheet. Your instructor will tell you whether to prepare the Statement of Changes in Owner's Equity manually or using Peachtree Complete® Accounting.

Step 7   Answer the Analyze question shown in your textbook.

Ending the Session

Step 8   Click the **Close Problem** button in the Glencoe Smart Guide window.

Computerized Accounting Using Spreadsheets

Problem 9-8   Preparing a Statement of Changes in Owner's Equity

Completing the Spreadsheet

Step 1   Read the instructions for Problem 9-8 in your textbook. This problem involves preparing a statement of changes in owner's equity.

Step 2   Open the Glencoe Accounting: Electronic Learning Center software.

Step 3   From the Program Menu, click on the **Peachtree Complete®** **Accounting Software and Spreadsheet Applications** icon.

Step 4   Log onto the Management System by typing your user name and password.

Step 5   Under the **Problems & Tutorials** tab, select template 9-8 from the Chapter 9 drop-down menu. The template should look like the one shown below.

```
PROBLEM 9-8
PREPARING A STATEMENT OF CHANGES IN OWNER'S EQUITY


(name)
(date)


                        OUTBACK GUIDE SERVICE
              STATEMENT OF CHANGES IN OWNER'S EQUITY
                 FOR THE MONTH ENDED SEPTEMBER 30, 20--
                                                                 0.00
Beginning Capital, September 1, 20--                   AMOUNT
Add: Investments by owner                              AMOUNT
     Net Income                                                  0.00
Total Increase in Capital                                        0.00
Subtotal
Less: Withdrawals by owner                                     AMOUNT
Ending Capital, September 30, 20--                            AMOUNT
```

Step 6   Key your name and today's date in the cells containing the *(name)* and *(date)* placeholders.

Step 7   Enter the investments by owner, net income, withdrawals by owner, and ending capital in the cells containing the AMOUNT placeholders. Remember, it is not necessary to add the decimal point and ending zeroes. The beginning capital and the total increase in capital will be automatically computed. When you have finished entering the amounts, ask your teacher to check your work.

Step 8   Save the spreadsheet using the **Save** option from the *File* menu. You should accept the default location of the save as this is handled by the management system.

Step 9   Print the completed spreadsheet.

Step 10   Exit the spreadsheet program.

Step 11   In the Close Options box, select the location where you would like to save your work.

Step 12   Answer the Analyze question from your textbook for this problem.

Notes

Remember, always save your work before performing What-If Analysis. Be sure to rename the spreadsheet if you save your work after performing What-If Analysis.

What-If Analysis

If the owner withdrew $3,200 during the period, what would beginning capital be?

Problem 9-7 Preparing Financial Statements
(1)

Kits and pups Grooming
Worksheet
For the month Ended Sept. 30, 2012

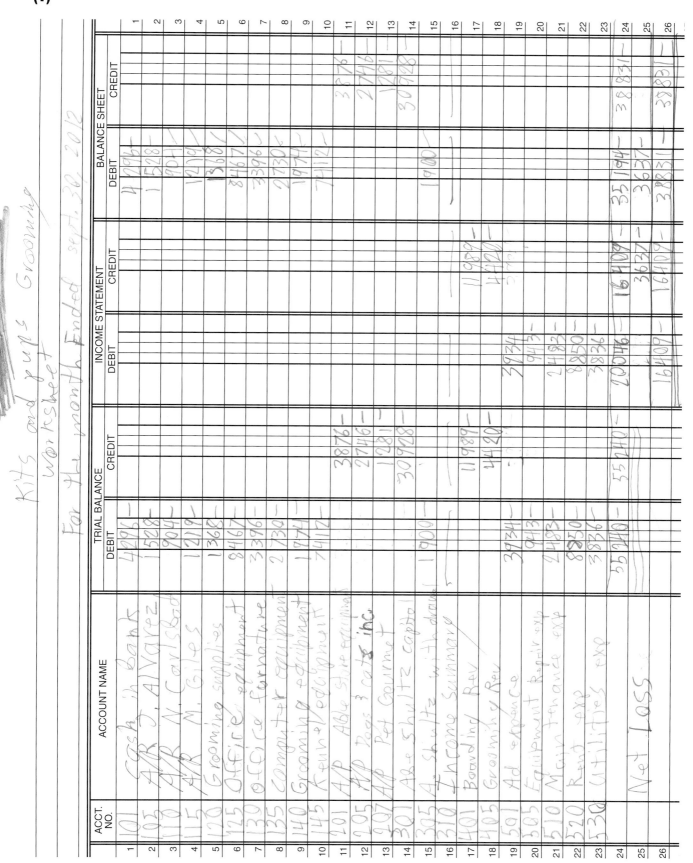

	ACCT. NO.	ACCOUNT NAME	TRIAL BALANCE DEBIT	TRIAL BALANCE CREDIT	INCOME STATEMENT DEBIT	INCOME STATEMENT CREDIT	BALANCE SHEET DEBIT	BALANCE SHEET CREDIT
1	101	Cash in Bank	4296				4296	
2	105	A/R J. Alvarez	1528				1528	
3	110	A/R N. Carlsbad	904				904	
4	115	A/R M. Giles	1219				1219	
5	120	Grooming supplies	1368				1368	
6	125	Office equipment	8467				8467	
7	130	Office furniture	3396				3396	
8	135	Computer equipment	2730				2730	
9	140	Grooming equipment	1974				1974	
10	145	Kennel equipment	7412				7412	
11	201	A/P Able Store equipment		3876				3876
12	205	A/P Dogs & Cats Inc		2746				2746
13	207	A/P Pet Gourmet		1281				1281
14	301	Abe Shultz Capital		30928				30928
15	305	Abe Shultz Withdrawal	1900				1900	
16	310	Income Summary						
17	401	Boarding Rev		11989		11989		
18	405	Grooming Rev		4420		4420		
19	501	Ad expense	3934		3934			
20	505	Equipment Repair exp	943		943			
21	510	Maintenance exp	2483		2483			
22	520	Rent exp	8850		8850			
23	530	Utilities exp	3836		3836			
24			55240	55240	20046	16409	35194	38831
25		Net Loss				3637	3637	
26					16409	16409	38831	38831

Problem 9-7   (continued)

(2)

Kits and Pups Grooming
income statement
For the month Ended

(3)

Problem 9-7 (concluded)

(4)

Analyze: _____

_____

Problem 9-8 Preparing a Statement of Changes in Owner's Equity

Analyze: _____

_____

CHAPTER  Financial Statements for a Sole Proprietorship

Self-Test

Part A   True or False

Directions: *Circle the letter* T *in the Answer column if the statement is true; circle the letter* F *if the statement is false.*

Answer

T   F   **1.** The Trial Balance section of the work sheet provides the information used in preparing the income statement.

T   F   **2.** The changes in the Cash in Bank account are reported in the statement of changes in owner's equity.

T   F   **3.** The balance sheet reports the final balances of the permanent accounts at the end of the fiscal period.

T   F   **4.** The balance sheet is prepared before the statement of changes in owner's equity.

T   F   **5.** The income statement represents the basic accounting equation.

T   F   **6.** A net income will increase the owner's capital account.

T   F   **7.** The heading is the same on all three financial statements.

T   F   **8.** The statement of changes in owner's equity summarizes the effects on the capital account of the various business transactions that occurred during the period.

T   F   **9.** The primary financial statements prepared for a sole proprietorship are the income statement and the balance sheet.

T   F   **10.** The information on the statement of changes in owner's equity is used in preparing the income statement.

Part B   Fill in the Missing Term

Directions: *In the Answer column, write the letter of the word or phrase that best completes the sentence. Some answers may be used more than once.*

A. balance sheet	**E.** Income Statement section of the work sheet	**H.** on a specific date
B. Balance Sheet section of the work sheet	**F.** heading	**I.** report form
C. financial statements	**G.** net income or net loss	**J.** statement of changes in owner's equity
D. income statement		**K.** work sheet

Answer

_____   **1.** The _____ is completed as a support document for the balance sheet.

_____   **2.** The balance sheet reports financial information _____.

_____   **3.** The information needed to prepare the income statement comes from the _____.

_____   **4.** _____ is reported on the income statement.

_____   **5.** The amount of net income or net loss reported on the income statement must match the amount shown on the _____.

_____   **6.** _____ summarize the changes resulting from business transactions that have occurred during a fiscal period.

_____   **7.** A(n) _____ is the financial statement that reports the final balances in all asset, liability, and owner's equity accounts at the end of the fiscal period.

_____   **8.** In the _____, the classification of balance sheet accounts are shown one under the other.

_____   **9.** The _____ reports a business's net income or net loss over an entire fiscal period.

_____   **10.** The source of information for completing the balance sheet comes from the work sheet and the _____.

_____   **11.** The _____ of a financial statement answers the questions: Who?, What?, and When?

Computerized Accounting Using Peachtree

Software Objectives

When you have completed this chapter, you will be able to use Peachtree to:
1. Close the current fiscal year.
2. Print a Post-Closing Trial Balance.

Problem 10-4   Preparing Closing Entries

INSTRUCTIONS

Beginning a Session

Step 1 Select the problem set: Wilderness Rentals (Prob. 10-4).
Step 2 Rename the company and set the system date to December 31, 2008.

Completing the Accounting Problem

Step 3 Perform the closing process.

To close the fiscal year:

1. Choose **System** from the *Tasks* menu and then choose **Year-End Wizard**.
2. **Year-End Wizard – Welcome** displays current open fiscal and payroll years. Click **Next** to continue.
3. **Year-End Wizard – Close Options** shows **Years to Close** for Fiscal and Payroll years. Click **Next** to continue.
4. Peachtree lets you print the General Ledger report before you continue the closing process if the problem has transactions recorded. **Year-End Wizard – Unprinted Item Warning** displays unprinted items. You do not have to print these reports unless instructed otherwise. Click **Next** to continue.
5. **Year-End Wizard – Reports** displays year-end reports. You do not have to print these reports. Deselect by removing check mark from Print box. Click **Next** to continue.
6. **Year-End Wizard – Back Up** is required in Peachtree. Save work to floppy or network. Click **Back Up** to continue.
7. **Back Up Company** will back up your company data and customized forms. Click **Back Up Now** to continue.
8. **Save Backup for Wilderness (Prob. 10-4) as:** displays the file name for backup and places it in the Wilderness (Prob. 10-4) folder as a backup (.ptb) file. Click **Save** to continue.
9. **Peachtree Accounting** alerts you to space requirements for the backup file. Click **OK** to continue.
10. **Year-End Wizard – Back Up** Peachtree recommends two backup files before closing the year. However, only one backup is required for this problem. Click **Next** to continue.
11. **Year-End Wizard – New Open Fiscal Years** displays the (12) periods in the new **Current Fiscal Year** (Jan. 1, 2009 to Dec. 31, 2009) and the (12) periods in the new **Next Fiscal Year**. Click **Next** to continue.
12. **Year-End Wizard – Confirm Close** displays the years that will close and the new open years. Click **Next** to continue.

DO YOU HAVE A QUESTION

Q. *Should you create a backup before closing the fiscal year?*
A. Yes. It is required that you make a backup of your work. You should do this because you cannot access transactions from a closed period. Also, the company data files may be damaged if the closing process is interrupted. Taking a few moments to make a backup could save you many hours of work in the event that your data files are accidentally damaged.
 If a problem occurs, you can always restore the original problem set.

Notes

 You do not have to manually enter the closing entries since the Peachtree software automatically performs the closing process.

13. **Year-End Wizard – Begin Close** allows you to close the years you have selected. **Important! You cannot cancel Year-End Wizard after this point!** Click **Begin Close** to continue. Peachtree creates the closing entries and automatically switches to the next fiscal year (Jan. 1 to Dec. 31, 2010).

14. **Year-End Wizard – Congratulations!** displays successfully closed years and newly opened years. Click **Finish** to exit the Year-End Wizard. (See Figure 10-4A.)

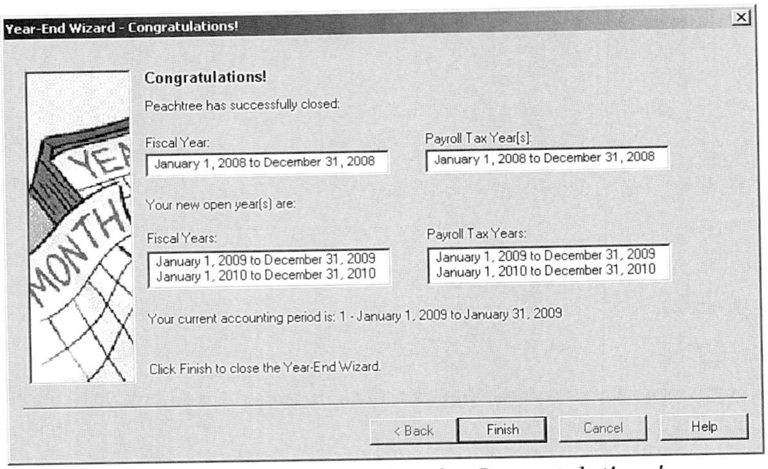

Figure 10-4A *Year-End Wizard—Congratulations!*

Notes

Display a General Ledger report or a General Ledger Trial Balance to view an account balance after you perform the closing process.

Step 4 Change the system date to January 1, 2009.
Step 5 Answer the Analyze question.

Ending the Session

Step 6 Click the **Close Problem** button in the Glencoe Smart Guide window.

Mastering Peachtree

Review the online help information to learn how to restore the files for the company using a backup floppy disk.

Problem 10-5 Preparing a Post-Closing Trial Balance

INSTRUCTIONS

Beginning a Session

Step 1 Select the problem set: Hot Suds Car Wash (Prob. 10-5).
Step 2 Rename the company and set the system date to January 1, 2009.

Preparing a Report

Step 3 Print a Post-Closing Trial Balance.
Step 4 Answer the Analyze question.

Ending the Session

Step 5 Click the **Close Problem** button in the Glencoe Smart Guide window.

Mastering Peachtree

Print a General Ledger Trial Balance, but change the report title to *Post-Closing Trial Balance*. Review the online help to learn how to change a report title.

DO YOU HAVE A QUESTION?

Q. *How do you print a post-closing trial balance since Peachtree does not include this report?*

A. Peachtree does not include a standard report called Post-Closing Trial Balance, but the standard General Ledger Trial Balance provides the same information. The only difference is the title of the report which you can change using the reports option. Click Options then Font tab and change the Title 1 Report Label.

Problem 10-6   Journalizing Closing Entries

INSTRUCTIONS

Beginning a Session

Step 1   Select the problem set: Kits & Pups Grooming (Prob. 10-6).

Step 2   Rename the company and set the system date to December 31, 2008.

Completing the Accounting Problem

Step 3   Perform the closing process.

TIP: Review the steps presented in Problem 10-4 if you need help closing the fiscal year.

Step 4   Change the system date to January 1, 2009.

Step 5   Answer the Analyze question.

Ending the Session

Step 6   Click the **Close Problem** button in the Glencoe Smart Guide window.

DO YOU HAVE A QUESTION

Q. *Can you undo the closing process?*

A. No. Once you close a fiscal year, you cannot undo the process and reset the company data files. For this reason, Peachtree requires that you backup your work. Before you close a fiscal year, you should proof the transactions, print summary reports, and print the financial statements for the current period.

Problem 10-7   Posting Closing Entries and Preparing a Post-Closing Trial Balance

INSTRUCTIONS

Beginning a Session

Step 1   Select the problem set: Outback Guide Service (Prob. 10-7).

Step 2   Rename the company and set the system date to December 31, 2008.

Completing the Accounting Problem

Step 3   Perform the closing process.

TIP: Review the steps presented in Problem 10-4 if you need help closing the fiscal year.

Step 4   Change the system date to January 1, 2009.

Step 5   Print a Post-Closing Trial Balance.

Step 6   Answer the Analyze question.

Ending the Session

Step 7   Click the **Close Problem** button in the Glencoe Smart Guide window.

DO YOU HAVE A QUESTION

Q. *Do you have to manually post the closing entries?*

A. No. When you choose to close the fiscal year, Peachtree generates the closing entries behind the scenes for you. The Peachtree software also automates the posting process just as it automatically posts the entries when you record a general journal transaction.

Notes

Choose to print the General Ledger Trial Balance whenever you need a Post-Closing Trial Balance.

Problem 10-8   Completing End-of-Period Activities

INSTRUCTIONS

Beginning a Session

Step 1   Select the problem set: Showbiz Video (Prob. 10-8).

Step 2   Rename the company and set the system date to December 31, 2008.

Completing the Accounting Problem

Step 3   Print a General Ledger Trial Balance.

Step 4   Print the financial statements.

Step 5   Perform the closing process.

Step 6   Change the system date to January 1, 2009.

Step 7   Print a Post-Closing Trial Balance.

Step 8   Answer the Analyze question.

Ending the Session

Step 9   Click the **Close Problem** button in the Glencoe Smart Guide window.

═════════════ FAQs ═════════════

What if you notice a mistake after you close the fiscal year?

You have two options if you notice a mistake after you close the fiscal year. If you made a backup before the closing, you can restore the company data, correct the error, and then close the fiscal year again. If you do not have the backup or you do not want to restore the company data, make a correcting journal entry in the new period.

Can you undo the closing process?

No. Once you close a fiscal year, you cannot undo the process and reset the company data files. For this reason, Peachtree requires that you backup your work. Before you close a fiscal year, you should proof the transactions, print summary reports, and print the financial statements for the current period.

How do you print a post-closing trial balance since Peachtree does not include this report?

Peachtree does not include a standard report called Post-Closing Trial Balance, but the standard General Ledger Trial Balance provides the same information. The only difference is the title of the report which you can change using the reports option. Click Options then Font tab and change the Title 1 Report Label.

Computerized Accounting Using Spreadsheets

Problem 10-9 Completing End-of-Period Activities

Completing the Spreadsheet

Step 1 Read the instructions for Problem 10-9 in your textbook. This problem involves preparing a six-column work sheet and the end-of-period financial statements for Job Connect.

Step 2 Open the Glencoe Accounting: Electronic Learning Center software.

Step 3 From the Program Menu, click on the **Peachtree Complete®** **Accounting Software and Spreadsheet Applications** icon.

Step 4 Log onto the Management System by typing your user name and password.

Step 5 Under the **Problems & Tutorials** tab, select template 10-9 from the Chapter 10 drop-down menu. The template should look like the one shown below.

```
PROBLEM 10-9
COMPLETING END-OF-PERIOD ACTIVITIES

(name)
(date)


JOB CONNECT
WORK SHEET
FOR THE MONTH ENDED DECEMBER 31, 20--
```

ACCOUNT NUMBER	ACCOUNT NAME	TRIAL BALANCE DEBIT	CREDIT	INCOME STATEMENT DEBIT	CREDIT	BALANCE SHEET DEBIT	CREDIT
101	Cash in Bank					0.00	
105	Accounts Receivable – CompuRite Systems					0.00	
110	Accounts Receivable – Marquez Manufacturing					0.00	
113	Accounts Receivable – Roaring Rivers Water Park					0.00	
115	Accounts Receivable – M. Spencer					0.00	
130	Office Equipment					0.00	
135	Office Furniture					0.00	
140	Computer Equipment					0.00	
201	Accounts Payable – Micro Solutions Inc.						0.00
205	Accounts Payable – Vega Internet Services						0.00
207	Accounts Payable – Wildwood Furniture Sales						0.00
301	Richard Tang, Capital						0.00
302	Richard Tang, Withdrawals					0.00	
303	Income Summary	----	----	----	----		
401	Placement Fees Revenue				0.00		
405	Technology Classes Revenue				0.00		
501	Advertising Expense			0.00			
505	Maintenance Expense			0.00			
510	Miscellaneous Expense			0.00			
520	Rent Expense			0.00			
530	Utilities Expense			0.00			
		0.00	0.00	0.00	0.00	0.00	0.00
	Net Income			0.00			0.00
				0.00	0.00	0.00	0.00

Step 6 Key your name and today's date in the cells containing the *(name)* and *(date)* placeholders.

Step 7 The first account, Cash in Bank, has a month-end balance of $6,000. Move the cell pointer to cell C14 and enter the account balance into the Trial Balance debit column of the work sheet: **6000**.

Step 8 Enter the remaining balances into the Trial Balance section of the work sheet. When you have entered all of the balances, move the cell pointer into the Income Statement and Balance Sheet sections of the work sheet. Notice that the amounts for the Income Statement and Balance Sheet are automatically entered. As you enter the data into the Trial Balance section of the work sheet, the program automatically calculates the remaining sections of the work sheet for you. The program also calculates the column totals and the net income for Job Connect.

Step 9 Now scroll down below the work sheet and look at the income statement, statement of changes in owner's equity, and balance sheet for Job Connect. Notice the financial statements are already completed. This is because the spreadsheet template includes formulas that automatically pull information from the filled-in work sheet to complete the financial statements.

Step 10 Now scroll down below the balance sheet and complete the closing entries in the general journal. The account names and posting references are given for you.

Step 11 Scroll down below the closing entries and look at the post-closing trial balance. The amounts have been automatically calculated using formulas.

Step 12 Save the spreadsheet using the **Save** option from the *File* menu. You should accept the default location for the save as this is handled by the management system.

Step 13 Print the completed spreadsheet.

TIP: When printing a long spreadsheet with multiple parts, you may want to insert page breaks between the sections so that each one begins printing at the top of a new page. Page breaks have already been entered into this spreadsheet template. Check your program's Help file for instructions on how to enter page breaks.

Step 14 Exit the spreadsheet program.

Step 15 In the Close Options box, select the location where you would like to save your work.

Step 16 Answer the Analyze question from your textbook for this problem.

What-If Analysis

TIP: Remember, always save your work before performing What-If Analysis. Be sure to rename the spreadsheet if you save your work after performing What-If Analysis.

If Cash in Bank were $5,000 and Advertising Expense were $4,000, what would Job Connect's net income be? What would Job Connect's ending capital be?

_____

_____

Problem 10-7 Posting Closing Entries and Preparing a Post-Closing Trial Balance

(1)

GENERAL LEDGER

ACCOUNT **Cash in Bank** ACCOUNT NO. **101**

DATE		DESCRIPTION	POST. REF.	DEBIT	CREDIT	BALANCE DEBIT	BALANCE CREDIT
20--							
Dec.	31	Balance	✓			1200000	

ACCOUNT **Accounts Receivable—Mary Johnson** ACCOUNT NO. **105**

DATE		DESCRIPTION	POST. REF.	DEBIT	CREDIT	BALANCE DEBIT	BALANCE CREDIT
20--							
Dec.	31	Balance	✓			60000	

ACCOUNT **Accounts Receivable—Feldman, Jones & Ritter** ACCOUNT NO. **110**

DATE		DESCRIPTION	POST. REF.	DEBIT	CREDIT	BALANCE DEBIT	BALANCE CREDIT
20--							
Dec.	31	Balance	✓			100000	

ACCOUNT **Accounts Receivable—Podaski Systems Inc.** ACCOUNT NO. **115**

DATE		DESCRIPTION	POST. REF.	DEBIT	CREDIT	BALANCE DEBIT	BALANCE CREDIT
20--							
Dec.	31	Balance	✓			90000	

ACCOUNT **Office Equipment** ACCOUNT NO. **130**

DATE		DESCRIPTION	POST. REF.	DEBIT	CREDIT	BALANCE DEBIT	BALANCE CREDIT
20--							
Dec.	31	Balance	✓			550000	

ACCOUNT **Office Furniture** ACCOUNT NO. **135**

DATE		DESCRIPTION	POST. REF.	DEBIT	CREDIT	BALANCE DEBIT	BALANCE CREDIT
20--							
Dec.	31	Balance	✓			740000	

Problem 10-7 (continued)

ACCOUNT __Computer Equipment_____ ACCOUNT NO. __140__

DATE		DESCRIPTION	POST. REF.	DEBIT	CREDIT	BALANCE	
						DEBIT	CREDIT
20--							
Dec.	31	Balance	✓			7 1 0 0 00	

ACCOUNT __Hiking Equipment_____ ACCOUNT NO. __145__

DATE		DESCRIPTION	POST. REF.	DEBIT	CREDIT	BALANCE	
						DEBIT	CREDIT
20--							
Dec.	31	Balance	✓			15 0 0 0 00	

ACCOUNT __Rafting Equipment_____ ACCOUNT NO. __150__

DATE		DESCRIPTION	POST. REF.	DEBIT	CREDIT	BALANCE	
						DEBIT	CREDIT
20--							
Dec.	31	Balance	✓			30 0 0 0 00	

ACCOUNT __Accounts Payable—A-1 Adventure Warehouse__ ACCOUNT NO. __201__

DATE		DESCRIPTION	POST. REF.	DEBIT	CREDIT	BALANCE	
						DEBIT	CREDIT
20--							
Dec.	31	Balance	✓				12 0 0 0 00

ACCOUNT __Accounts Payable—Peak Equipment Inc.__ ACCOUNT NO. __205__

DATE		DESCRIPTION	POST. REF.	DEBIT	CREDIT	BALANCE	
						DEBIT	CREDIT
20--							
Dec.	31	Balance	✓				9 0 0 0 00

ACCOUNT __Accounts Payable—Premier Processors__ ACCOUNT NO. __207__

DATE		DESCRIPTION	POST. REF.	DEBIT	CREDIT	BALANCE	
						DEBIT	CREDIT
20--							
Dec.	31	Balance	✓				6 0 0 0 00

Problem 10-7   (continued)

ACCOUNT __Juanita Ortega, Capital__   ACCOUNT NO. ___301___

DATE		DESCRIPTION	POST. REF.	DEBIT	CREDIT	BALANCE DEBIT	BALANCE CREDIT
20--							
Dec.	31	Balance	✓				5020000

ACCOUNT __Juanita Ortega, Withdrawals__   ACCOUNT NO. ___302___

DATE		DESCRIPTION	POST. REF.	DEBIT	CREDIT	BALANCE DEBIT	BALANCE CREDIT
20--							
Dec.	31	Balance	✓			400000	

ACCOUNT __Income Summary__   ACCOUNT NO. ___310___

DATE	DESCRIPTION	POST. REF.	DEBIT	CREDIT	BALANCE DEBIT	BALANCE CREDIT

ACCOUNT __Guide Service Revenue__   ACCOUNT NO. ___401___

DATE		DESCRIPTION	POST. REF.	DEBIT	CREDIT	BALANCE DEBIT	BALANCE CREDIT
20--							
Dec.	31	Balance	✓				1630000

ACCOUNT __Advertising Expense__   ACCOUNT NO. ___501___

DATE		DESCRIPTION	POST. REF.	DEBIT	CREDIT	BALANCE DEBIT	BALANCE CREDIT
20--							
Dec.	31	Balance	✓			300000	

ACCOUNT __Maintenance Expense__   ACCOUNT NO. ___505___

DATE		DESCRIPTION	POST. REF.	DEBIT	CREDIT	BALANCE DEBIT	BALANCE CREDIT
20--							
Dec.	31	Balance	✓			110000	

Problem 10-7 (concluded)

ACCOUNT __Rent Expense__ ACCOUNT NO. __515__

DATE		DESCRIPTION	POST. REF.	DEBIT	CREDIT	BALANCE DEBIT	BALANCE CREDIT
20--							
Dec.	31	Balance	✓			400000	

ACCOUNT __Utilities Expense__ ACCOUNT NO. __525__

DATE		DESCRIPTION	POST. REF.	DEBIT	CREDIT	BALANCE DEBIT	BALANCE CREDIT
20--							
Dec.	31	Balance	✓			190000	

(2)

Analyze: _____

Problem 10-8   Completing End-of-Period Activities

(1)

ACCT. NO.	ACCOUNT NAME	TRIAL BALANCE		INCOME STATEMENT		BALANCE SHEET	
		DEBIT	CREDIT	DEBIT	CREDIT	DEBIT	CREDIT
1							
2							
3							
4							
5							
6							
7							
8							
9							
10							
11							
12							
13							
14							
15							
16							
17							
18							
19							
20							
21							
22							
23							
24							
25							
26							
27							

Problem 10-8 (continued)

(2)

Problem 10-9   Completing End-of-Period Activities

(1)

ACCT. NO.	ACCOUNT NAME	TRIAL BALANCE DEBIT	TRIAL BALANCE CREDIT	INCOME STATEMENT DEBIT	INCOME STATEMENT CREDIT	BALANCE SHEET DEBIT	BALANCE SHEET CREDIT
1							
2							
3							
4							
5							
6							
7							
8							
9							
10							
11							
12							
13							
14							
15							
16							
17							
18							
19							
20							
21							
22							
23							
24							
25							

Problem 10-9   (continued)

(2)

CHAPTER 10 Completing the Accounting Cycle for a Sole Proprietorship

Self-Test

Part A   True or False

Directions: *Circle the letter* T *in the Answer column if the statement is true; circle the letter* F *if the statement is false.*

Answer

T   F   **1.** Revenue and expense accounts must be closed out because their balances apply to only one fiscal period.

T   F   **2.** Closing entries transfer the net income or net loss to the withdrawals account.

T   F   **3.** To close a revenue account, debit it for the amount of its credit balance.

T   F   **4.** When expense accounts are closed, the Income Summary account is credited.

T   F   **5.** Before closing entries are journalized and posted, the Income Summary account in the general ledger has a normal credit balance.

T   F   **6.** The Income Summary account is a simple income statement in the ledger.

T   F   **7.** After the closing entries have been posted, the balance in the capital account reflects the net income or net loss and the withdrawals for the period.

T   F   **8.** The Income Summary account is located in the owner's equity section of the general ledger.

T   F   **9.** Closing the revenue account is the second closing entry.

T   F   **10.** If a business reports a net loss for the period, the journal entry to close the Income Summary account would be a debit to capital and a credit to Income Summary.

T   F   **11.** The last step in the accounting cycle is the preparation of the post-closing trial balance.

T   F   **12.** To close the withdrawals account, the amount of its balance is debited to the capital account and credited to the withdrawals account.

Part B Multiple Choice

Directions: *Only one of the choices given with each of the following statements is correct. Write the letter of the correct answer in the Answer column.*

Answer

_____ 1. Which of the following accounts does *not* require a closing entry?
(A) Fees.
(B) Income Summary.
(C) Maintenance Expense.
(D) Klaus Braun, Capital.

_____ 2. Transferring the expense account balances to the Income Summary account is the
(A) first closing entry.
(B) second closing entry.
(C) third closing entry.
(D) fourth closing entry.

_____ 3. Accounts that start each new fiscal period with zero balances are
(A) permanent.
(B) assets.
(C) liabilities.
(D) temporary.

_____ 4. Which of the following is a true statement?
(A) The Income Summary account is located in the owner's equity section of the general ledger.
(B) The Income Summary account has a normal balance on the debit side.
(C) The Income Summary account is a permanent account.
(D) The Income Summary account is used throughout the accounting period.

_____ 5. The balance of the revenue account is transferred to the
(A) debit side of the Cash in Bank account.
(B) credit side of the owner's capital account.
(C) credit side of the Income Summary account.
(D) debit side of the owner's withdrawals account.

_____ 6. If a business has a net income for the period, the journal entry to close the balance of the Income Summary account is
(A) a debit to owner's capital, a credit to Income Summary.
(B) a debit to Fees, a credit to owner's capital.
(C) a debit to Income Summary, a credit to owner's capital.
(D) a debit to owner's capital, a credit to Fees.

CHAPTER 11 Cash Control and Banking Activities

Study Plan

Check Your Understanding

Section 1	Read Section 1 on pages 258–263 and complete the following exercises on page 264.
	❏ Thinking Critically
	❏ Communicating Accounting
	❏ Problem 11-1 *Preparing a Deposit Slip and Writing Checks*
Section 2	Read Section 2 on pages 265–272 and complete the following exercises on page 273.
	❏ Thinking Critically
	❏ Computing in the Business World
	❏ Problem 11-2 *Analyzing a Source Document*
Summary	Review the Chapter 11 Summary on page 275 in your textbook.
	❏ Key Concepts
Review and Activities	Complete the following questions and exercises on pages 276–277 in your textbook.
	❏ Using Key Terms
	❏ Understanding Accounting Concepts and Procedures
	❏ Case Study
	❏ Conducting an Audit with Alex
	❏ Internet Connection
	❏ Workplace Skills
Computerized Accounting	Read the Computerized Accounting information on page 278 in your textbook.
	❏ Making the Transition from a Manual to a Computerized System
	❏ Reconciling the Bank Statement in Peachtree
Problems	Complete the following end-of-chapter problems for Chapter 11 in your textbook.
	❏ Problem 11-3 *Handling Deposits*
	❏ Problem 11-4 *Maintaining the Checkbook*
	❏ Problem 11-5 *Reconciling the Bank Statement*
	❏ Problem 11-6 *Reconciling the Bank Statement*
	❏ Problem 11-7 *Reconciling the Bank Statement*
Challenge Problem	❏ Problem 11-8 *Reconciling the Bank Statement Using the Account Form*
Chapter Reviews and Working Papers	Complete the following exercises for Chapter 11 in your Chapter Reviews and Working Papers.
	❏ Chapter Review
	❏ Self-Test

CHAPTER 11 REVIEW
Cash Control and Banking Activities

Part 1 Accounting Vocabulary (21 points)

Total Points 40
Student's Score

Directions: *Using terms from the following list, complete the sentences below. Write the letter of the term you have chosen in the space provided.*

A. bank service charge	**G.** deposit slip	**M.** NSF check	**R.** restrictive endorsement
B. bank statement	**H.** drawee	**N.** outstanding checks	**S.** signature card
C. canceled checks	**I.** drawer	**O.** outstanding deposits	**T.** stop payment order
D. check	**J.** endorsement	**P.** payee	**U.** voiding a check
E. checking account	**K.** external controls	**Q.** reconciling the	**V.** electronic funds transfer
F. depositor	**L.** internal controls	bank statement	

_____*I*_____ **0.** The person who signs a check is the _____.

_____ **1.** The process of determining any differences between the balance shown on the bank statement and the checkbook balance is known as _____.

_____ **2.** Checks that are paid by the bank, deducted from the depositor's account, and returned with the bank statement are called _____.

_____ **3.** The bank on which a check is written is called the _____.

_____ **4.** The person or business to whom a check is written is the _____.

_____ **5.** A(n) _____ is a bank form on which the currency (bills and coins) and checks to be deposited are listed.

_____ **6.** A(n) _____ is an authorized signature that is written or stamped on the back of a check.

_____ **7.** A(n) _____ limits how a check may be handled and protects a check from being cashed by anyone except the payee.

_____ **8.** A written order from a depositor telling the bank to pay cash to a person or business is a(n) _____.

_____ **9.** A(n) _____ is a demand by the depositor that the bank not honor a certain check.

_____ **10.** A bank account that allows a bank customer to deposit cash and to write checks against the account balance is a(n) _____.

_____ **11.** Checks that have been written but not yet presented to the bank for payment are called _____.

_____ **12.** A fee charged to the depositor by the bank for maintaining bank records and for processing bank statement items is a(n) _____.

_____ **13.** Deposits that have been made and recorded in the checkbook but that do not appear on the bank statement are called _____.

_____ **14.** A person or business that has cash on deposit in a bank is a(n) _____.

_____ **15.** The card containing the signature(s) of the person(s) authorized to write checks on the bank account is called a(n) _____.

_____ **16.** Those steps the business takes to protect cash are called _____.

_____ **17.** A check returned by the bank because there are not enough funds in the drawer's checking account to cover the amount of the check is a(n) _____.

_____ **18.** Those controls on cash from outside the business are called _____.

_____ **19.** Writing the word "VOID" across the front of a check is known as _____.

_____ **20.** A(n) _____ is an itemized record of all the transactions occurring in a depositor's account over a given period, usually a month.

_____ **21.** _____ allows banks to transfer funds between accounts without the exchange of paper checks.

Computerized Accounting Using Peachtree

Software Objectives

When you have completed this chapter, you will be able to use Peachtree to:

1. Record a payment.
2. Use the **Account Reconciliation** feature to reconcile a bank statement.
3. Adjust an account.
4. Print the Account Reconciliation reports.

Problem 11-4   Maintaining the Checkbook

INSTRUCTIONS

Beginning a Session

Step 1   Select the problem set: Hot Suds Car Wash (Prob. 11-4).

Step 2   Rename the company and set the system date to October 31, 2008.

Completing the Accounting Problem

Step 3   Record Check 504 issued on October 3.

> ***Issued Check 504 for $868.45 to Custom Construction for construction supplies.***

To record a payment:

- Choose **Payments** from the *Tasks* menu.
- Type **CC** (the account ID for Custom Construction) in the *Vendor ID* field.

  If you do not know the account code for a payee, click the button, highlight the vendor name, and click the **OK** button.

- Type **504** in the *Check Number* field.
- Type **10/3/08** in the *Date* field. Make sure that Peachtree shows 2008 for the year. If not, change the system date.

TIP: As a shortcut, you can just type the day of the month in most date fields. Peachtree will automatically format the date.

- Type **Construction Supplies** in the *Memo* field.
- Move to the *Description* field on the **Apply to Expenses** tab and type **Construction Supplies.**
- Skip the *GL Account* field since the default account code, **Maintenance Expense**, is correct.
- Type **868.45** in the *Amount* field to record the amount of the check.

  When you enter the amount, Peachtree automatically displays the amount in the upper portion of the entry window.

DO YOU HAVE A QUESTION

Q. *Do you have to use Peachtree to print checks?*

A. No. You do not have to use Peachtree to print checks, although Peachtree does offer this feature. If you prefer, you can manually write checks and then use Peachtree to record these payments.

Notes

Lookup fields, such as the Vendor ID *field, are case-sensitive.*

- Proof the information you just recorded. Check all of the information you entered. If you notice a mistake, move to that field and make the correction. Compare the information on your screen to the completed transaction shown in Figure 11-4A.

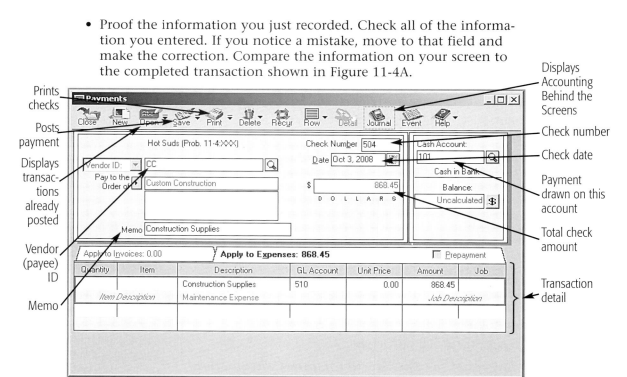

Prints checks
Posts payment
Displays transactions already posted
Vendor (payee) ID
Memo

Displays Accounting Behind the Screens
Check number
Check date
Payment drawn on this account
Total check amount
Transaction detail

Figure 11-4A *Completed Payment Transaction (October 3, Check 504)*

- Click [Journal] to display the Accounting Behind the Screens window.
  Use the information in this window to check your work again. (See Figure 11-4B.)

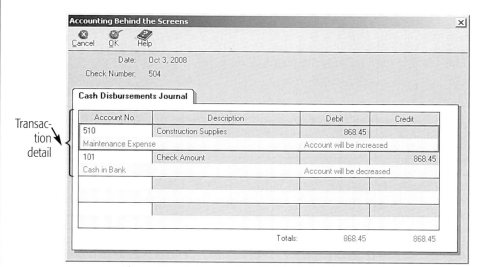

Transaction detail

Figure 11-4B *Accounting Behind the Screens (Payment)*

Notes

Viewing the Accounting Behind the Screens window is not required each time you enter a transaction. This information, however, is useful to understand how Peachtree records a transaction such as a payment.

The Accounting Behind the Screens window shows you the accounts and the debit/credit amounts that Peachtree will record for this transaction. For the payment, the window should show the following: **Maintenance Expense** ($868.45 DR) and **Cash in Bank** ($868.45 CR).

- Click [Cancel] to close the Accounting Behind the Screens window.

- Click [Save] to post the payment.

Although not required, you could use Peachtree to print real checks either on pre-printed checks or duplicate checks on blank paper. If you choose to print a real check, you would leave the check-number blank and click [Print]. Select a check form (e.g., *AP Preprint 2 Stub*). Then, click the **Real** button to print the check. If you enter a check number, Peachtree prints *Duplicate* on the check. The software automatically posts the transaction and is ready for the next payment after the check is printed.

TIP: If you notice an error after you post an entry, click [Open] and choose the transaction you want to edit. Make the corrections and post the transaction again to record the changes.

Step 4   Record the remaining checks issued during the month: Checks 505–507.

For each of the payments, do not change the default GL Account. These default accounts have already been setup for you in the problem set. For example, whenever you choose to pay Union Utilities, Peachtree displays account **530 Utilities Expense**. You will learn more about payments in a later chapter.

Step 5   Record the deposits—October 3 ($601.35—Tape 303) and October 10 ($342.80—Tape 304).

Use the **General Journal Entry** option to enter the deposits. For both deposits, debit **101 Cash** and credit **401 Wash Revenue**.

Step 6   Print an Account Register report.

To print an Account Register report:

- Choose **Account Reconciliation** from the **Reports** menu.
- Make sure the *Account Reconciliation* report area is selected.
- Select **Account Register** in the report list and click .
- Click **OK** to accept the standard option settings.

> **Notes**
>
> *The Account Register report shows all of the account activity (withdrawals and deposits) for the current period.*

Step 7   Proof the information on the report. If you notice an error, select the appropriate task option (**Payments** or **General Journal Entry**) and choose to edit the incorrect entry. Print a revised Account Register report.

Step 8   Answer the Analyze question.

Ending the Session

Step 9   Click the **Close Problem** button in the Glencoe Smart Guide window.

Mastering Peachtree

On a separate sheet of paper, answer the following questions:
How would you purchase pre-printed checks if you wanted to use Peachtree to prepare real checks? What security features are available on Peachtree checks? Use the online help to learn how to order pre-printed checks.

Problem 11-5   Reconciling the Bank Statement

INSTRUCTIONS

Beginning a Session

Step 1   Select the problem set: Kits & Pups Grooming (Prob. 11-5).
Step 2   Rename the company and set the system date to October 31, 2008.

Completing the Accounting Problem

Step 3   Reconcile the bank statement.

The problem set includes all of the transaction data needed to perform the account reconciliation. Payments and deposits made throughout the month have already been recorded for you. Bank service charges and other fees have not been recorded.

To reconcile the bank statement:

- Choose **Account Reconciliation** from the ***Tasks*** menu.
- Type **101** (the **Cash in Bank** account ID) in the *Account to Reconcile* field.

You must identify the account you want to reconcile. Some companies may have more than one checking account. For example, a company may have one account for regular payments and another account for payroll.

- Change the statement date to **10/30/08**.
- Identify the checks that have cleared (those that would appear on the bank statement). Click the *Clear* box for all of the checks except Check 768 and Check 772 since these two are outstanding.
- Identify which deposits have cleared. If a deposit was made, but does not appear on the bank statement, do not check the *Clear* box.
- Type **1380.00** in the *Statement Ending Balance* field.
- Compare the Account Reconciliation window on your screen to the one shown in Figure 11-5A. Outstanding checks should total $835.00 and deposits in transit should be $405.00. The GL (System) Balance is the same as the checkbook balance.

The unreconciled difference should show −$10.00 because you have yet to account for the bank service charge. When this amount is zero ($0.00), the account will be reconciled.

DO YOU HAVE A QUESTION ?

Q. *When you use the **Account Reconciliation** option and record an adjustment, do you have to make a separate entry to update the general ledger?*

A. No. Peachtree automatically creates a general journal entry behind the scenes when you enter an adjustment (e.g., bank service charge) using the **Account Reconciliation** option.

Notes

Peachtree displays the checks and deposits in the Account Reconciliation window after you identify the account you want to reconcile.

Records
account
recon-
ciliation
data

Click here
to mark
check as
cleared

Do not
mark out-
standing
checks

Allows you to enter
adjustments such as a
bank service charge

Bank statement date

Bank statement balance

Checkbook balance

Unreconciled
difference

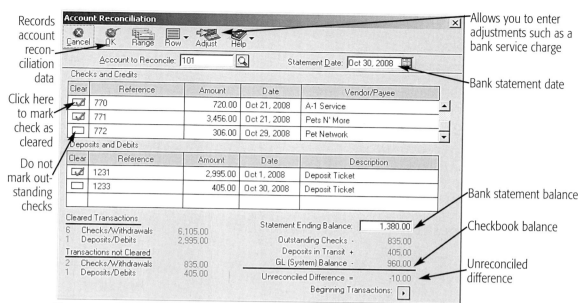

Figure 11-5A *Account Reconciliation Window (Checks/Deposits Marked)*

- Click ▣ to enter the bank service charge.
- Record the bank service charge as a withdrawal in the Additional
  Transactions window. Enter **10.00** for the amount, type **Bank
  Service Charge** for the description, and record **512** (Miscellaneous
  Expense) for the account. (See Figure 11-5B.) Click **OK** to record the
  additional transaction.

Record bank
service charge
here

GL account

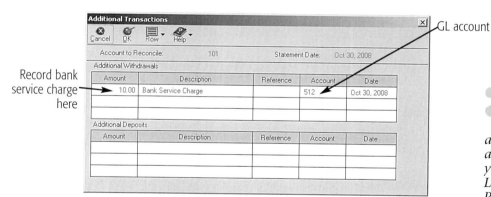

Figure 11-5B *Additional Transactions Window*

- Scroll the *Checks and Credits* list to display the first entry in
  the Account Reconciliation window. You should see the bank
  service charge you just entered.
- Click the *Clear* box next to the service charge to mark it as cleared.
- Verify that the unreconciled difference is zero ($0.00). Also, notice
  that the GL (System) Balance is $950.00. Peachtree automatically
  updated the **Cash in Bank** GL account to reflect the service
  charge. If these figures do not match your screen, verify which
  checks/deposits you marked as cleared. Also, check the bank
  statement amount and the bank service charge amount.
- Click ▣ to complete the reconciliation process.

> **Notes**
>
> *When you enter an
> additional charge such
> as the bank service charge,
> you must enter a General
> Ledger account number.
> Peachtree automatically
> generates a general journal
> entry for you behind
> the scenes.*

Step 4 Print the following Account Reconciliation reports: Account Register, Account Reconciliation, Deposits in Transit, and Outstanding Checks.

Step 5 Answer the Analyze question.

Ending the Session

Step 6 Click the **Close Problem** button in the Glencoe Smart Guide window.

Mastering Peachtree

Do you know how to change the layout of reports? Adjust the Account Reconciliation report layout to widen the date field to show the entire date (month, day, year). Print an updated Account Reconciliation report.

Problem 11-7   Reconciling the Bank Statement

INSTRUCTIONS

Beginning a Session

Step 1 Select the problem set: Showbiz Video (Prob. 11-7).

Step 2 Rename the company and set the system date to October 31, 2008.

Completing the Accounting Problem

Step 3 Reconcile the bank statement.

TIP: Use the account reconciliation **Adjust** option to record a bank service charge. Apply the charge to the **Miscellaneous Expense** account.

Step 4 Print the following Account Reconciliation reports: Account Register, Account Reconciliation, Deposits in Transit, and Outstanding Checks.

Step 5 Answer the Analyze question.

Ending the Session

Step 6 Click the **Close Problem** button in the Glencoe Smart Guide window.

> **Notes**
>
> *If you make a mistake when you enter an additional (adjustment) transaction, you must close the Account Reconciliation window and then use **General Journal Entry** option to edit or delete the entry.*

> **Notes**
>
> *You do not have to make a separate general journal entry to record a bank service charge. Peachtree automatically creates this entry when you make an adjustment to reconcile an account.*

Problem 11-8   Reconciling the Bank Statement Using the Account Form

INSTRUCTIONS

Beginning a Session

Step 1   Select the problem set: Job Connect (Prob. 11-8).

Step 2   Rename the company and set the system date to October 20, 2008.

Completing the Accounting Problem

Step 3   Reconcile the bank statement.

> **Note:** The $200 check issued to Fontenot Inc. was issued to pay for maintenance and repairs.

TIP: Use the account reconciliation **Adjust** option to record returned checks, bank service fees, and stop payment charges to the **Miscellaneous Expense** account.

Step 4   Print the following Account Reconciliation reports: Account Register, Account Reconciliation, Deposits in Transit, and Outstanding Checks.

Step 5   Answer the Analyze question.

Ending the Session

Step 6   Click the **Close Problem** button in the Glencoe Smart Guide window.

FAQs

How do you edit/delete an additional (adjustment) transaction entered using the Account Reconciliation feature?

Once you enter an additional transaction when reconciling an account, Peachtree will not allow you to edit the transaction. You must close the Account Reconciliation window and then choose the **General Journal Entry** option. Click the **Open** button and select the entry you want to change. Edit or delete the transaction and then choose the **Account Reconciliation** option to continue where you left off.

When you use the Account Reconciliation option and record an adjustment, do you have to make a separate entry to update the general ledger?

No. Peachtree automatically creates a general journal entry behind the scenes when you enter an adjustment (e.g., bank service charge) using the **Account Reconciliation** option.

Computerized Accounting Using Spreadsheets

Problem 11-6   Reconciling the Bank Statement

Completing the Spreadsheet

Step 1 Read the instructions for Problem 11-6 in your textbook. This problem involves reconciling a bank statement for Outback Guide Service.

Step 2 Open the Glencoe Accounting: Electronic Learning Center software.

Step 3 From the Program Menu, click on the **Peachtree Complete® Accounting Software and Spreadsheet Applications** icon.

Step 4 Log onto the Management System by typing your user name and password.

Step 5 Under the **Problems & Tutorials** tab, select template 11-6 from the Chapter 11 drop-down menu. The template should look like the one shown below.

```
PROBLEM 11-6
RECONCILING THE BANK STATEMENT

(name)
(date)


OUTBACK GUIDE SERVICE
BANK RECONCILIATION
OCTOBER 30, 20--

Balance on bank statement                                      AMOUNT

Deposits in transit:
                        30-Oct                   AMOUNT
TOTAL DEPOSITS                                                    0.00

Outstanding checks:
             Check #872                          AMOUNT
             Check #881                          AMOUNT
             Check #883                          AMOUNT
             Check #887                          AMOUNT
TOTAL OUTSTANDING CHECKS                                          0.00
ADJUSTED BANK BALANCE                                             0.00

Balance in checkbook                                           AMOUNT

Additions:
      Interest earned                            AMOUNT
TOTAL ADDITIONS                                                   0.00

Deductions:
         Bank service charge                     AMOUNT
         NSF check                               AMOUNT
TOTAL DEDUCTIONS                                                  0.00
ADJUSTED CHECKBOOK BALANCE                                        0.00
```

Step 6 Key your name and today's date in the cells containing the *(name)* and *(date)* placeholders.

Step 7 The balance shown on the bank statement is $2,272.36. Move the cell pointer to cell E12 and enter the bank statement balance: **2272.36**. (Remember, it is not necessary to include a comma as part of the entry.)

Step 8 A deposit was not reflected on the bank statement. Move the cell pointer to cell D15 and enter the amount of the deposit.

Step 9 Beginning in cell D19, enter the amounts for the outstanding checks. The spreadsheet template will automatically calculate the adjusted bank balance.

Step 10 Move the cell pointer to cell E26 and enter the checkbook balance.

Step 11 No interest was earned for the period, so there are no additions to the checkbook balance. Move the cell pointer to cell D29 and enter **0** as the amount of interest earned.

Step 12 Move the cell pointer to cell D33 and enter the amount of the bank service charge.

Step 13 Move the cell pointer to cell D34 and enter the amount of the NSF check. The spreadsheet template will automatically calculate the adjusted checkbook balance.

Step 14 The adjusted bank balance and adjusted checkbook balance should be equal. If they are not equal, find the error(s) and make the necessary corrections.

Step 15 Save the spreadsheet using the **Save** option from the *File* menu. You should accept the default location for the save as this is handled by the management system.

Step 16 Print the completed spreadsheet.

Step 17 Exit the spreadsheet program.

Step 18 In the Close Options box, select the location where you would like to save your work.

Step 19 Answer the Analyze question from your textbook for this problem.

Problem 11-8   Reconciling the Bank Statement Using the Account Form

SNB *Security National Bank*
3001 Porterfield Street, Durham, NC 27704

Job Connect
405 McLocklin Drive
Durham, NC 27713

FDIC

Account Number: 555113

Statement Date: 10/18/20––

Balance Last Statement	Deposits & Other Credits		Checks & Other Debits		Balance This Statement
	No.	Amount	No.	Amount	
452.46	1	288.66	4	396.54	344.58

Description	Checks & Other Debits	Deposits & Other Credits	Date	Balance
Balance Forward			09/15	452.46
Check	105.00		09/18	347.46
Check	59.71		09/22	287.75
Check	87.00		09/29	200.75
Check	45.41		09/29	155.34
Deposit		288.66	10/06	444.00
Returned Check	68.42		10/13	375.58
Service Charge	7.00		10/13	368.58
Service Charge	10.00		10/15	358.58
Service Charge	14.00		10/18	344.58

PLEASE EXAMINE YOUR STATEMENT CAREFULLY. DIRECT ALL INQUIRIES IMMEDIATELY.

Problem 11-8 (concluded)

Analyze: _____

CHAPTER **11** Cash Control and Banking Activities

Self-Test

Part A Fill in the Missing Term

Directions: *In the Answer column, write the letter of the word or phrase that best completes the sentence.*

A. bank service charge	**E.** NSF check	**H.** reconciling the bank statement
B. bank statement	**F.** outstanding checks	
C. canceled checks	**G.** outstanding deposits	**I.** restrictive
D. internal controls		**J.** stop payment

Answer

_____ 1. An itemized record of all the transactions occurring in a depositor's account over a given period is a(n) _____.

_____ 2. A(n) _____ is a fee charged by the bank for maintaining bank records and for processing bank statement items for the depositor.

_____ 3. A(n) _____ is a check returned by the bank because there are not sufficient funds in the drawer's checking account to cover the amount of the check.

_____ 4. _____ are checks that are paid by the bank, deducted from the depositor's account, and returned with the bank statement.

_____ 5. _____ are checks that have been written but not yet presented to the bank for payment.

_____ 6. _____ are steps that a business takes to protect its cash and other assets.

_____ 7. A demand by the depositor that the bank not honor a certain specific check is a(n) _____.

_____ 8. _____ is the process of determining any differences between the balance shown on the bank statement and the checkbook balance.

_____ 9. _____ are deposits that have been made and recorded in the checkbook but that do not appear on the bank statement.

_____ 10. A type of endorsement that limits how a check may be handled to protect the check from being cashed by anyone except the payee is a(n) _____ endorsement.

Part B   True or False

Directions: _Circle the letter_ T _in the Answer column if the statement is true;_
circle the letter F _if the statement is false._

Answer

T	F	**1.** A check should be written before the check stub is filled out.
T	F	**2.** Cash is the most liquid asset of a business.
T	F	**3.** The ending balance on the bank statement seldom agrees with the balance in the checkbook.
T	F	**4.** Bank service charges should be recorded in the checkbook before reconciling the bank statement.
T	F	**5.** An example of an internal control is the daily deposit of cash receipts in the bank.
T	F	**6.** A stop payment order and voiding a check mean the same thing.
T	F	**7.** Prompt reconciliation of the bank statement is a good way to guard against disorderly cash records or cash loss.
T	F	**8.** Checks written in pencil are acceptable.
T	F	**9.** When a business receives a check in payment for a product or service, it acquires the right to that check.
T	F	**10.** Outstanding checks and voided checks are the most frequent causes for differences between the bank statement balance and the checkbook balance.

MINI PRACTICE SET 2

Fast Track Tutoring Service

CHART OF ACCOUNTS
ASSETS
101   Cash in Bank
110   Accounts Receivable—Carla DiSario
120   Accounts Receivable—George McGarty
140   Office Supplies
150   Office Equipment
155   Instructional Equipment
LIABILITIES
210   Accounts Payable—Educational Software
215   Accounts Payable—T & N School Equip.
OWNER'S EQUITY
301   Jennifer Rachael, Capital
305   Jennifer Rachael, Withdrawals
310   Income Summary
REVENUE
401   Group Lessons Fees
405   Private Lessons Fees
EXPENSES
505   Maintenance Expense
510   Miscellaneous Expense
515   Rent Expense
525   Utilities Expense

Mini Practice Set 2   Source Documents

Instructions: *Use the following source documents to record the transactions for this practice set.*

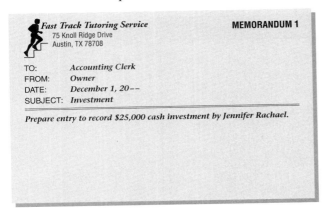

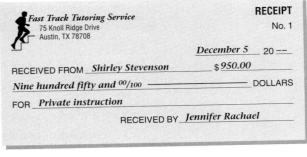

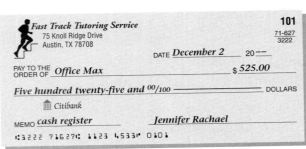

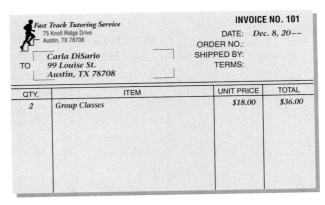

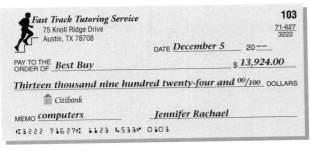

Mini Practice Set 2   (continued)

Fast Track Tutoring Service
75 Knoll Ridge Drive
Austin, TX 78708

INVOICE NO. 102

DATE: *Dec. 10, 20--*
ORDER NO.:
SHIPPED BY:
TERMS:

TO
George McGarty
31 Vale Street
Austin, TX 78705

QTY.	ITEM	UNIT PRICE	TOTAL
5	*Special Group Classes*	$55.00	$275.00

T&N School Equipment
111 Stratford Drive, #2A
Rollingwood, TX 77081

INVOICE NO. 5495

DATE: *Dec. 10, 20--*
ORDER NO.:
SHIPPED BY:
TERMS:

TO
Fast Track Tutoring Service
75 Knoll Ridge Drive
Austin, TX 78708

QTY.	ITEM	UNIT PRICE	TOTAL
	Microcomputer System		$2,375.00

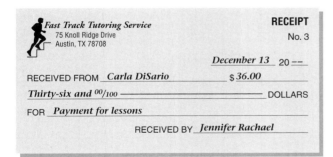

Fast Track Tutoring Service
75 Knoll Ridge Drive
Austin, TX 78708

RECEIPT
No. 2

December 11 20 --

RECEIVED FROM *Cash customers* $695.00

Six hundred ninety-five and 00/100 ————————— DOLLARS

FOR *20 private lessons between 12/1 and 12/10*

RECEIVED BY *Jennifer Rachael*

Fast Track Tutoring Service
75 Knoll Ridge Drive
Austin, TX 78708

RECEIPT
No. 3

December 13 20 --

RECEIVED FROM *Carla DiSario* $ 36.00

Thirty-six and 00/100 ————————— DOLLARS

FOR *Payment for lessons*

RECEIVED BY *Jennifer Rachael*

Fast Track Tutoring Service
75 Knoll Ridge Drive
Austin, TX 78708

105
71-627
3222

DATE *December 14* 20 --

PAY TO THE ORDER OF *Educational Software* $ 200.00

Two hundred and 00/100 ————————— DOLLARS

🏛 *Citibank*

MEMO *software*   *Jennifer Rachael*

⑆3222 ⑆⑆627⑆ ⑆123 4533⑈ 0105

Fast Track Tutoring Service
75 Knoll Ridge Drive
Austin, TX 78708

106
71-627
3222

DATE *December 15* 20 --

PAY TO THE ORDER OF *Union Painting Service* $ 750.00

Seven hundred fifty and 00/100 ————————— DOLLARS

🏛 *Citibank*

MEMO *painting*   *Jennifer Rachael*

⑆3222 ⑆⑆627⑆ ⑆123 4533⑈ 0106

Fast Track Tutoring Service
75 Knoll Ridge Drive
Austin, TX 78708

107
71-627
3222

DATE *December 18* 20 --

PAY TO THE ORDER OF *Jennifer Rachael* $ 500.00

Five hundred and 00/100 ————————— DOLLARS

🏛 *Citibank*

MEMO *personal withdrawal*   *Jennifer Rachael*

⑆3222 ⑆⑆627⑆ ⑆123 4533⑈ 0107

Fast Track Tutoring Service
75 Knoll Ridge Drive
Austin, TX 78708

108
71-627
3222

DATE *December 20* 20 --

PAY TO THE ORDER OF *Edison Electric* $ 183.00

One hundred eighty-three and 00/100 ————————— DOLLARS

🏛 *Citibank*

MEMO *electricity bill*   *Jennifer Rachael*

⑆3222 ⑆⑆627⑆ ⑆123 4533⑈ 0108

Fast Track Tutoring Service
75 Knoll Ridge Drive
Austin, TX 78708

109
71-627
3222

DATE *December 24* 20 --

PAY TO THE ORDER OF *U.S. Postal Service* $ 45.00

Forty-five and 00/100 ————————— DOLLARS

🏛 *Citibank*

MEMO *stamps*   *Jennifer Rachael*

⑆3222 ⑆⑆627⑆ ⑆123 4533⑈ 0109

Computerized Accounting Using Peachtree

Mini Practice Set 2

INSTRUCTIONS

Beginning a Session

Step 1 Open the Glencoe Accounting: Electronic Learning Center software.

Step 2 From the Program Menu, click on the **Peachtree Complete®
Accounting Software and Spreadsheet Applications** icon.

Step 3 Log onto the Management System by typing your user name and
password.

Step 4 Under the **Problems & Tutorials** tab, select problem set: Fast
Track Tutoring Service (MP-2) from the drop-down menu.

Step 5 Rename the company by adding your initials, e.g., Fast Track
(MP-2: XXX).

Step 6 Set the system date to December 31, 2008.

Completing the Accounting Problem

Step 7 Analyze each business transaction shown in your textbook for Fast
Track Tutoring Service.

Step 8 Record all of the transactions using the **General Journal Entry** option.

TIP: Proof each general journal entry before you post it. Check the
account numbers, descriptions, and amounts.

Step 9 Use the **Account Reconciliation** option to reconcile the bank
statement.

TIP: If you use the account reconciliation feature, Peachtree automatically inserts the general journal entry behind the scenes.

Preparing Reports and Proofing Your Work

Step 10 Print a General Journal report.

Step 11 Proof your work. Make any corrections as needed and print a revised
report, if necessary.

TIP: While viewing a General Journal report, you can double-click on
an entry to display it in the General Journal Entry window. You
can edit the transaction and then close the window to see an
updated report.

Step 12 Print the Account Register and Account Reconciliation reports.

Step 13 Print a General Ledger report.

Step 14 Print a General Ledger Trial Balance.

Peachtree Guide

Preparing Financial Statements

Step 15   Print an Income Statement and a Statement of Changes in Owner's Equity.

Step 16   Print a Balance Sheet.

Checking Your Work

IMPORTANT: Save your work for the mini practice set before you perform the closing process.

Step 17   Click the **Save Pre-closing Balances** button in the Glencoe Smart Guide window.

Note: When this button is clicked for the *first time*, balances will be saved automatically. When this button is clicked subsequent times, a dialog box appears asking you if you want to overwrite previously saved pre-closing balances.

Close the Accounting Period

Step 18   Close the fiscal year.

Step 19   Print a Post-Closing Trial Balance.

Analyzing Your Work

Step 20   Answer the Analyze question.

Step 21   Complete the Audit Test.

Ending the Session

Step 22   Click the **Close Problem** button in the Glencoe Smart Guide window.

Continuing from a Previous Session

- If you were previously directed to save your work on the network, select the problem from the scrolling menu and click **OK.** The system will retrieve your files from your last session.

- If you were previously directed to save your work on a floppy disk, insert the floppy, select the corresponding problem from the scrolling menu, and click **OK.** The system will retrieve your files from the floppy disk.

Notes

When you print the financial statements, use the custom Income Statement, custom Statement of Changes in Owner's Equity, and <Standard> Balance Sheet report.

Notes

Use the reports you prepared to answer the Analyze question and to complete the Audit Test.

Mini Practice Set 2

GENERAL JOURNAL

PAGE _____

	DATE	DESCRIPTION	POST. REF.	DEBIT	CREDIT	
1						1
2						2
3						3
4						4
5						5
6						6
7						7
8						8
9						9
10						10
11						11
12						12
13						13
14						14
15						15
16						16
17						17
18						18
19						19
20						20
21						21
22						22
23						23
24						24
25						25
26						26
27						27
28						28
29						29
30						30
31						31
32						32
33						33
34						34
35						35
36						36
37						37

Mini Practice Set 2   (continued)

GENERAL JOURNAL

	DATE	DESCRIPTION	POST. REF.	DEBIT	CREDIT	
1						1
2						2
3						3
4						4
5						5
6						6
7						7
8						8
9						9
10						10
11						11
12						12
13						13
14						14
15						15
16						16
17						17
18						18
19						19
20						20
21						21
22						22
23						23
24						24
25						25
26						26
27						27
28						28
29						29
30						30
31						31
32						32
33						33
34						34
35						35
36						36
37						37

Mini Practice Set 2 (continued)

GENERAL LEDGER

ACCOUNT _____ ACCOUNT NO. _____

DATE	DESCRIPTION	POST. REF.	DEBIT	CREDIT	BALANCE	
					DEBIT	CREDIT

ACCOUNT _____ ACCOUNT NO. _____

DATE	DESCRIPTION	POST. REF.	DEBIT	CREDIT	BALANCE	
					DEBIT	CREDIT

Mini Practice Set 2   (continued)

ACCOUNT _____ ACCOUNT NO. _____

DATE	DESCRIPTION	POST. REF.	DEBIT	CREDIT	BALANCE	
					DEBIT	CREDIT

ACCOUNT _____ ACCOUNT NO. _____

DATE	DESCRIPTION	POST. REF.	DEBIT	CREDIT	BALANCE	
					DEBIT	CREDIT

ACCOUNT _____ ACCOUNT NO. _____

DATE	DESCRIPTION	POST. REF.	DEBIT	CREDIT	BALANCE	
					DEBIT	CREDIT

ACCOUNT _____ ACCOUNT NO. _____

DATE	DESCRIPTION	POST. REF.	DEBIT	CREDIT	BALANCE	
					DEBIT	CREDIT

ACCOUNT _____ ACCOUNT NO. _____

DATE	DESCRIPTION	POST. REF.	DEBIT	CREDIT	BALANCE	
					DEBIT	CREDIT

ACCOUNT _____ ACCOUNT NO. _____

DATE	DESCRIPTION	POST. REF.	DEBIT	CREDIT	BALANCE	
					DEBIT	CREDIT

Mini Practice Set 2   (continued)

ACCOUNT _____ ACCOUNT NO. _____

DATE	DESCRIPTION	POST. REF.	DEBIT	CREDIT	BALANCE	
					DEBIT	CREDIT

ACCOUNT _____ ACCOUNT NO. _____

DATE	DESCRIPTION	POST. REF.	DEBIT	CREDIT	BALANCE	
					DEBIT	CREDIT

ACCOUNT _____ ACCOUNT NO. _____

DATE	DESCRIPTION	POST. REF.	DEBIT	CREDIT	BALANCE	
					DEBIT	CREDIT

ACCOUNT _____ ACCOUNT NO. _____

DATE	DESCRIPTION	POST. REF.	DEBIT	CREDIT	BALANCE	
					DEBIT	CREDIT

ACCOUNT _____ ACCOUNT NO. _____

DATE	DESCRIPTION	POST. REF.	DEBIT	CREDIT	BALANCE	
					DEBIT	CREDIT

Mini Practice Set 2   (continued)

ACCOUNT _____ ACCOUNT NO. _____

DATE	DESCRIPTION	POST. REF.	DEBIT	CREDIT	BALANCE	
					DEBIT	CREDIT

ACCOUNT _____ ACCOUNT NO. _____

DATE	DESCRIPTION	POST. REF.	DEBIT	CREDIT	BALANCE	
					DEBIT	CREDIT

ACCOUNT _____ ACCOUNT NO. _____

DATE	DESCRIPTION	POST. REF.	DEBIT	CREDIT	BALANCE	
					DEBIT	CREDIT

ACCOUNT _____ ACCOUNT NO. _____

DATE	DESCRIPTION	POST. REF.	DEBIT	CREDIT	BALANCE	
					DEBIT	CREDIT

Mini Practice Set 2   (continued)

(5)

BANK RECONCILIATION FORM

PLEASE EXAMINE YOUR STATEMENT AT ONCE. ANY DISCREPANCY SHOULD BE REPORTED TO THE BANK IMMEDIATELY.

1. Record any transactions appearing on this statement but not listed in your checkbook.

2. List any checks still outstanding in the space provided to the right.

3. Enter the balance shown on this statement here.

4. Enter deposits recorded in your checkbook but not shown on this statement.

5. Total Lines 3 and 4 and enter here.

6. Enter total checks outstanding here.

7. Subtract Line 6 from Line 5. This adjusted bank balance should agree with your checkbook balance.

CHECKS OUTSTANDING		
Number	Amount	
TOTAL		

(7)

	No. 114
$ _____	
Date _____ 20 ___	
To _____	
For _____	

	Dollars	Cents
Balance brought forward		
Add deposits		
Total		
Less this check		
Balance carried forward		

Fast Track Tutoring Service
75 Knoll Ridge Drive
Austin, TX 78708

114

71-627
3222

DATE _____ 20 ___

PAY TO THE
ORDER OF _____ $ _____

_____ DOLLARS

🏛 Citibank

MEMO _____

⑆3222 7⑈627⑆ ⑈⑈23 4533⑆ 0⑈⑈4

Mini Practice Set 2   (continued)

(8)

ACCT. NO.	ACCOUNT NAME	TRIAL BALANCE DEBIT	TRIAL BALANCE CREDIT	INCOME STATEMENT DEBIT	INCOME STATEMENT CREDIT	BALANCE SHEET DEBIT	BALANCE SHEET CREDIT

Mini Practice Set 2   (continued)

(9)

(10)

Mini Practice Set 2 (concluded)

(11)

Analyze: _____

_____

MINI PRACTICE SET 2

Fast Track Tutoring Service

Audit Test

Directions: *Use your completed solutions to answer the following questions. Write the answer in the space to the left of each question.*

_____ **1.** Did the transaction on December 18 increase or decrease owner's capital?

_____ **2.** What was the balance in the Private Lessons Fees account on December 23?

_____ **3.** Did the transaction on December 8 increase or decrease accounts receivable?

_____ **4.** What was the amount of office supplies purchased during the month?

_____ **5.** What was the checkbook balance after the bank service charge was recorded on the check stub?

_____ **6.** What account was debited to record the bank service charge amount?

_____ **7.** What was the total amount of outstanding checks listed on the bank reconciliation statement?

_____ **8.** To which creditor did Fast Track Tutoring Service owe the most money on December 31?

_____ **9.** What was the balance of the owner's capital account reported on the trial balance?

_____ **10.** To what section of the work sheet was the balance of the Jennifer Rachael, Withdrawals account extended?

_____ **11.** What was the total of the Income Statement Debit column of the work sheet before the net income or net loss was determined?

_____ **12.** What was the amount of net income or net loss for December?

_____ **13.** What was the amount of total revenue for the period?

_____ **14.** From what source did Fast Track Tutoring Service earn most of its revenue?

_____ **15.** What were the total expenses for the month?

_____ **16.** Did all the temporary capital accounts appear on the income statement?

_____ **17.** How many asset accounts were listed on the balance sheet?

_____ **18.** What were Fast Track Tutoring Service's total liabilities at the end of the month?

_____ **19.** How many closing entries were needed to close the temporary capital accounts?

_____ **20.** To close Rent Expense, was the account debited or credited?

_____ **21.** How many accounts in the general ledger were closed?

_____ **22.** The final closing entry closed which account?

_____ **23.** How many accounts were listed on the post-closing trial balance?

_____ **24.** What was the balance in the Jennifer Rachael, Capital account reported on the post-closing trial balance?

_____ **25.** What were the debit and credit totals of the post-closing trial balance?

Problem 12-4   Preparing a Payroll Check

PAYROLL REGISTER

PAY PERIOD ENDING _March 23_, 20--   DATE OF PAYMENT _March 23, 20--_

EMPLOYEE NUMBER	NAME	MAR. STATUS	ALLOW.	TOTAL HOURS	RATE	EARNINGS			DEDUCTIONS							NET PAY	CK. NO.
						REGULAR	OVERTIME	TOTAL	SOC. SEC. TAX	MED. TAX	FED. INC. TAX	STATE INC. TAX	HOSP. INS.	OTHER	TOTAL		
18	Burns, Janice	S	1	42	7.80	312 00	23 40	335 40	20 79	4 86	35 00	6 71	4 10	—	71 46	263 94	79

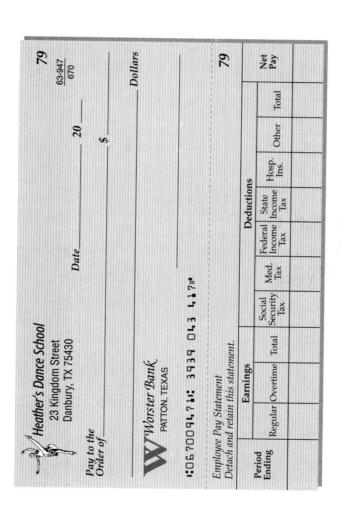

Heather's Dance School
23 Kingdom Street
Danbury, TX 75430

No. 79
63-947 / 670

Date _____ 20 ____

Pay to the
Order of _____ $ _____

_____ Dollars

Worster Bank
PATTON, TEXAS

⑈06 7009 47 ⑈: 3939 043 4 17⑈

Employee Pay Statement
Detach and retain this statement.   No. 79

Period Ending	Earnings			Deductions						Net Pay
	Regular	Overtime	Total	Social Security Tax	Med. Tax	Federal Income Tax	State Income Tax	Hosp. Ins.	Other	Total	

Computerized Accounting Using Peachtree

Software Objectives

When you have completed this chapter, you will be able to use Peachtree to:

1. Update an employee record.
2. Calculate gross earnings.
3. Print payroll checks.
4. Calculate and record payroll deductions.
5. Print a Payroll Register report.
6. Print a Current Earnings report.

Problem 12-5   Calculating Gross Pay

INSTRUCTIONS

Beginning a Session

Step 1   Select the problem set: Wilderness Rentals (Prob. 12-5).

Step 2   Rename the company by adding your initials, e.g., Wilderness (Prob. 12-5: XXX).

Step 3   Set the system date to February 1, 2008.

Completing the Accounting Problem

Step 4   Review the payroll information provided in your textbook.

Step 5   Update the employee record for John Gilmartin to include his regular and overtime hourly rate.

All of the employee records have already been set up for you, except that John Gilmartin's record does not include the pay rates. Follow the instructions provided here to record this information.

To update an employee's record:

- Choose **Employees/Sales Reps** from the **Maintain** menu.
- Type **GIL** in the *Employee ID* field.
- Click the **Pay Info** tab.
- Type **6.80** for the regular hourly rate and **10.20** for the overtime rate.
- Compare the information you entered to the employee record shown in Figure 12-5A.
- Click 🖫 to save the changes and then click 🗖 to close the data entry window.

> **DO YOU HAVE A QUESTION**
>
> **Q.** *What kinds of pay methods does Peachtree support?*
>
> **A.** Peachtree supports hourly and salary pay methods. The program will automatically calculate gross pay based on the employees' pay rates. Peachtree will also support other pay methods such as a sales commission. For these "special" pay methods, you have to manually enter the payroll amounts.

> **Notes**
>
> *You can use the **Employee/Sales Reps** option to update any of the following employee record fields: address, social security number, hire date, and pay rate.*

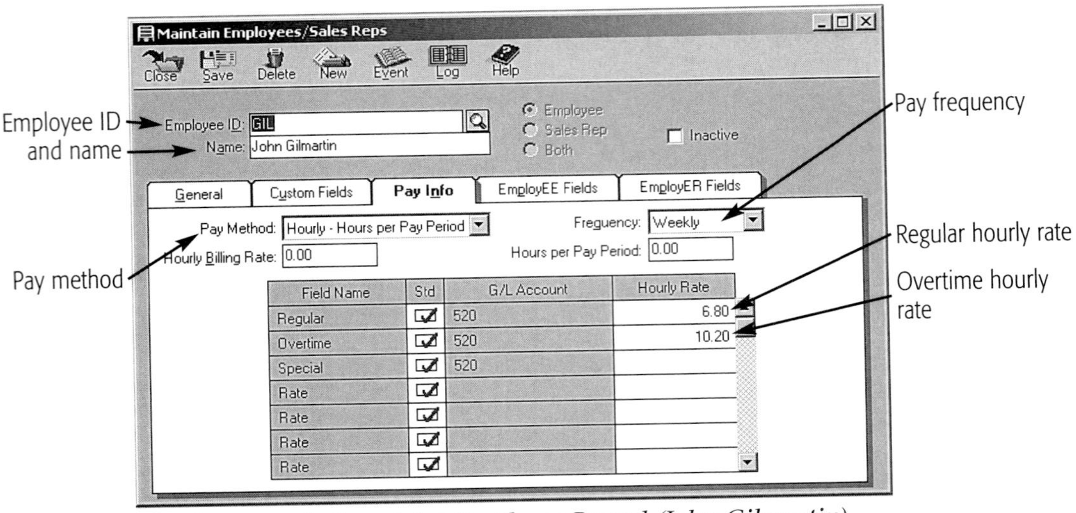

Employee ID and name →

Pay method →

Pay frequency

Regular hourly rate

Overtime hourly rate

Figure 12-5A *Completed Employee Record (John Gilmartin)*

Step 6   Record the payroll for John Gilmartin (43 hours).

To record a payroll entry:

- Choose **Payroll Entry** from the ***Tasks*** menu to display the Payroll Entry window.
- Type **GIL** in the *Employee ID* field.

TIP: Click 🔍 or press **SHIFT+?** when the cursor is in the *Employee ID* field to display a list of employees and their IDs.

- Type **40** in the *Regular Hours* field.
- Type **3** in the *Overtime Hours* field and press **ENTER**.

  As you enter the regular and overtime hours (if any), Peachtree automatically calculates the gross pay based on the rate/salary information stored in the employee's record. As you can see, Peachtree also includes fields to record the payroll taxes and other deductions. You will learn how to record information in these fields in a later problem.

- Proof the information you just recorded. Check all of the information you entered. If you notice a mistake, move to that field and make the correction. Compare the information on your screen to the completed transaction shown in Figure 12-5B.

- Click 💾 to record the payroll entry.

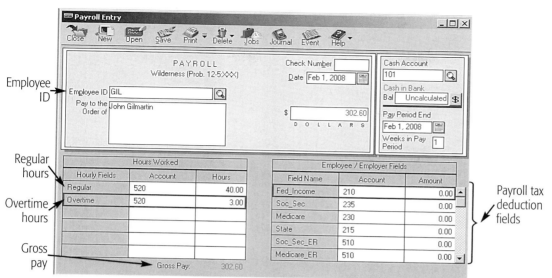

Figure 12-5B *Completed Payroll Entry (John Gilmartin)*

Step 7   Record a payroll entry for each of the following employees: Arlene Stone, Tom Driscoll, and Ann Ryan.

TIP: To record the payroll entry for an employee who earned a commission, you must calculate the commission amount and manually enter it in the *Commission* field.

Step 8   Print a Payroll Register and proof your work.

To print a Payroll Register report:

• Choose **Payroll** from the ***Reports*** menu to display the Select a Report window. (See Figure 12-5C.)

TIP: As a shortcut, you can double-click a report title to go directly to the report screen.

     • Select Payroll Register in the report list.
     • Click [Preview] and then click **OK** to display the report.
     • Click [Print] to print the report.
     • Click [Close] to close the report window.

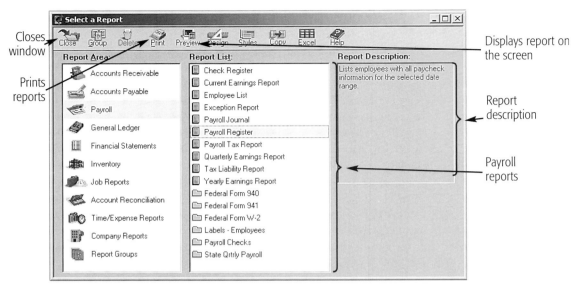

Closes window

Prints reports

Displays report on the screen

Report description

Payroll reports

Figure 12-5C *Select a Report Window*

Step 9 If you notice any errors, choose the **Payroll Entry** command again, click ![Open], select the record you want to edit, and post the changes.

Step 10 Answer the analyze question.

Ending the Session

Step 11 Click the **Close Problem** button in the Glencoe Smart Guide window. Select the save option as directed by your teacher.

Continuing a Problem from a Previous Session

If you want to continue working on a problem that you did not complete in a previous session, follow step 1. The management system will retrieve your files from your last session.

Mastering Peachtree

Can you customize the employee record to store extra information such as position/title, review date, and benefit plan? Explain your answer.

Problem 12-7   Preparing Payroll Checks and Earnings Records

INSTRUCTIONS

Beginning a Session

Step 1 Select the problem set: Kits & Pups Grooming (Prob. 12-7).

TIP: Make sure that you set the system date before you print any payroll checks.

Step 2 Rename the company and set the system date to October 18, 2008.

> **DO YOU HAVE A QUESTION**
>
> **Q.** *Why don't the Peachtree payroll checks look like real checks?*
>
> **A.** When you choose to print payroll checks, Peachtree prints only the pertinent information since it assumes that you are using pre-printed checks. A pre-printed check is a real check with the company name, routing numbers, and other information. Peachtree simply prints the date, employee name, amount, and deductions. If you are printing on plain paper, only these items appear.

Completing the Accounting Problem

Step 3 Review the payroll information provided in your textbook.

Step 4 Print a payroll check for Mildred Hurd.

The payroll information (hours and deductions) for the pay period ending October 17 has already been recorded for you. Normally, you would enter the earnings data and deductions yourself before printing payroll checks.

To print a payroll check:

- Choose **Payroll Entry** from the *Tasks* menu.

- Click [Open] and choose the payroll entry for Mildred Hurd.

- Review the payroll information for Mildred Hurd. Her net pay should be $219.48.

- Click [Print] to print a payroll check.

- Choose **PR MultiP Chks 2 Stub** or an equivalent form.

By choosing the payroll check form, you let Peachtree know how it should organize the information it is about to print. The layout for the employee name, check amount, and deductions depend on the particular form.

- Click **Real** to print a real check and enter **92** for the first check number.

Step 5 Print payroll checks for José Montego, Amanda Pilly, and Margaret Steams.

Step 6 Print a Payroll Register for this pay period.

To print a Payroll Register report for a specific pay period:

- Choose **Payroll** from the *Reports* menu.
- Select Payroll Register in the report list.
- Click [Preview].
- Choose **This Week-to-Date** for date range on the Filter options tab and then press **OK**.

The date range should show October 12, 2008 to October 18, 2008. If the report is already on the screen, click [Options] to select a pay period.

- Click [Print] to print the report.

- Click [Close] to close the report window.

Step 7 Print a Current Earnings report for José Montego and Amanda Pilly by choosing it from the Payroll Report list.

> **Notes**
>
> *The payroll earnings and deductions have already been entered for the employees.*

TIP: Use the report filter options to select which employee you want to appear on the Current Earnings report.

Step 8 Answer the analyze question.

Peachtree Guide

Ending the Session

Step 9   Click the **Close Problem** button in the Glencoe Smart Guide window. Select the save option as directed by your teacher.

Mastering Peachtree

Where can you obtain pre-printed payroll checks? Are there different styles from which to choose? Explain.

Problem 12-8   Preparing the Payroll

INSTRUCTIONS

Beginning a Session

Step 1   Select the problem set: Outback Guide Service (Prob. 12-8).

> **TIP:** Make sure that you set the system date before you print any payroll checks.

Step 2   Rename the company and set the system date to October 18, 2008.

Completing the Accounting Problem

Step 3   Review the payroll information provided in your textbook. Round actual time recorded to the nearest quarter hour.

Step 4   Record the payroll and print a payroll check for each of the employees.

To process the payroll and print checks:

- Choose **Payroll Entry** from the **Tasks** menu.
- Manually enter the payroll information (rate, taxes, and other deductions) for an employee. Make sure that you enter the deductions as a negative number (See Figure 12-8A.)
- Review the payroll information.
- Click ⬚ to print a payroll check.
- Choose **PR MultiP Chks 2 Stub** or an equivalent form.

   By choosing the payroll check form, you let Peachtree know how it should organize the information it is about to print. The layout for the employee name, check amount, and deductions depend on the particular form.

- Click **Real** to print a real check and enter **82** for the first check number.

Step 5   Print a Payroll Register for the pay period.

To print a Payroll Register report for a specific pay period:

- Choose **Payroll** from the **Reports** menu.
- Select Payroll Register in the report list.
- Click ⬚.
- Choose **This Week-to-Date** for date range on the Filter options tab and then press **OK**.

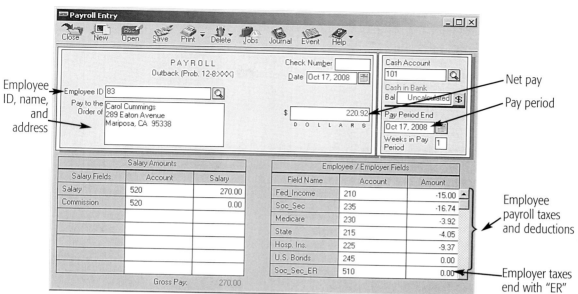

Employee ID, name, and address →

Net pay

Pay period

Employee payroll taxes and deductions

Employer taxes end with "ER"

Figure 12-8A _Payroll Entry Window_

The date range should show October 12, 2008 to October 18, 2008.

If the report is already on the screen, click [Options] to select a pay period.

- Click [Print] to print the report.
- Click [Close] to close the report window.

Step 6 Print a Current Earnings report for all of the employees.
Step 7 Answer the Analyze question.

Ending the Session

Step 8 Click the **Close Problem** button in the Glencoe Smart Guide window. Select the save option as directed by your teacher.

Problem 12-9 Preparing the Payroll Register

INSTRUCTIONS

Beginning a Session

Step 1 Select the problem set: Showbiz Video (Prob. 12-9).
Step 2 Rename the company and set the system date to October 25, 2008.

Completing the Accounting Problem

Step 3 Review the payroll information provided in your textbook. Round actual time recorded to the nearest quarter hour.
Step 4 Process the payroll information for all of the employees, but do not print payroll checks.

TIP: Remember that you must manually enter the payroll taxes and other deductions. Be sure to enter the deductions as negative amounts.

Step 5 Print a Payroll Register for the pay period October 19, 2008 to October 25, 2008.

Step 6 Proof your work. If necessary, use the **Payroll Entry** option to edit any employee pay records. Print revised reports if you make any changes.

Step 7 Answer the Analyze question.

Checking Your Work and Ending the Session

Step 8 Click the **Close Problem** button in the Glencoe Smart Guide window.

Step 9 If your teacher has asked you to check your solution, select *Check my answer to this problem*. Review, print, and close the report.

Please note that the Solution Software does not check the employee account balances (gross pay, net pay, etc.). Rather it checks the general ledger balances for the accounts affected by the payroll entries (e.g., Cash, Payroll Tax Expense, Salaries Expense, Federal Income Tax Payable, etc.).

Step 10 Click the **Close Problem** button and select the save option as directed by your teacher.

Problem 12-10   Calculating Gross Earnings

INSTRUCTIONS

Beginning a Session

Step 1 Select the problem set: Job Connection (Prob. 12-10).

Step 2 Rename the company and set the system date to October 25, 2008.

Completing the Accounting Problem

Step 3 Review the payroll information provided in your textbook.

Step 4 Process the payroll information for all employees to compute gross earnings. **Note:** You do **not** have to record the payroll taxes and other deductions.

Step 5 Print a Payroll Register for the pay period.

Step 6 Proof your work.

Step 7 Answer the Analyze question.

Notes
You must manually compute and record commission earnings.

Ending the Session

Step 8 Click the **Close Problem** button in the Glencoe Smart Guide window.

Notes
Leave the check number field blank.

Mastering Peachtree

Explain the step necessary to update an employee's marital status and number of exemptions.

FAQs

Why doesn't Peachtree automatically calculate the payroll taxes?

For a real company, Peachtree will automatically calculate all of the payroll taxes. To perform these calculations, a company must have the tax tables for the current year. For the payroll problems in your text, the base year is 2008. To avoid potential software conflicts, the automatic payroll tax calculation feature was turned off.

Spreadsheet Guide

Computerized Accounting Using Spreadsheets

Problem 12-6   Preparing a Payroll Register

Completing the Spreadsheet

Step 1   Read the instructions for Problem 12-6 in your textbook. This problem involves preparing a payroll register.

Step 2   Open the Glencoe Accounting: Electronic Learning Center software.

Step 3   From the Program Menu, click on the **Peachtree Complete® Accounting Software and Spreadsheet Applications** icon.

Step 4   Log onto the Management System by typing your user name and password.

Step 5   Under the **Problems & Tutorials** tab, select template 12-6 from the Chapter 12 drop-down menu. The template should look like the one shown below.

```
PROBLEM 12-6
PREPARING A PAYROLL REGISTER


(name)
(date)


PAYROLL REGISTER
PAY PERIOD ENDING OCTOBER 9, 20--
```

EMPLOYEE NUMBER	NAME	MARITAL STATUS	ALLOW.	TOTAL HOURS	HOURLY RATE	EARNINGS REGULAR	OVERTIME	><	NET PAY
108	Dumser, James					0.00	0.00	><	0.00
112	Job, Gail					0.00	0.00	><	0.00
102	Liptak, James					0.00	0.00	><	0.00
109	Stern, Bruce					0.00	0.00	><	0.00
TOTAL						0.00	0.00	><	0.00

Step 6   Key your name and today's date in the cells containing the *(name)* and *(date)* placeholders.

Step 7   Enter the marital status, number of allowances, total hours worked, and hourly rate for each employee in the appropriate cells of the spreadsheet template. The spreadsheet template will automatically calculate the regular earnings, overtime earnings, total earnings, Social Security tax, Medicare tax, and state income tax for each employee.

TIP: The cells for Social Security tax, Medicare tax, and state income tax are set up to round these numbers to two decimal places. When you are entering data in this spreadsheet template, always round numbers to two decimal places when rounding is necessary.

Step 8   Use the tax tables in your textbook to determine the federal income tax for each employee. Enter the federal income tax for each employee.

Step 9   Enter the hospital insurance deduction of $6.75 for the employees who have health and hospital insurance.

Step 10 Enter the union dues of $4.50 for the employees who are union members. The spreadsheet template automatically calculates the total deductions and net pay for each employee.

Step 11 Save the spreadsheet using the **Save** option from the *File* menu. You should accept the default location for the save as this is handled by the management system.

Step 12 Print the completed spreadsheet.

TIP: If your spreadsheet is too wide to fit on an 8.5-inch wide piece of paper, you can change your print settings to print the worksheet *landscape*. Landscape means that the worksheet will be printed broadside on the page. Some spreadsheet applications also allow you to choose a "fit to page" option. This function will reduce the width and/or depth of the worksheet to fit on one page.

Step 13 Exit the spreadsheet program.

Step 14 In the Close Options box, select the location where you would like to save your work.

Step 15 Answer the Analyze question from your textbook for this problem.

What-If Analysis

If James Dumser worked 43 hours, what would his net pay be?

_____

TIP: Remember to update the federal withholding tax to reflect Mr. Dumser's gross pay.

Problem 12-10   Source Documents

Instructions: *Use the following source documents to record the transactions for this problem.*

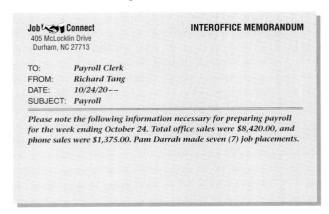

Job Connect
405 McLocklin Drive
Durham, NC 27713

INTEROFFICE MEMORANDUM

TO:	*Payroll Clerk*
FROM:	*Richard Tang*
DATE:	*10/24/20--*
SUBJECT:	*Payroll*

Please note the following information necessary for preparing payroll for the week ending October 24. Total office sales were $8,420.00, and phone sales were $1,375.00. Pam Darrah made seven (7) job placements.

NO. **15**

NAME **Doris Franco**

SOC. SEC. NO.

WEEK ENDING **10/24/20--**

DAY	IN	OUT	IN	OUT	IN	OUT	TOTAL
M	9:00	12:01	12:35	5:35			8
T	9:01	12:02	12:37	5:36			8
W	8:00	11:59	12:28	5:31			9
Th	8:58	12:01	12:29	5:46			8¼
F	9:02	11:58	12:31	5:29			8
S							
S							
						TOTAL HOURS	41¼

	HOURS	RATE	AMOUNT
REGULAR			
OVERTIME			
	TOTAL EARNINGS		

SIGNATURE _____ DATE _____

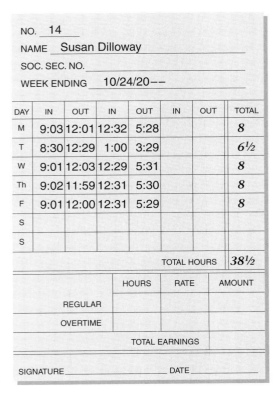

NO. **14**

NAME **Susan Dilloway**

SOC. SEC. NO.

WEEK ENDING **10/24/20--**

DAY	IN	OUT	IN	OUT	IN	OUT	TOTAL
M	9:03	12:01	12:32	5:28			8
T	8:30	12:29	1:00	3:29			6½
W	9:01	12:03	12:29	5:31			8
Th	9:02	11:59	12:31	5:30			8
F	9:01	12:00	12:31	5:29			8
S							
S							
						TOTAL HOURS	38½

	HOURS	RATE	AMOUNT
REGULAR			
OVERTIME			
	TOTAL EARNINGS		

SIGNATURE _____ DATE _____

NO. **17**

NAME **David Facini**

SOC. SEC. NO.

WEEK ENDING **10/24/20--**

DAY	IN	OUT	IN	OUT	IN	OUT	TOTAL
M	3:00	8:01					5
T	2:59	8:02					5
W	3:01	8:02					5
Th	3:00	8:00					5
F	3:00	6:01					3
S							
S							
						TOTAL HOURS	23

	HOURS	RATE	AMOUNT
REGULAR			
OVERTIME			
	TOTAL EARNINGS		

SIGNATURE _____ DATE _____

Problem 12-10   Calculating Gross Earnings

Name	Gross Earnings
Austin, Lynn	
Darrah, Pam	
Dilloway, Susan	
Facini, David	
Franco, Doris	
Miller, Barbara	
Womack, Charlene	
Total Gross Earnings	

Analyze: _____

_____

CHAPTER 12   Payroll Accounting

Self-Test

Part A   True or False

Directions: *Circle the letter* T *in the Answer column if the statement is true; circle the letter* F *if the statement is false.*

Answer

T	F	**1.** A payroll register is prepared for each pay period.
T	F	**2.** Most businesses use computers to prepare the payroll.
T	F	**3.** The social security tax and Medicare tax are both part of the FICA system.
T	F	**4.** Form W-4 lists the marital status and the number of exemptions claimed by each employee.
T	F	**5.** A person can change their social security number at any time.
T	F	**6.** The number of hours worked multiplied by the hourly wage gives the net earnings for the pay period.
T	F	**7.** Most employees claim more than eight exemptions on their W-4 form.
T	F	**8.** Overtime is paid to most employees after 44 hours of work in a pay week.
T	F	**9.** The payroll register summarizes information about employees' earnings for the pay period.
T	F	**10.** The social security system was established by the Federal Insurance Contributions Act.
T	F	**11.** Payroll is not a major expense of most businesses.
T	F	**12.** The account Medicare Tax Payable is a liability of the business.

Part B   Fill in the Missing Term

Directions: *In the Answer column, write the letter of the word or phrase that best completes the sentence. Some answers may be used more than once.*

A. accumulated earnings	**D.** direct deposit	**H.** pay period	**K.** salary
	E. gross earnings	**I.** payroll	**L.** timecard
B. commission	**F.** net pay	**J.** payroll register	**M.** wage
C. deduction	**G.** overtime rate		

Answer

_____ **1.** An amount paid to employees at a specific rate per hour is a(n) _____.

_____ **2.** The amount of money actually received by the employee after all deductions are subtracted is called the _____.

_____ **3.** _____ are the year-to-date gross earnings of an employee.

_____ **4.** A(n) _____ is paid to the employee as a percentage of the employee's sales.

_____ **5.** The amount of time for which an employee is paid is called the _____.

_____ **6.** Most businesses use a(n) _____ to keep track of an hourly wage employee's hours.

_____ **7.** The _____ is the total amount earned by the employee in the pay period.

_____ **8.** A(n) _____ is an amount subtracted from gross earnings.

_____ **9.** A list of employees in a business and the earnings due each employee for a specific period of time is a(n) _____.

_____ **10.** A(n) _____ is the deposit by the employer of an employee's net pay into a personal bank account.

Computerized Accounting Using Peachtree

Software Objectives

When you have completed this chapter, you will be able to use Peachtree to:

1. Record the employer's payroll taxes.
2. Record the payment of tax liabilities.
3. Print General Journal and General Ledger reports to verify the payroll entries.

Problem 13-7   Recording the Payment of the Payroll

INSTRUCTIONS

Beginning a Session

Step 1   Select the problem set: Kits & Pups Grooming (Prob. 13-7).
Step 2   Rename the company and set the system date to December 31, 2008.

Completing the Accounting Problem

Step 3   Review the payroll information provided in your textbook.
Step 4   Record the payroll entry using the **General Journal Entry** option.

TIP: You can use the Row, Add and Row, Remove buttons in the General Journal Entry window to edit a multi-part entry.

Step 5   Print a General Journal report and a General Ledger report.
Step 6   Proof your work.
Step 7   Answer the Analyze question.

Checking Your Work and Ending the Session

Step 8   Click the **Close Problem** button in the Glencoe Smart Guide window.
Step 9   If your teacher has asked you to check your solution, select *Check my answer to this problem*. Review, print, and close the report.
Step 10   Click the **Close Problem** button and select the save option as directed by your teacher.

Problem 13-8   Journalizing Payroll Transactions

INSTRUCTIONS

Beginning a Session

Step 1   Select the problem set: Outback Guide Service (Prob. 13-8).
Step 2   Rename the company and set the system date to December 31, 2008.

Completing the Accounting Problem

Step 3   Review the payroll information provided in your textbook.
Step 4   Add a new General Ledger account—**242 Union Dues Payable**. Make sure that you identify the account type as *Other Current Liabilities*.
Step 5   Record the entry for the payment of the payroll using the **General Journal Entry** option.

DO YOU HAVE A QUESTION

Q. *Can Peachtree automatically generate the journal entries to record the payment of the payroll and the employer's payroll tax liabilities?*

A. Yes, Peachtree will automatically record the payroll journal entries if you use the payroll features to record each employee's earnings and deductions. You must also enter the employer's tax liabilities for each employee. When you post a payroll entry, Peachtree updates the necessary general ledger accounts.

In this chapter, however, the focus is on manually recording the payroll entries. You will use the general journal to manually record the entries.

Step 6   Record the employer's payroll taxes.
Step 7   Print a General Journal report and a General Ledger report.
Step 8   Proof your work.
Step 9   Answer the Analyze question.

Ending the Session

Step 10   Click the **Close Problem** button in the Glencoe Smart Guide window and select the save option as directed by your teacher.

Problem 13-9   Recording and Posting Payroll Transactions

INSTRUCTIONS

Beginning a Session

Step 1   Select the problem set: Showbiz Video (Prob. 13-9).
Step 2   Rename the company and set the system date to December 31, 2008.

Completing the Accounting Problem

Step 3   Review the transactions provided in your textbook.
Step 4   Record the transactions.
Step 5   Print a General Journal report and a General Ledger report.
Step 6   Proof your work.
Step 7   Answer the Analyze question.

Checking Your Work and Ending the Session

Step 8   Click the **Close Problem** button in the Glencoe Smart Guide window.
Step 9   If your teacher has asked you to check your solution, select *Check my answer to this problem*. Review, print, and close the report.
Step 10   Click the **Close Problem** button and select the appropriate save option.

Problem 13-10   Recording and Posting Payroll Transactions

INSTRUCTIONS

Beginning a Session

Step 1   Select the problem set: Job Connect (Prob. 13-10).
Step 2   Rename the company and set the system date to December 31, 2008.

Completing the Accounting Problem

Step 3   Review the payroll information provided in your textbook.
Step 4   Record the transactions.
Step 5   Prepare Federal Tax Deposit-Form 8109 for each of the two federal tax deposits. Use the forms provided in the working papers.
Step 6   Print a General Journal report and a General Ledger report.
Step 7   Proof your work.
Step 8   Answer the Analyze question.

Ending the Session

Step 9   Click the **Close Problem** button in the Glencoe Smart Guide window.

FAQs

Can you print a Federal Tax Deposit-Form 8109 using the Peachtree software?

No, Peachtree does not include a Federal Tax Deposit-Form 8109 report. You must manually prepare this form when you submit a payroll tax deposit.

Computerized Accounting Using Spreadsheets

Problem 13-6   Calculating Employer's Payroll Taxes

Completing the Spreadsheet

Step 1   Read the instructions for Problem 13-6 in your textbook. This problem involves calculating employer's payroll taxes.

Step 2   Open the Glencoe Accounting: Electronic Learning Center software.

Step 3   From the Program Menu, click on the **Peachtree Complete®** **Accounting Software and Spreadsheet Applications** icon.

Step 4   Log onto the Management System by typing your user name and password.

Step 5   Under the **Problems & Tutorials** tab, select template 13-6 from the Chapter 13 drop-down menu. The template should look like the one shown below.

```
PROBLEM 13-6
CALCULATING EMPLOYER'S PAYROLL TAXES

(name)
(date)
```

Total Gross Earnings	Social Security Tax	Medicare Tax	Federal Unemployment Tax	State Unemployment Tax
	$0.00	$0.00	$0.00	$0.00
	$0.00	$0.00	$0.00	$0.00
	$0.00	$0.00	$0.00	$0.00
	$0.00	$0.00	$0.00	$0.00
	$0.00	$0.00	$0.00	$0.00

Step 6   Key your name and today's date in the cells containing the *(name)* and *(date)* placeholders.

Step 7   Enter the total gross earnings for the first employee in cell A11 of the spreadsheet template: **914.80**. Remember, it is not necessary to enter a dollar sign. The spreadsheet template will automatically calculate the Social Security tax, Medicare tax, federal unemployment tax, and state unemployment tax for the first employee using the rates stated in your textbook.

 TIP: The cells for Social Security tax, Medicare tax, federal unemployment tax, and state unemployment tax are set up to round these numbers to two decimal places. When you are entering data in this spreadsheet template, always round numbers to two decimal places when rounding is necessary.

Step 8   Enter the total gross earnings for the remaining employees. The Social Security tax, Medicare tax, federal unemployment tax, and state unemployment tax will be automatically calculated for each employee.

Spreadsheet Guide

TIP: To check your work, multiply the total gross earnings of each employee by the tax rates given in your textbook.

Step 9 Save the spreadsheet using the **Save** option from the **File** menu. You should accept the default location for the save as this is handled by the management system.

Step 10 Print the completed spreadsheet.

Step 11 Exit the spreadsheet program.

Step 12 In the Close Options box, select the location where you would like to save your work.

Step 13 Answer the Analyze question from your textbook for this problem.

What-If Analysis

What would the employer's payroll taxes be on total gross earnings of $1,891.02, assuming the employee had not reached the taxable earnings limit?

Problem 13-10   (concluded)

$ _____ No. 1602
Date _____ 20 ___
To _____
For _____

	Dollars	Cents
Balance brought forward	29,441	61
Add deposits		
Total		
Less this check		
Balance carried forward		

Job Connect 1602
4457 Market Street
Kingstown, NC 28150

4-58 / 810

DATE _____ 20 ___

PAY TO THE ORDER OF _____ $ _____

_____ DOLLARS

SNB Security National Bank
KINGSTOWN, NC

MEMO _____

⑈0810 0058⑈ 4163 697⑈ 1602

$ _____ No. 1603
Date _____ 20 ___
To _____
For _____

	Dollars	Cents
Balance brought forward		
Add deposits		
Total		
Less this check		
Balance carried forward		

Job Connect 1603
4457 Market Street
Kingstown, NC 28150

4-58 / 810

DATE _____ 20 ___

PAY TO THE ORDER OF _____ $ _____

_____ DOLLARS

SNB Security National Bank
KINGSTOWN, NC

MEMO _____

⑈0810 0058⑈ 4163 697⑈ 1603

$ _____ No. 1604
Date _____ 20 ___
To _____
For _____

	Dollars	Cents
Balance brought forward		
Add deposits		
Total		
Less this check		
Balance carried forward		

Job Connect 1604
4457 Market Street
Kingstown, NC 28150

4-58 / 810

DATE _____ 20 ___

PAY TO THE ORDER OF _____ $ _____

_____ DOLLARS

SNB Security National Bank
KINGSTOWN, NC

MEMO _____

⑈0810 0058⑈ 4163 697⑈ 1604

$ _____ No. 1605
Date _____ 20 ___
To _____
For _____

	Dollars	Cents
Balance brought forward		
Add deposits		
Total		
Less this check		
Balance carried forward		

Job Connect 1605
4457 Market Street
Kingstown, NC 28150

4-58 / 810

DATE _____ 20 ___

PAY TO THE ORDER OF _____ $ _____

_____ DOLLARS

SNB Security National Bank
KINGSTOWN, NC

MEMO _____

⑈0810 0058⑈ 4163 697⑈ 1605

CHAPTER **13** Payroll Liabilities and Tax Records

Self-Test

Part A   True or False

Directions: *Circle the letter* T *in the Answer column if the statement is true; circle the letter* F *if the statement is false.*

Answer

T   F   **1.** The Cash in Bank account is debited for the amount the employees actually earn in the pay period.

T   F   **2.** The only taxes that the employer must pay are unemployment taxes.

T   F   **3.** The amount entered for Payroll Tax Expense is the total gross earnings of the pay period.

T   F   **4.** Social security tax is paid by the employer and the employees.

T   F   **5.** Amounts deducted from employees' earnings and held for payment by the employer become liabilities to the business.

T   F   **6.** The entry for Salaries Expense is recorded each pay period but the entry for Payroll Tax Expense is recorded only at the end of the year.

T   F   **7.** Federal withholding taxes and social security taxes are normally paid once a year by the employer.

T   F   **8.** The payroll register is the source of information for preparing the journal entry for payroll.

T   F   **9.** Form W-2 must be prepared and given to each employee by January 31 of the year following that in which the taxes were deducted.

T   F   **10.** The employer's payroll taxes are operating expenses of the business.

Part B   Debit or Credit

Directions: *Each of the following statements can be completed with the word* Debit *or* Credit. *Circle the correct word in the Answer column.*

Answer

Debit   Credit   **1.** The total amount of gross earnings is entered on the _____ side of the Salaries Expense account.

Debit   Credit   **2.** The account Employees' Federal Income Tax Payable would normally have a _____ balance.

Debit   Credit   **3.** Salaries Expense would normally have a _____ balance.

Debit   Credit   **4.** The total net pay for the period is entered on the _____ side of the Cash in Bank account.

Debit   Credit   **5.** The Medicare Tax Payable account would normally have a _____ balance.

Debit   Credit   **6.** When a payment is made to the state government for the employees' state income tax, an entry is made on the _____ side of the Employees' State Income Tax Payable account.

Debit   Credit   **7.** The Payroll Tax Expense account normally has a _____ balance.

Debit   Credit   **8.** A decrease in U.S. Savings Bonds Payable would be a _____.

Debit   Credit   **9.** An increase in State Unemployment Tax Payable would be a _____.

Debit   Credit   **10.** The account Insurance Premiums Payable would normally have a _____ balance.

MINI PRACTICE SET **3**

Green Thumb Plant Service

Instructions: *Complete the following time cards and use to record payroll information in the payroll register.*

(1)

NO. __019__
NAME __Michael Alter__
SOC. SEC. NO. __049-71-8436__
WEEK ENDING __7/25/20－－__

DAY	IN	OUT	IN	OUT	IN	OUT	TOTAL
M			2:00	5:00			
T			2:00	6:00			
W			3:00	5:00			
Th			2:00	6:00			
F			2:00	6:00			
S			9:00	2:00			
S							
					TOTAL HOURS		

	HOURS	RATE	AMOUNT
REGULAR			
OVERTIME			
TOTAL EARNINGS			

SIGNATURE _____ DATE _____

NO. __018__
NAME __Christine Cuddy__
SOC. SEC. NO. __223-56-0992__
WEEK ENDING __7/25/20－－__

DAY	IN	OUT	IN	OUT	IN	OUT	TOTAL
M	9:00	12:00	12:30	5:00			
T	9:00	11:30	12:00	5:00			
W	9:00	1:00					
Th	9:00	12:00	12:30	4:00			
F	8:30	1:00	1:30	3:00			
S	9:00	1:30					
S							
					TOTAL HOURS		

	HOURS	RATE	AMOUNT
REGULAR			
OVERTIME			
TOTAL EARNINGS			

SIGNATURE _____ DATE _____

NO. __013__
NAME __Joclyn Filley__
SOC. SEC. NO. __042-97-3814__
WEEK ENDING __7/25/20－－__

DAY	IN	OUT	IN	OUT	IN	OUT	TOTAL
M	9:00	12:00	1:00	3:00			
T	9:00	12:00	1:00	5:00			
W	8:00	12:00	1:00	5:00			
Th	9:00	12:00	1:00	3:30			
F	9:00	12:00	1:00	4:00			
S	9:00	12:00					
S							
					TOTAL HOURS		

	HOURS	RATE	AMOUNT
REGULAR			
OVERTIME			
TOTAL EARNINGS			

SIGNATURE _____ DATE _____

NO. __016__
NAME __Daniel Ripp__
SOC. SEC. NO. __011-79-2118__
WEEK ENDING __7/25/20－－__

DAY	IN	OUT	IN	OUT	IN	OUT	TOTAL
M	9:00	12:00	12:30	5:00			
T	9:00	12:30	1:00	6:00			
W	9:00	12:00	1:00	4:30			
Th	8:30	12:30	1:00	5:00			
F	9:00	11:30	12:00	5:00			
S	9:00	1:00					
S							
					TOTAL HOURS		

	HOURS	RATE	AMOUNT
REGULAR			
OVERTIME			
TOTAL EARNINGS			

SIGNATURE _____ DATE _____

Mini Practice Set 3

Federal Income Tax Table

					SINGLE Persons—WEEKLY Payroll Period							
					(For Wages Paid in 20--)							
If the wages are—		And the number of withholding allowances claimed is—										
At least	But less than	0	1	2	3	4	5	6	7	8	9	10
		The amount of income tax to be withheld is—										
125	130	11	4	0	0	0	0	0	0	0	0	0
130	135	12	5	0	0	0	0	0	0	0	0	0
135	140	13	5	0	0	0	0	0	0	0	0	0
140	145	14	6	0	0	0	0	0	0	0	0	0
145	150	14	7	0	0	0	0	0	0	0	0	0
150	155	15	8	0	0	0	0	0	0	0	0	0
155	160	16	8	1	0	0	0	0	0	0	0	0
160	165	17	9	1	0	0	0	0	0	0	0	0
165	170	17	10	2	0	0	0	0	0	0	0	0
170	175	18	11	3	0	0	0	0	0	0	0	0
175	180	19	11	4	0	0	0	0	0	0	0	0
180	185	20	12	4	0	0	0	0	0	0	0	0
185	190	20	13	5	0	0	0	0	0	0	0	0
190	195	21	14	6	0	0	0	0	0	0	0	0
195	200	22	14	7	0	0	0	0	0	0	0	0
200	210	23	15	8	0	0	0	0	0	0	0	0
210	220	25	17	9	2	0	0	0	0	0	0	0
220	230	26	18	11	3	0	0	0	0	0	0	0
230	240	28	20	12	5	0	0	0	0	0	0	0
240	250	29	21	14	6	0	0	0	0	0	0	0
250	260	31	23	15	8	0	0	0	0	0	0	0
260	270	32	24	17	9	2	0	0	0	0	0	0
270	280	34	26	18	11	3	0	0	0	0	0	0
280	290	35	27	20	12	5	0	0	0	0	0	0
290	300	37	29	21	14	6	0	0	0	0	0	0
300	310	38	30	23	15	8	0	0	0	0	0	0
310	320	40	32	24	17	9	1	0	0	0	0	0
320	330	41	33	26	18	11	3	0	0	0	0	0
330	340	43	35	27	20	12	4	0	0	0	0	0
340	350	44	36	29	21	14	6	0	0	0	0	0

Mini Practice Set 3   (continued)

Federal Income Tax Table

MARRIED Persons—WEEKLY Payroll Period											
(For Wages Paid in 20--)											
If the wages are—		And the number of withholding allowances claimed is—										
At least	But less than	0	1	2	3	4	5	6	7	8	9	10
		The amount of income tax to be withheld is—										
340	350	33	26	18	10	3	0	0	0	0	0	0
350	360	35	27	19	12	4	0	0	0	0	0	0
360	370	36	29	21	13	6	0	0	0	0	0	0
370	380	38	30	22	15	7	0	0	0	0	0	0
380	390	39	32	24	16	9	1	0	0	0	0	0
390	400	41	33	25	18	10	2	0	0	0	0	0
400	410	42	35	27	19	12	4	0	0	0	0	0
410	420	44	36	28	21	13	5	0	0	0	0	0
420	430	45	38	30	22	15	7	0	0	0	0	0
430	440	47	39	31	24	16	8	1	0	0	0	0
440	450	48	41	33	25	18	10	2	0	0	0	0
450	460	50	42	34	27	19	11	4	0	0	0	0
460	470	51	44	36	28	21	13	5	0	0	0	0
470	480	53	45	37	30	22	14	7	0	0	0	0
480	490	54	47	39	31	24	16	8	1	0	0	0
490	500	56	48	40	33	25	17	10	2	0	0	0
500	510	57	50	42	34	27	19	11	4	0	0	0
510	520	59	51	43	36	28	20	13	5	0	0	0
520	530	60	53	45	37	30	22	14	7	0	0	0
530	540	62	54	46	39	31	23	16	8	0	0	0
540	550	63	56	48	40	33	25	17	10	2	0	0
550	560	65	57	49	42	34	26	19	11	3	0	0
560	570	66	59	51	43	36	28	20	13	5	0	0
570	580	68	60	52	45	37	29	22	14	6	0	0
580	590	69	62	54	46	39	31	23	16	8	0	0
590	600	71	63	55	48	40	32	25	17	9	2	0
600	610	72	65	57	49	42	34	26	19	11	3	0
610	620	74	66	58	51	43	35	28	20	12	5	0
620	630	75	68	60	52	45	37	29	22	14	6	0
630	640	77	69	61	54	46	38	31	23	15	8	0
640	650	78	71	63	55	48	40	32	25	17	9	2
650	660	80	72	64	57	49	41	34	26	18	11	3
660	670	81	74	66	58	51	43	35	28	20	12	5
670	680	83	75	67	60	52	44	37	29	21	14	6
680	690	84	77	69	61	54	46	38	31	23	15	8
690	700	86	78	70	63	55	47	40	32	24	17	9
700	710	87	80	72	64	57	49	41	34	26	18	11
710	720	89	81	73	66	58	50	43	35	27	20	12
720	730	90	83	75	67	60	52	44	37	29	21	14
730	740	92	84	76	69	61	53	46	38	30	23	15

Mini Practice Set 3 (continued) **(2), (3), (4), (5), (6)**

PAYROLL REGISTER

PAY PERIOD ENDING 20____ DATE OF PAYMENT _____

EMPLOYEE NUMBER	NAME	MAR. STATUS	ALLOW.	TOTAL HOURS	RATE	EARNINGS			DEDUCTIONS							NET PAY	CK. NO.
						REGULAR	OVERTIME	TOTAL	SOC. SEC. TAX	MED. TAX	FED. INC. TAX	STATE INC. TAX	HOSP. INS.	OTHER	TOTAL		
1																	
2																	
3																	
4																	
5																	
6																	
7																	
8																	
9																	
10																	
11																	
12																	
13																	
14																	
15																	
16																	
17																	
18																	
19																	
20																	
21																	
22																	
23																	
24																	
25																	
	TOTALS																

Other Deductions: Write the appropriate code letter to the left of the amount: B—U.S. Savings Bonds; C—Credit Union; UD—Union Dues; UW—United Way.

Mini Practice Set 3   (continued)

(7)

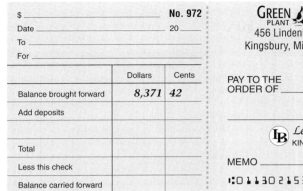

	Dollars	Cents
Balance brought forward	8,371	42
Add deposits		
Total		
Less this check		
Balance carried forward		

$ _____ No. 972
Date _____ 20 ____
To _____
For _____

GREEN THUMB PLANT SERVICE
456 Lindenhurst Street
Kingsbury, Michigan 03855
972
53-215
113
DATE _____ 20 ____
PAY TO THE ORDER OF _____ $ _____
_____ DOLLARS
Lexington Bank
KINGSBURY, MICHIGAN
MEMO _____
⑈011302153⑈ 331 234 9⑈ 0972

(12)

	Dollars	Cents
Balance brought forward		
Add deposits		
Total		
Less this check		
Balance carried forward		

$ _____ No. 973
Date _____ 20 ____
To _____
For _____

GREEN THUMB PLANT SERVICE
456 Lindenhurst Street
Kingsbury, Michigan 03855
973
53-215
113
DATE _____ 20 ____
PAY TO THE ORDER OF _____ $ _____
_____ DOLLARS
Lexington Bank
KINGSBURY, MICHIGAN
MEMO _____
⑈011302153⑈ 331 234 9⑈ 0973

(14)

	Dollars	Cents
Balance brought forward		
Add deposits		
Total		
Less this check		
Balance carried forward		

$ _____ No. 974
Date _____ 20 ____
To _____
For _____

GREEN THUMB PLANT SERVICE
456 Lindenhurst Street
Kingsbury, Michigan 03855
974
53-215
113
DATE _____ 20 ____
PAY TO THE ORDER OF _____ $ _____
_____ DOLLARS
Lexington Bank
KINGSBURY, MICHIGAN
MEMO _____
⑈011302153⑈ 331 234 9⑈ 0974

(7)

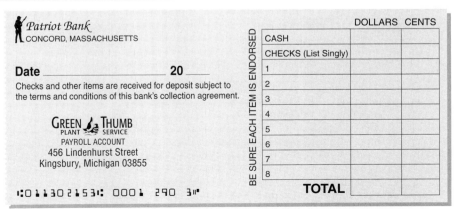

Patriot Bank
CONCORD, MASSACHUSETTS

Date _____ 20 ____

Checks and other items are received for deposit subject to the terms and conditions of this bank's collection agreement.

GREEN THUMB PLANT SERVICE
PAYROLL ACCOUNT
456 Lindenhurst Street
Kingsbury, Michigan 03855

⑈011302153⑈ 0001 290 3⑈

BE SURE EACH ITEM IS ENDORSED

	DOLLARS	CENTS
CASH		
CHECKS (List Singly)		
1		
2		
3		
4		
5		
6		
7		
8		
TOTAL		

Mini Practice Set 3 (continued)
(8), (11), (13), (14)

GENERAL JOURNAL   PAGE _____

	DATE		DESCRIPTION	POST. REF.	DEBIT	CREDIT	
1							1
2							2
3							3
4							4
5							5
6							6
7							7
8							8
9							9
10							10
11							11
12							12
13							13
14							14
15							15
16							16
17							17
18							18
19							19
20							20
21							21
22							22
23							23
24							24
25							25
26							26
27							27
28							28
29							29
30							30
31							31
32							32
33							33
34							34
35							35
36							36

Mini Practice Set 3 (continued)

GENERAL LEDGER (PARTIAL)

ACCOUNT **Cash in Bank**   ACCOUNT NO. **101**

DATE		DESCRIPTION	POST. REF.	DEBIT	CREDIT	BALANCE DEBIT	BALANCE CREDIT
20--							
July	18	Balance	✓			8 3 7 1 42	

ACCOUNT **Employees' Federal Income Tax Payable**   ACCOUNT NO. **205**

DATE		DESCRIPTION	POST. REF.	DEBIT	CREDIT	BALANCE DEBIT	BALANCE CREDIT
20--							
July	18	Balance	✓				1 8 3 00

ACCOUNT **Employees' State Income Tax Payable**   ACCOUNT NO. **210**

DATE		DESCRIPTION	POST. REF.	DEBIT	CREDIT	BALANCE DEBIT	BALANCE CREDIT
20--							
July	18	Balance	✓				2 4 5 74

ACCOUNT **Social Security Tax Payable**   ACCOUNT NO. **215**

DATE		DESCRIPTION	POST. REF.	DEBIT	CREDIT	BALANCE DEBIT	BALANCE CREDIT
20--							
July	18	Balance	✓				2 1 7 96

Mini Practice Set 3   (continued)

ACCOUNT __Medicare Tax Payable__   ACCOUNT NO. __220__

DATE		DESCRIPTION	POST. REF.	DEBIT	CREDIT	BALANCE DEBIT	BALANCE CREDIT
20--							
July	18	Balance	✓				5444

ACCOUNT __Insurance Premiums Payable__   ACCOUNT NO. __225__

DATE		DESCRIPTION	POST. REF.	DEBIT	CREDIT	BALANCE DEBIT	BALANCE CREDIT
20--							
July	18	Balance	✓				17100

ACCOUNT __Federal Unemployment Tax Payable__   ACCOUNT NO. __235__

DATE		DESCRIPTION	POST. REF.	DEBIT	CREDIT	BALANCE DEBIT	BALANCE CREDIT
20--							
July	4	Balance	✓				3071
	11		G18		1643		4714
	18		G18		1436		6150

ACCOUNT __State Unemployment Tax Payable__   ACCOUNT NO. __240__

DATE		DESCRIPTION	POST. REF.	DEBIT	CREDIT	BALANCE DEBIT	BALANCE CREDIT
20--							
July	4	Balance	✓				20620
	11		G18		10416		31036
	18		G18		9679		40715

Mini Practice Set 3   (continued)

ACCOUNT __U.S. Savings Bonds Payable__   ACCOUNT NO. ___245___

DATE		DESCRIPTION	POST. REF.	DEBIT	CREDIT	BALANCE DEBIT	BALANCE CREDIT
20--							
July	4	Balance	✓				4000
	11		G18		2000		6000
	18		G18		2000		8000

ACCOUNT __United Way Payable__   ACCOUNT NO. ___250___

DATE		DESCRIPTION	POST. REF.	DEBIT	CREDIT	BALANCE DEBIT	BALANCE CREDIT
20--							
July	4	Balance	✓				1200
	11		G18		1200		2400
	18		G18		1200		3600

ACCOUNT __Payroll Tax Expense__   ACCOUNT NO. ___620___

DATE		DESCRIPTION	POST. REF.	DEBIT	CREDIT	BALANCE DEBIT	BALANCE CREDIT
20--							
July	4	Balance	✓			568920	
	11		G18	36941		605861	
	18		G18	36617		642478	

ACCOUNT __Salaries Expense__   ACCOUNT NO. ___630___

DATE		DESCRIPTION	POST. REF.	DEBIT	CREDIT	BALANCE DEBIT	BALANCE CREDIT
20--							
July	11	Balance	✓			4394739	
	18		G18	296314		4691053	

Mini Practice Set 3   (continued)

(9)

GREEN THUMB PLANT SERVICE
PAYROLL ACCOUNT
456 Lindenhurst Street
Kingsbury, Michigan 03855

310

53-215
113

Date _____ 20 _____

Pay to the
Order of _____ $ _____

_____ Dollars

Patriot Bank
CONCORD, MASSACHUSETTS

⑆011302153⑆ 0001 290 3⑈ 0310

Employee Pay Statement
Detach and retain this statement.

310

Period Ending	Earnings			Deductions							Net Pay
	Regular	Overtime	Total	Social Security Tax	Med. Tax	Federal Income Tax	State Income Tax	Hosp. Ins.	Other	Total	

GREEN THUMB PLANT SERVICE
PAYROLL ACCOUNT
456 Lindenhurst Street
Kingsbury, Michigan 03855

311

53-215
113

Date _____ 20 _____

Pay to the
Order of _____ $ _____

_____ Dollars

Patriot Bank
CONCORD, MASSACHUSETTS

⑆011302153⑆ 0001 290 3⑈ 0311

Employee Pay Statement
Detach and retain this statement.

311

Period Ending	Earnings			Deductions							Net Pay
	Regular	Overtime	Total	Social Security Tax	Med. Tax	Federal Income Tax	State Income Tax	Hosp. Ins.	Other	Total	

Mini Practice Set 3   (continued)

GREEN THUMB
PLANT SERVICE
PAYROLL ACCOUNT
456 Lindenhurst Street
Kingsbury, Michigan 03855

312
53-215
113

Date _____ 20 ____

Pay to the
Order of _____ $ _____

_____ Dollars

Patriot Bank
CONCORD, MASSACHUSETTS

⑆011302153⑆ 0001 290 3⑈ 0312

Employee Pay Statement
Detach and retain this statement.

312

Period Ending	Earnings			Deductions							Net Pay
	Regular	Overtime	Total	Social Security Tax	Med. Tax	Federal Income Tax	State Income Tax	Hosp. Ins.	Other	Total	

GREEN THUMB
PLANT SERVICE
PAYROLL ACCOUNT
456 Lindenhurst Street
Kingsbury, Michigan 03855

313
53-215
113

Date _____ 20 ____

Pay to the
Order of _____ $ _____

_____ Dollars

Patriot Bank
CONCORD, MASSACHUSETTS

⑆011302153⑆ 0001 290 3⑈ 0313

Employee Pay Statement
Detach and retain this statement.

313

Period Ending	Earnings			Deductions							Net Pay
	Regular	Overtime	Total	Social Security Tax	Med. Tax	Federal Income Tax	State Income Tax	Hosp. Ins.	Other	Total	

Mini Practice Set 3 (continued)

GREEN THUMB
PLANT SERVICE
PAYROLL ACCOUNT
456 Lindenhurst Street
Kingsbury, Michigan 03855

314

53-215
113

Date_____ 20 ____

Pay to the
Order of _____ $ _____

_____ Dollars

Patriot Bank
CONCORD, MASSACHUSETTS

⑆011302153⑆ 0001 290 3⑈ 0314

Employee Pay Statement
Detach and retain this statement.

314

Period Ending	Earnings			Deductions							Net Pay
	Regular	Overtime	Total	Social Security Tax	Med. Tax	Federal Income Tax	State Income Tax	Hosp. Ins.	Other	Total	

GREEN THUMB
PLANT SERVICE
PAYROLL ACCOUNT
456 Lindenhurst Street
Kingsbury, Michigan 03855

315

53-215
113

Date_____ 20 ____

Pay to the
Order of _____ $ _____

_____ Dollars

Patriot Bank
CONCORD, MASSACHUSETTS

⑆011302153⑆ 0001 290 3⑈ 0315

Employee Pay Statement
Detach and retain this statement.

315

Period Ending	Earnings			Deductions							Net Pay
	Regular	Overtime	Total	Social Security Tax	Med. Tax	Federal Income Tax	State Income Tax	Hosp. Ins.	Other	Total	

Mini Practice Set 3   (continued)

GREEN THUMB
PLANT SERVICE
PAYROLL ACCOUNT
456 Lindenhurst Street
Kingsbury, Michigan 03855

316

53-215
113

Date _____ 20 _____

Pay to the
Order of _____ $ _____

_____ Dollars

Patriot Bank
CONCORD, MASSACHUSETTS

⑆011302153⑆ 0001 290 3⑈ 0316

Employee Pay Statement
Detach and retain this statement.

316

Period Ending	Earnings			Deductions							Net Pay
	Regular	Overtime	Total	Social Security Tax	Med. Tax	Federal Income Tax	State Income Tax	Hosp. Ins.	Other	Total	

GREEN THUMB
PLANT SERVICE
PAYROLL ACCOUNT
456 Lindenhurst Street
Kingsbury, Michigan 03855

317

53-215
113

Date _____ 20 _____

Pay to the
Order of _____ $ _____

_____ Dollars

Patriot Bank
CONCORD, MASSACHUSETTS

⑆011302153⑆ 0001 290 3⑈ 0317

Employee Pay Statement
Detach and retain this statement.

317

Period Ending	Earnings			Deductions							Net Pay
	Regular	Overtime	Total	Social Security Tax	Med. Tax	Federal Income Tax	State Income Tax	Hosp. Ins.	Other	Total	

Mini Practice Set 3   (continued)

(10)

EMPLOYEE'S EARNINGS RECORD FOR QUARTER ENDING **September 30, 20--**

Alter
Last Name

Michael   Initial
First

Address
479 Lindon Street

Kingsbury, Michigan

EMPLOYEE NO. **019**

POSITION **Supply Clerk**

RATE OF PAY **7.10**

MARITAL STATUS **S**

ALLOWANCES **1**

SOC. SEC. NO. **049-71-8436**

PAY PERIOD NO.	ENDED	EARNINGS REGULAR	OVERTIME	TOTAL	DEDUCTIONS SOC. SEC. TAX	MED. TAX	FED. INC. TAX	STATE INC. TAX	HOSP. INS.	OTHER	TOTAL	NET PAY	ACCUMULATED EARNINGS
													2,572 10
1	7/4	158 60		158 60	9 83	2 30	8 00	3 17	6 00	5 00	34 30	124 30	2,730 70
2	7/11	154 10		154 10	9 55	2 23	8 00	3 08	6 00	5 00	33 86	120 24	2,884 80
3	7/18	147 30		147 30	9 13	2 13	6 00	2 94	6 00	5 00	31 20	116 10	3,032 10
4													
5													
6													
7													
8													
9													
10													
11													
12													
13													
QUARTERLY TOTALS													

Other Deductions: B—U.S. Savings Bonds; C—Credit Union; UD—Union Dues; UW—United Way.

Mini Practice Set 3   (continued)

EMPLOYEE'S EARNINGS RECORD FOR QUARTER ENDING _September 30, 20--_

Last Name _Millette_
First _Greg_   Initial
Address _86 Meadow Road_
Kingsbury, Michigan

EMPLOYEE NO. _011_
POSITION _Salesperson_
RATE OF PAY _$450/week + 10%_

MARITAL STATUS _M_
SOC. SEC. NO. _046-29-8403_
ALLOWANCES _2_

PAY PERIOD		EARNINGS			DEDUCTIONS							NET PAY	ACCUMULATED EARNINGS
NO.	ENDED	REGULAR	OVERTIME	TOTAL	SOC. SEC. TAX	MED. TAX	FED. INC. TAX	STATE INC. TAX	HOSP. INS.	OTHER	TOTAL		
													9,349 20
1	7/4	526 10		526 10	32 62	7 63	44 00	10 52	9 00	5 00	108 77	417 33	9,875 30
2	7/11	519 60		519 60	32 22	7 53	42 00	10 39	9 00	5 00	106 14	413 46	10,394 90
3	7/18	584 20		584 20	36 22	8 47	51 00	11 68	9 00	5 00	121 37	462 83	10,979 10
4													
5													
6													
7													
8													
9													
10													
11													
12													
13													
QUARTERLY TOTALS													

Other Deductions: B—U.S. Savings Bonds; C—Credit Union; UD—Union Dues; UW—United Way.

Mini Practice Set 3   (concluded)

(12)

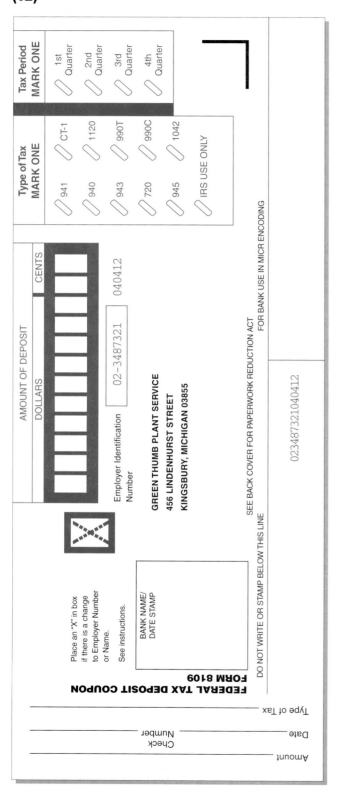

Computerized Accounting Using Peachtree

Mini Practice Set 3

INSTRUCTIONS

Beginning a Session

Step 1 Open the Glencoe Accounting: Electronic Learning Center software.

Step 2 Click on the **Peachtree Complete® Accounting Software and Spreadsheet Applications** icon.

Step 3 Log onto the Management System by typing your user name and password.

Step 4 Under the **Problems & Tutorials** tab, select problem set: Green Thumb Plant Service (MP-3) from the drop-down menu.

Step 5 Rename the company by adding your initials, e.g., Green Thumb (MP-3: XXX).

Step 6 Set the system date to July 25, 2008.

Completing the Accounting Problem

Step 7 Review the payroll information shown in your textbook for Green Thumb Plant Service.

Step 8 Change employee record 022 using the **Employees/Sales Reps** option. Enter your name in the *Name* field. Do not change any of the other employee information.

TIP: All of the employee information is already recorded in the problem set for you.

Step 9 Record the weekly payroll and print checks for all employees using the **Payroll Entry** option.

IMPORTANT: You must manually enter each employee's payroll tax deductions and other deductions. Use the federal tax tables found in your workbook. Remember to enter the employee deductions (Fed_Income, Soc_Sec, etc.) as **negative** amounts.

TIP: Select *PR MultiP Chks 2 Stub* (or equivalent) for the payroll check format, and use 310 for the first payroll check number.

TIP: Peachtree automatically generates the entries to record the payroll, but it does not record the employer's payroll tax expense.

Step 10 Print a Payroll Register report for the current week.

TIP: While viewing a Payroll Register report, you can double-click a report item to go directly to the Payroll Entry window.

Step 11 Proof your work. Make any corrections as needed and print a revised report, if necessary.

Step 12 Print a Payroll Journal for the current week.

Step 13 Calculate and record the employer's payroll tax expense (Social Security, Medicare, FUTA, and SUTA) using the **General Journal Entry** option.

Step 14 Record the deposit for the taxes owed to the federal government, and enter the monthly insurance premium ($228) using the **General Journal Entry** option. Manually complete Federal Tax Deposit Coupon Form 8109.

TIP: Display a General Ledger report to determine the balances in the payroll tax liability accounts.

Step 15 Print a General Journal report for July 25, 2008, and proof your work.

Step 16 Print a General Ledger report.

Step 17 Print Quarterly Earnings reports for Michael Alter and Greg Millette for July 25, 2008, from the Payroll Report List.

Analyzing Your Work

Step 18 Answer the Analyze questions.

Checking Your Work and Ending the Session

Step 19 Click the **Close Problem** button in the Glencoe Smart Guide window.

Step 20 If your teacher has asked you to check your solution, select *Check my answer to this problem.* Review, print, and close this report.

Step 21 Click the **Close Problem** button and select the appropriate save option.

MINI PRACTICE SET 3

Green Thumb Plant Service

Audit Test

Instructions: *Use your completed solutions to answer the following questions. Write the answer in the space to the left of each question.*

Answer

_____   **1.** What rate is used to compute employee state income tax?

_____   **2.** What rate is used to compute the employer's Federal unemployment tax?

_____   **3.** What commission amount did Greg Millette earn?

_____   **4.** How many employees reached the maximum taxable amount for the social security tax this period?

_____   **5.** What was the total net pay for the pay period ending July 25, 20--?

_____   **6.** How many payroll checks were issued for this period?

_____   **7.** What was the amount of check 972?

_____   **8.** What was the amount paid to American Insurance Company for employee hospital insurance?

_____   **9.** What accounts were debited when check 973 was recorded in the general ledger?

_____   **10.** What is the ending balance of the Employees' Federal Income Tax Payable account at July 31?

_____ **11.** What is the ending balance of the Cash in Bank account at month end?

_____ **12.** What is the amount of payroll check 312?

_____ **13.** What is the amount of accumulated earnings for Michael Alter after the July 25 paycheck?

_____ **14.** What amount was remitted with the Federal Tax Deposit Coupon Form 8109 on July 25?

_____ **15.** What total amount of federal income tax was withheld from the paycheck of Greg Millette at the July 25 pay period?

_____ **16.** What is the total of payroll liabilities at July 25?

_____ **17.** What amount was debited to Salaries Expense on July 25?

_____ **18.** What is the balance in the Green Thumb Plant Service regular bank account after check 974 is recorded?

_____ **19.** What is the next pay period date for this business?

CHAPTER 14   Accounting for Sales and Cash Receipts

Study Plan

Check Your Understanding

Section 1	*Read Section 1 on pages 354–356 and complete the following exercises on page 357.* ❑ Thinking Critically ❑ Communication Accounting ❑ Problem 14-1 *Recording Merchandising Transactions*
Section 2	*Read Section 2 on pages 358–365 and complete the following exercises on page 366.* ❑ Thinking Critically ❑ Analyzing Accounting ❑ Problem 14-2 *Recording Sales on Account and Sales Returns and Allowances Transactions*
Section 3	*Read Section 3 on pages 367–375 and complete the following exercises on page 376.* ❑ Thinking Critically ❑ Analyzing Accounting ❑ Problem 14-3 *Analyzing a Source Document* ❑ Problem 14-4 *Recording Cash Receipts*
Summary	*Review the Chapter 14 Summary on page 377 in your textbook.* ❑ Key Concepts
Review and Activities	*Complete the following questions and exercises on pages 378–379 in your textbook.* ❑ Using Key Terms ❑ Understanding Accounting Concepts and Procedures ❑ Case Study ❑ Conducting an Audit with Alex ❑ Internet Connection ❑ Workplace Skills
Computerized Accounting	*Read the Computerized Accounting information on page 380 in your textbook.* ❑ *Making the Transition from a Manual to a Computerized System* ❑ *Entering Sales and Cash Receipts in Peachtree*
Problems	*Complete the following end-of-chapter problems for Chapter 14.* ❑ Problem 14-5 *Recording Sales and Cash Receipts* ❑ Problem 14-6 *Posting Sales and Cash Receipts* ❑ Problem 14-7 *Recording Sales and Cash Receipts* ❑ Problem 14-8 *Recording Sales and Cash Receipts Transactions*
Challenge Problem	❑ Problem 14-9 *Recording and Posting Sales and Cash Receipts*
Chapter Reviews and Working Papers	*Complete the following exercises for Chapter 14 in your Chapter Reviews and Working Papers.* ❑ Chapter Review ❑ Self-Test

CHAPTER 14 REVIEW — Accounting for Sales and Cash Receipts

Part 1 Accounting Vocabulary (24 points)

Total Points	52
Student's Score	

Directions: *Using terms from the following list, complete the sentences below. Write the letter of the term you have chosen in the space provided.*

A. accounts receivable subsidiary ledger	**G.** contra account	**N.** merchandising business	**T.** sales on account
B. bankcard	**H.** controlling account	**O.** receipt	**U.** sales return
C. cash discount	**I.** credit card	**P.** retailer	**V.** sales slip
D. cash receipt	**J.** credit memorandum	**Q.** sales	**W.** sales tax
E. cash sale	**K.** credit terms	**R.** sales allowance	**X.** subsidiary ledger
F. charge customer	**L.** inventory	**S.** sales discount	**Y.** wholesaler
	M. merchandise		

___X___   **0.** A(n) _____ is a book that is summarized in a controlling account in the general ledger.

_____   **1.** The amount a customer may deduct if the payment for merchandise is made within a certain time is a(n) _____.

_____   **2.** An account whose balance decreases another account's balance is a(n) _____.

_____   **3.** A(n) _____ is a credit card issued by a bank and honored by many businesses.

_____   **4.** A(n) _____ is a form that serves as a record of cash received.

_____   **5.** A(n) _____ is a business that sells to the final user.

_____   **6.** A(n) _____ is a transaction that occurs when a business receives full payment for the merchandise sold at the time of the sale.

_____   **7.** A form that lists the details of a sale and is used as a record of the transaction is a(n) _____.

_____   **8.** The _____ maintains a list in alphabetical order of the charge customers of a business.

_____   **9.** The cash received by a business is referred to as a(n) _____.

_____   **10.** The goods a business buys for resale to customers are known as _____.

_____   **11.** A(n) _____ is a customer to whom a sale on account is made.

_____   **12.** Merchandise returned to the seller for full credit is a(n) _____.

_____   **13.** Accounts Receivable is a(n) _____ because its balance must equal the total of the accounts receivable subsidiary ledger.

_____   **14.** Many states and some cities require that a business add a(n) _____, or a percentage of the selling price of the goods, at the time of the sale.

_____   **15.** The _____ of a sale set out the time allowed for payment.

_____   **16.** A(n) _____ is a business that sells to retailers.

_____   **17.** A charge customer is entitled to charge merchandise with a(n) _____.

_____   **18.** The items of merchandise a business has in stock are referred to as _____.

_____   **19.** A(n) _____ permits a customer to pay for the merchandise at a later date.

_____   **20.** An example of a(n) _____ is Wal-Mart.

_____   **21.** A(n) _____ is a document prepared by the seller granting credit for damaged or returned merchandise.

_____   **22.** A price reduction granted by a business for damaged goods kept by the customer is called a(n) _____.

_____   **23.** The revenue account for a merchandising business is _____.

_____   **24.** A cash discount issued by the seller is a(n) _____.

Computerized Accounting Using Peachtree

Software Objectives

When you have completed this chapter, you will be able to use Peachtree to:

1. Record sales on account transactions using the **Sales/Invoicing** option.
2. Enter cash receipts using the **Receipts** option.
3. Record credit memorandums.
4. Print a Sales Journal report and a Cash Receipts Journal report.
5. Print a General Ledger report.

Problem 14-5 Recording Sales and Cash Receipts

Review the transactions listed in your textbook for Sunset Surfwear, a California-based merchandising store. As you will learn, the process for recording the sales and cash receipts transactions using the Peachtree accounting software is different than the method you learned in your text. Instead of recording these transactions in a general journal, you will record sales on account using the **Sales/Invoicing** option and process cash receipts with the **Receipts** option.

INSTRUCTIONS

Beginning a Session

Step 1 Select the problem set: Sunset Surfwear (Prob. 14-5).
Step 2 Rename the company by adding your initials, e.g., Sunset (Prob. 14-5: XXX).
Step 3 Set the system date to January 31, 2008.

Completing the Accounting Problem

The instructions in this section explain how to enter sales on account, credit memorandums, and cash receipts using the Peachtree software. To simplify the data entry process, you will batch similar transactions by type and then enter them using the appropriate software task option.

Entering Sales on Account

Step 4 Review the transactions for Sunset Surfwear and identify the sales on account.
Step 5 Enter the sale on account transaction for January 1.

January 1, Sold $300 in merchandise plus a sales tax of $18 on account to Martha Adams, Sales Slip 777.

To enter the sale on account:

- Choose **Sales/Invoicing** from the *Tasks* menu to display the Sales/Invoicing window.
- Click [Template] and then select <Predefined> Service to display a simplified entry form. (See Figure 14-5A.) **Note:** You must perform this step only once per session.

  Note: You can use the simplified form to enter the sales on account for Sunset Surfwear since you do not have to complete the following fields: Quantity, Item, Unit Price, and Freight Amt. Using a simplified form makes the data entry process easier.

DO YOU HAVE A QUESTION

Q. *Why can't you use the Peachtree **General Journal Entry** option to record sales on account and cash receipts from customers?*

A. Unlike a manual general journal, the Peachtree software does not provide a way for you to identify the corresponding **Accounts Receivable** subsidiary ledger account in the general journal. When you use the **Sales/Invoicing** option, for example, you must enter a Customer ID. Peachtree uses this information to update both the subsidiary ledger account and the **Accounts Receivable** balance.

Peachtree Guide

- Type **ADA** (the account code for Martha Adams) in the *Customer ID* field.

  If you do not know the account code for a customer, click the button, highlight the customer name and click the **OK** button.

TIP: As a shortcut, remember that you can press **SHIFT+?** when you are in a Lookup field to display a list of choices.

- Type **777** for the sales slip number in the *Invoice #* field.

- Type **1/1/08** to record the transaction date in the *Date* field. Make sure that Peachtree shows 2008 for the year. If not, change the system date.

- Move to the *Description* field by pressing **ENTER** or by clicking in the field.

  You can skip over the *Ship To, Customer PO, Ship Via, Ship Date, Terms,* and *Sales Rep ID* fields. Some of these fields, such as the *Ship To* field, are optional. You do not need to complete these fields when you are entering transactions from your text, but a company using Peachtree to keep its books might need to enter this information. For other fields, such as the *Terms* field, default information is already recorded in the field. You do not need to change this field unless different terms apply.

- Type **Sale on account** for the description.

- Skip the *GL Account* field. The **Sales** account number already appears in this field, Skip the Tax field.

- Type **300.00** in the *Amount* field to record the amount of the sale.

  When you enter the amount, Peachtree automatically calculates the sales tax based on the default sales tax code established for the customer. For example, if the customer was set up with a tax exempt status, Peachtree would not include any sales tax. Peachtree also calculates and displays the invoice total and amount due.

- Tab past the Job field.

Notes

Peachtree uses the term "invoice" instead of "sales slip" for sales on account.

TIP: Remember to enter the decimal point when you enter an amount. If you enter 300, Peachtree will automatically format the amount to $3.00 unless the decimal entry option is set to manual.

- Proof the information you just recorded. Check all of the information you entered. If you notice a mistake, move to that field and make the correction. Compare the information on your screen to the completed transaction shown in Figure 14-5A.

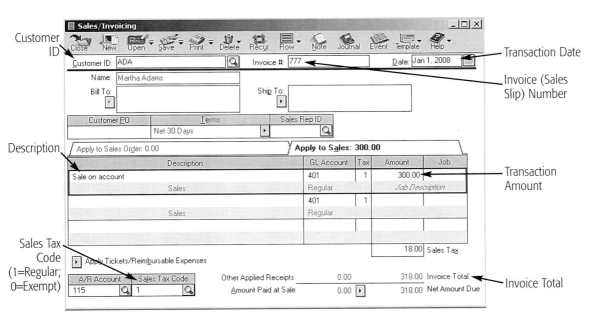

Figure 14-5A *Completed Sales/Invoicing Transaction (January 1, Sales Slip 777)*

- Click [Journal] to display the Accounting Behind the Screens window. Use the information in this window to check your work again. (See Figure 14-5B.)

The Accounting Behind the Screens window shows you the accounts and the debit/credit amounts that Peachtree will record for this transaction. Although Peachtree does not show the **Accounts Receivable** subsidiary ledger information, it will automatically update the customer account balance, too. As you can see, Peachtree uses the data you entered to make a transaction "behind the scenes" that is equivalent to the transactions you learned about in your text.

For the sale on account transaction, the window should show the following: **Accounts Receivable** ($318.00 DR), **Sales** ($300.00 CR), and **Sales Tax Payable** ($18.00 CR).

- Click [Cancel] to close the Accounting Behind the Screens window.

- Click [Save] to post the transaction.

> **Notes**
>
> *Viewing the Accounting Behind the Screens window is not required each time you enter a transaction. This information, however, is useful to understand how Peachtree records a transaction such as a sale on account.*

If you need a printed copy of the invoice (sales slip), you could choose the **Print** option. Peachtree lets you print on plain paper or on pre-printed forms. If you choose to print an invoice, select the *Invoice Plain* form, then click **OK**. Click the **Real** button, verify the correct invoice number is displayed, then click **OK** to print the invoice. After Peachtree prints the invoice, the software automatically posts the transaction and is ready for the next invoice.

TIP: If you notice an error after you post an entry, click [Open] and choose the transaction you want to edit. Make the corrections and post the transaction again to record the changes.

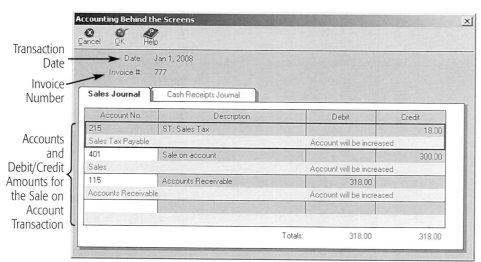

Transaction Date

Invoice Number

Accounts and Debit/Credit Amounts for the Sale on Account Transaction

Figure 14-5B *Accounting Behind the Screens (Sales/Invoicing)*

Step 6 Enter the remaining sales on account transactions for the month: January 5 and January 25.

TIP: If you enter a sale on account for a customer who is tax exempt, make sure that the sales tax code is set to **0** (zero).

Recording Credit Memorandums/Sales Returns

Step 7 Review the transactions shown in your textbook and identify those where the company issued a credit memorandum to a customer.

Step 8 Record the credit memo issued on January 10.

> ***January 10, Issued Credit Memo 102 to Martha Adams for $318 covering $300 in returned merchandise plus $18 sales tax.***

To enter the credit memorandum:

- Choose **Sales/Invoicing** from the *Tasks* menu to display the Sales/Invoicing window if it is not already on your screen.
- Click [Template] and then select <Predefined> Service to display a simplified entry form, if necessary.
- Type **ADA** (the account code for Martha Adams) in the *Customer ID* field.
- Type **CM102** for the credit memo number in the *Invoice #* field.
- Type **1/10/08** to record the transaction date in the *Date* field.
- Move to the *Description* field and enter **Sales return** as the description.
- Type **410** (the **Sales Returns and Allowances** account number) in the *GL Account* field.

   For Peachtree to apply the credit to the appropriate general ledger account, you must enter the **Sales Returns and Allowances** account number in the *GL Account* field. If you do not enter this information, the program will apply the credit to sales instead of the contra-sales account.

TIP: If the *GL Account* field does not appear, choose **Global** from the **Options** menu. Change the setting to show the general ledger accounts for Accounts Receivable. You will have to close the Sales/Invoicing window and then choose this option again for the new setting to take effect.

- Type **-300.00** in the *Amount* field to record the credit amount.
- Tab past the Job field.
- Proof the transaction. Verify that you entered the customer account code, credit memo reference, and credit amount correctly. The invoice total should show -318.00. Compare the information on your screen to the completed transaction shown in Figure 14-5C.

Notes

To record a credit memo, you must enter a negative amount in the Amount *field. The Peachtree software will also calculate the sales tax credit, if applicable.*

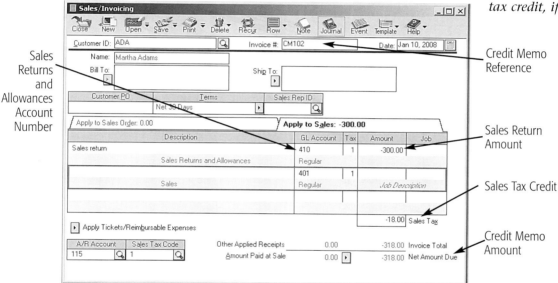

Figure 14-5C *Completed Sales Return Transaction (January 10, Credit Memo 102)*

- *Optional:* Click [icon] to display the Accounting Behind the Screens window. Compare the information shown to the following: **Sales Tax Payable** ($18.00 DR), **Sales Returns and Allowances** ($300.00 DR), and **Accounts Receivable** ($318.00 CR). Close the Accounting Behind the Screens window when you finish.
- Click [icon] to post the credit memo.
- Click [icon] to close the Sales/Invoicing window.

To apply a credit memo to an invoice (sales slip):

- Choose **Receipts** from the *Tasks* menu to display the Receipts window so that you can apply the credit memo to a specific invoice.
- Click **OK** to select **Cash in Bank** as the default cash account, if prompted.
- Type **1/10/08** in the *Deposit ticket ID* field.
- Type **ADA** as the customer ID for Martha Adams.
- Type **CM102** in the *Reference* field.

- Tab past the Receipt Number field.

- Record the transaction date, **1/10/08,** in the *Date* field.

  You do not need to change the payment method or the cash account in the top portion on the window. These fields do not apply to a credit memo transaction.

- Click the **Pay** check box for the credit memo shown in the Apply to Invoices tab located in the lower portion of the window.

- Find the invoice to which you want to apply the credit memo—Invoice 777. Click the **Pay** check box next to the invoice since the credit is for the entire amount.

> **TIP:** If a discount amount appears when you apply a credit memo, move to the *Discount* field and delete the amount shown.

> **Notes**
>
> *If the credit is the same as the invoice amount, just click the **Pay** check box. You do not have to change the amount. If the credit is for only a portion of the total invoice, you must specify in the Amount Paid column the amount of the credit to apply to the invoice.*

After you mark both the credit memo and the invoice, the *Receipt Amount* field should show 0.00 since no cash was actually exchanged. Applying the credit memo removes the original invoice (sales slip) from the customer's record. (See Figure 14-5D.)

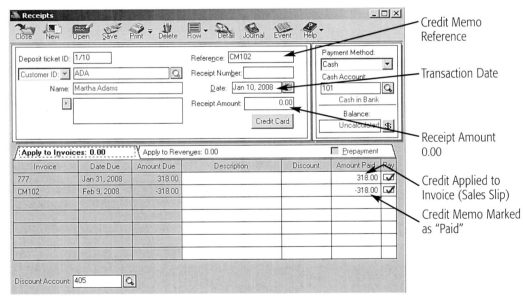

Figure 14-5D *Receipts Window with Applied Credit Memo (January 10, Credit Memo 102)*

- *Optional:* Click [Journal] to display the Accounting Behind the Screens window.

  Check the information shown. Notice that entries show a debit and a credit to **Accounts Receivable** for the same amount. The net effect is $0.00 on **Accounts Receivable,** but the subsidiary ledger is now up-to-date.

  Click [Cancel] to close the Accounting Behind the Screens window.

- Click to post the transaction.
- Click to close the Receipts window.

Step 9   Record the other credit memos issued on January 28.

Recording Cash Receipts

Step 10   Review the transactions and identify the cash receipt transactions for Sunset Surfwear.

Step 11   Enter the January 7 cash receipt on account.

> ***January 7, Received $400 from Alex Hamilton on account, Receipt 345.***

To enter the cash receipt on account:

- Choose **Receipts** from the ***Tasks*** menu to display the Receipts window if it is not already on your screen.
- Type **1/7/08** in the *Deposit ticket ID* field.
- Type **HAM** (the customer code for Alex Hamilton) in the *Customer ID* field, or click and select the code from the Lookup list.
- Type **345** in the *Reference* field.
- Tab past the Receipt Number field.
- Type **1/7/08** for the transaction date in the *Date* field.

   Since the cash you received will go into the **Cash in Bank** account, you do not need to change the payment method or the cash account in the top, right portion of the window.

- In the Apply to Invoices tab, identify the invoice to which the cash receipt should be applied. Click the **Pay** check box next to that invoice to indicate that the customer paid it. After you mark the invoice, the *Receipt Amount* field should show $400.00, the amount of cash received on account.

Notes

The Deposit ticket ID field is useful when you reconcile the bank statement. For our purposes, however, you can use the transaction date.

TIP: If a cash receipt on account applies to more than one invoice, you can mark several invoices to be paid in one transaction. If a customer sends only a partial payment, you can mark an invoice for payment and then change the amount to reflect the actual cash received.

- Proof the information you just recorded. If you notice a mistake, move to that field and make the correction. Compare the information on your screen to the completed transaction shown in Figure 14-5E.
- Click to post the transaction.

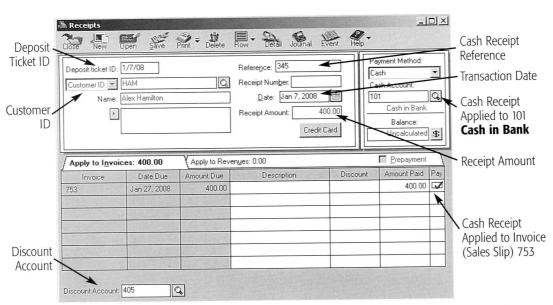

Figure 14-5E *Completed Cash Receipt on Account Transaction (January 7, Receipt 345)*

Step 12   Enter the cash sale on January 15.

> ### January 15, Recorded cash sales of $800 plus $48 in sales tax, Tape 39.

To enter the cash/bankcard sales:

- Choose **Receipts** from the **Tasks** menu to display the Receipts window, if necessary.
- Type **1/15/08** in the *Deposit ticket ID* field.
- Leave the *Customer ID* field empty since this is a cash sale, but enter **CASH** in the *Name* field. **Note:** Peachtree requires that you complete this field even if there is not a specific customer.
- Type **T39** in the *Reference* field.
- Tab past the Receipt Number field.
- Type **1/15/08** for the transaction date in the *Date* field.
- Do not change the payment method or cash account. These fields should already be set to Cash and G/L Account 101 **(Cash in Bank),** respectively.

    Peachtree lets you change the general ledger account if a company has more than one cash account to which it deposits cash receipts. For the problems in your textbook, there is only one bank account and it is set up as the default cash account.

- In the Apply to Revenues tab, move to the first detail line and enter **800.00** in the *Amount* field.

    For a cash sale, you do not have to include the Quantity, Item, Description, Unit Price, and Job information.

- If a sales tax code is not shown at the bottom of the Receipts window, enter **1** for the standard sales tax. Verify that Peachtree calculates the correct sales tax amount.

> ## Notes
>
> *The steps to enter bankcard sales are the same as the steps shown here to enter cash sales. Use BANKCARD instead of CASH for the customer name.*

- Proof the information you just recorded. The total receipt amount should be $848.00. (See Figure 14-5F.) If you notice a mistake, move to that field and make the correction.
- Click [Save] to post the transaction.

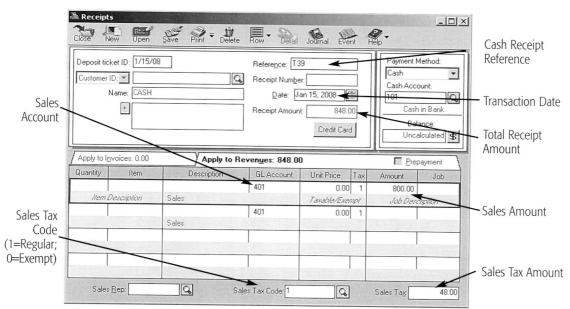

Figure 14-5F *Completed Cash Sale Transaction (January 15, Tape 39)*

Step 13   Record the remaining cash receipt transactions: January 15, January 20, and January 30.

Preparing Reports

Step 14   Print the Sales Journal report.

    To print a Sales Journal report:

- Choose **Accounts Receivable** from the **Reports** menu to display the Select a Report window. (See Figure 14-5G.)

TIP: As an alternative, you can click the ⬚ Sales ⬚ navigation aid and then choose the Sales Journal report to go directly to the report.

- Select Sales Journal in the report list.
- Click [Preview] and then click **OK** to display the Sales Journal report.

TIP: As a shortcut, remember that you can double-click on a report in the report list to automatically display a report on the screen.

- Click [Print] to print the report.
- Click [Close] to close the report window.

Peachtree Guide

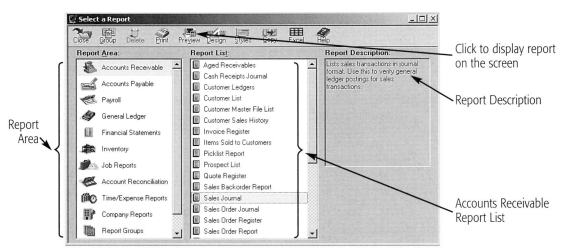

Figure 14-5G *Select a Report Window with the Accounts Receivable reports.*

Step 15 Print the Cash Receipts Journal report.

Step 16 Proof the information shown on the reports. If there are any corrections needed, choose the corresponding task option—**Sales/Invoicing** or **Receipts.** Click [Open] and select the transaction to edit. Make the necessary changes and post the transaction to record your work.

Analyzing Your Work

Step 17 Choose **General Ledger** from the *Reports* menu to display the Select a Report window.

Step 18 Double-click the General Ledger report title to display the report.

Step 19 Locate the **Sales Returns and Allowances** account on the General Ledger report. What is the sum of the debits to this account during January?

Checking Your Work and Ending the Session

Step 20 Click the **Close Problem** button in the Glencoe Smart Guide window.

Step 21 If your teacher has asked you to check your solution, click *Check my answer to this problem.*

Step 22 Click **OK**.

Continuing a Problem from a Previous Session

- If you were previously directed to save your work on the network, select the problem from the scrolling menu and click **OK.** The system will retrieve your files from your last session.
- If you were previously directed to save your work on a floppy disk, insert the floppy, select the corresponding problem from the scrolling menu, and click **OK.** The system will retrieve your files from the floppy disk.

Mastering Peachtree

Why does the Sales/Invoicing form include the *Sales Rep ID* field? Although you did not need to complete this field when you entered the transactions for this problem, a company might find it useful to record this information. Explore the uses a company may have for this information. Write your findings on a separate sheet of paper.

TIP: Search the Help index to learn more about the *Sales Rep ID* field.

Problem 14-6   Posting Sales and Cash Receipts

Review the General Journal entries listed in your textbook for InBeat CD Shop. Then, print the General Ledger and Customer Ledgers reports to see how Peachtree posts the transactions to the corresponding accounts.

INSTRUCTIONS

Beginning a Session

Step 1   Select the problem set: InBeat CD Shop (Prob. 14-6).

Step 2   Rename the company and set the system date to January 31, 2008.

Completing the Accounting Problem

Step 3   Print a General Ledger report.

Step 4   Print a Customer Ledgers report from the Accounts Receivable Report List.

TIP: When a report list appears, you can double-click a report title to go directly to a report, skipping the report options window.

Step 5   Review the reports. Compare the General Journal entries shown in your textbook to the information shown on the reports.

TIP: Credit balances appear as negative amounts on a General Ledger report.

DO YOU HAVE A QUESTION

Q. *Why doesn't Peachtree require you to manually post transactions to the General Ledger?*

A. Peachtree does not require you to post transactions to the General Ledger since the software automatically performs this step for you each time you record a transaction. When you record a Sales/Invoicing transaction, for example, Peachtree posts the transaction amounts to the corresponding General Ledger accounts. Automatic updates to the General Ledger is one of the advantages of using a computerized accounting system.

Analyzing Your Work

Step 6   Review the General Ledger report to answer the **Analyze** question shown in your textbook.

Ending the Session

Step 7   Click the **Close Problem** button in the Glencoe Smart Guide window and select the appropriate save option.

Mastering Peachtree

Although the General Ledger report you printed for this problem is essentially the same as a manual General Ledger report, there are a few minor differences. What are the differences between a Peachtree General Ledger report and the one that you have learned to prepare manually? Write your answer on a separate sheet of paper.

Problem 14-7   Recording Sales and Cash Receipts

INSTRUCTIONS

Beginning a Session

Step 1   Select the problem set: Shutterbug Cameras (Prob. 14-7).

Step 2   Rename the company and set the system date to January 31, 2008.

Completing the Accounting Problem

Step 3   Review the transactions listed in your textbook and group them by type—sales on account, credit memos, cash/bankcard sales, and other cash receipts.

Step 4   Use the **Sales/Invoicing** option to record the sales on account transactions.

TIP: Refer to the instructions for Problem 14-5 if you need help entering the transactions for this problem.

When you enter the January 12 transaction for the merchandise sold on account to FastForward Productions, notice that the credit terms **2/10, Net 30 Days** appears in the *Terms* field. The terms for FastForward Productions have already been set up for you in the problem set.

Step 5   Enter the credit memos issued by Shutterbug Cameras in January.

Remember that there are two steps to record credit memos. First, use the **Sales/Invoicing** option to enter the credit memo. Be sure to change the general ledger account number to **410 Sales Returns and Allowances** and enter a negative invoice amount. Next, apply the credit memo to a specific invoice using the **Receipts** option. Mark both the credit memo and the invoice to which it applies as "Paid." The receipt amount should be $0.00.

Step 6   Enter the cash receipts transaction for the sale of supplies to Betty's Boutique on January 3.

The process for recording a cash receipt for the sale of an asset (e.g., supplies or office equipment) is similar to the steps required to record the sale of merchandise. Use the **Receipts** option, but change the *GL Account* from the default, **Sales,** account to the asset account affected by the transaction. Follow the steps listed below to record the transaction.

January 3, Received $50 in cash from the sale of supplies to Betty's Boutique, Receipt 201.

To enter the cash receipt for the sale of an asset (e.g., supplies):

- Choose **Receipts** from the *Tasks* menu to display the Receipts window, if necessary.
- Type **1/3/08** in the *Deposit ticket ID* field.
- Enter **Betty's Boutique** in the *Name* field. **Note:** A customer account is not set up for this company, but you must still complete the *Name* field.

- Type **201** in the *Reference* field.
- Tab past the Receipt Number field.
- Type **1/3/08** for the transaction date in the *Date* field.

TIP: To record the sale of an asset (e.g., supplies), you must change the *GL Account* field so that the cash receipt is applied to the appropriate general ledger account, not the default account—**Sales**.

- Do not change the payment method or cash account. These fields should already be set to Cash and G/L Account 101 **(Cash in Bank),** respectively.
- In the Apply to Revenues tab, move to the first detail line and enter **Sale of supplies** in the *Description* field.
- Change the general ledger account to **130** (Supplies) in the *GL Account* field.
- Enter **50.00** in the *Amount* field.
- Make sure that the sales tax amount is $0.00 since this is not a taxable transaction.
- Proof the information you just recorded. The total receipt amount should be $50.00. (See Figure 14-7A.) If you notice a mistake, move to that field and make the correction.

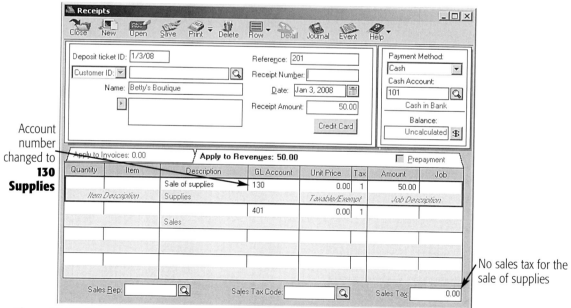

Account number changed to **130 Supplies**

No sales tax for the sale of supplies

Figure 14-7A *Completed Cash Receipt for the Sale of an Asset (January 3, Receipt 201)*

- *Optional:* Click [Journal] to display the Accounting Behind the Screens window, and then review the accounts and amounts to check your work again. (See Figure 14-7B.) The entry should show the following: **Cash** ($50.00 DR) and **Supplies** ($50.00 CR). Close the window when you finish.
- Click [Save] to post the transaction.

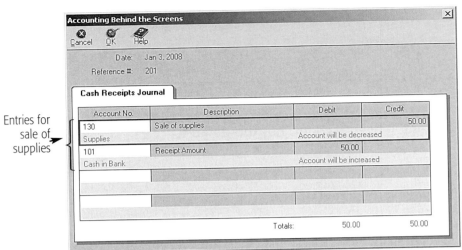

Figure 14-7B *Accounting Behind the Screens Window (January 3, Receipt 201)*

Entries for sale of supplies

Step 7 Record the remaining cash receipts (sales on account, cash sales, and bankcard sales).

To record a cash receipt on account involving a discount, follow the steps you already learned to record the transaction. After you mark the invoice (sales slip) as "Paid," move to the *Discount* field and enter the sales discount if Peachtree does not calculate it correctly. Verify that the *Discount* and *Amount Paid* fields are correct and then post.

Step 8 Print a Sales Journal report and a Cash Receipts Journal report.
Step 9 Proof your work. Update the transactions and print revised reports if you identify any errors.
Step 10 Print the General Ledger for account 215 **(Sales Tax Payable)**, and answer the Analyze question shown in your textbook.

Checking Your Work and Ending the Session

Step 11 Click the **Close Problem** button in the Glencoe Smart Guide window.
Step 12 If your teacher has asked you to check your solution, click *Check my answer to this problem.* Review, print, and close the report.
Step 13 Click the **Close Problem** button and select the appropriate save option.

Mastering Peachtree

How do you set up the standard (default) customer payment terms for a company? Can you set up different payment terms for one or two customers? If so, explain how. Use a separate sheet of paper to record your answer.

Problem 14-8 Recording Sales and Cash Receipts Transactions

INSTRUCTIONS

Beginning a Session

Step 1 Select the problem set: River's Edge Canoe & Kayak (Prob. 14-8).
Step 2 Rename the company and set the system date to January 31, 2008.

Completing the Accounting Problem

Step 3   Review the transactions listed in your textbook and group them by type—sales on account, credit memos, cash/bankcard sales, and other cash receipts.

TIP: Refer to the instructions for Problem 14-5 if you need help entering the transactions for this problem.

Step 4   Record the sales on account transactions.
Step 5   Enter the credit memos.

   When you apply a credit memo to a specific invoice and the customer is set up to receive a sales discount, you will most likely have to change the amounts that automatically appear. Set the discount amounts to $0.00 after you mark the credit memo and invoice as "Paid." Also, if a credit is for an amount less than the original invoice (sales slip) amount, you will have to change the default amount paid to reflect the amount of the credit.

TIP: Remember that when you apply a credit memo to an invoice using the **Receipts** options, the receipt amount must equal $0.00.

Step 6   Enter the cash receipts transactions.

TIP: If Peachtree does not calculate a sales discount correctly, you can manually enter the discount amount in the *Discount* field in the Cash Receipts (Apply to Invoices) window.

Step 7   Print a Sales Journal report and a Cash Receipts Journal report.
Step 8   Proof your work. Update the transactions and print revised reports if you identify any errors.
Step 9   Print the General Ledger for account 405 (Sales Discounts), and answer the Analyze question.

Ending the Session

Step 10   Click the **Close Problem** button in the Glencoe Smart Guide window.

Mastering Peachtree

A business may have customers who are exempt from paying sales tax. Peachtree lets you set up customer accounts so that you can identify those customers who do not have to pay sales tax. When you enter a sales on account transaction, the sales tax code will automatically appear. How do you change the default sales tax code for a customer? Record your answer on a separate sheet of paper.

> **DO YOU HAVE A QUESTION**
>
> **Q.** *Why do you have to apply a credit memo to a specific invoice?*
>
> **A.** Suppose a customer purchases merchandise on account for $100 and then receives a $25 credit. When you enter the $25 credit, you record it as a "negative" invoice amount. The customer's account correctly shows a balance of $75, but there are now two outstanding invoices—one for $100 and another for -$25. When you apply the credit, you mark the -$25 credit memo (invoice) as paid and then record a $25 cash receipt against the original invoice. Now there is only one outstanding invoice for $75 and the customer's account balance remains at $75 because the net cash received for this part of the transaction was $0.

Problem 14-9   Recording and Posting Sales and Cash Receipts

INSTRUCTIONS

Beginning a Session

Step 1   Select the problem set: Buzz Newsstand (Prob. 14-9).

Step 2   Rename the company and set the system date to January 31, 2008.

Completing the Accounting Problem

Step 3   Review the transactions listed in your textbook for Buzz Newsstand.

 TIP: To save time entering the transactions, group them by type—sales on account, credit memos, cash/bankcard sales, and other cash receipts.

Step 4   Record the transactions.

Step 5   Print the following reports: Sales Journal, Cash Receipts Journal, Customer Ledgers, and General Ledger.

Step 6   Proof your work.

Step 7   Answer the Analyze question.

Ending the Session

Step 8   Click the **Close Problem** button in the Glencoe Smart Guide window and select the appropriate save option.

Mastering Peachtree

If a tax authority changes the sales tax rate, you would have to change the tax rate in the Peachtree software so that it would apply the latest rate for each sales transaction. How do you change the default sales tax? Record your answer on a separate sheet of paper.

FAQs

What steps are required to correct a sales invoice with the wrong customer ID code or sale amount?

If you enter the wrong information for a sales invoice, you can correct it even if you have already posted the transaction. Choose the **Sales/Invoicing** option and click the **Open** button. Select the invoice (sales slip) you want to change. When the invoice appears in the Sales/Invoicing data entry window, make the necessary corrections and then post the updated transaction.

What steps are required to correct a cash receipt entry?

You can change any information on a cash receipt unless you have closed the current period. To make a change, choose the **Receipts** option. Click the **Open** button and select the receipt you want to update. Make the changes and post the corrected transaction. Peachtree automatically applies the corrections.

Is it required to use the <Predefined> Service (simplified invoice) in the Sales/Invoicing window?

No. You can use the standard invoice to enter sales on account. Just leave the *Quantity* and *Item* fields empty when you complete the data entry form.

Problem 14-8 Recording Sales and Cash Receipts Transactions

GENERAL JOURNAL   PAGE _____

	DATE		DESCRIPTION	POST. REF.	DEBIT	CREDIT	
1							1
2							2
3							3
4							4
5							5
6							6
7							7
8							8
9							9
10							10
11							11
12							12
13							13
14							14
15							15
16							16
17							17
18							18
19							19
20							20
21							21
22							22
23							23
24							24
25							25
26							26
27							27
28							28
29							29
30							30
31							31
32							32
33							33
34							34
35							35
36							36
37							37

Problem 14-8   (concluded)

GENERAL JOURNAL   PAGE _____

	DATE	DESCRIPTION	POST. REF.	DEBIT	CREDIT	
1						1
2						2
3						3
4						4
5						5
6						6
7						7
8						8
9						9
10						10
11						11
12						12
13						13
14						14
15						15
16						16
17						17

Analyze: _____

_____

Problem 14-9   Source Documents

Instructions: *Use the following source documents to record the transactions for this problem.*

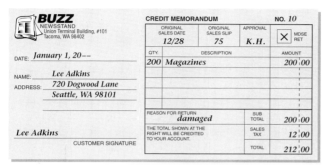

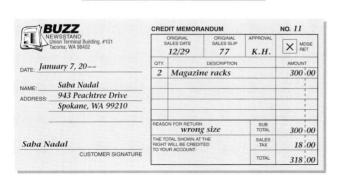

Problem 14-9   (continued)

BUZZ
NEWSSTAND
Union Terminal Building, #101
Tacoma, WA 98402

RECEIPT
No. 77

January 20   20 --

RECEIVED FROM   *Janson Lee* $ *40.00*

Forty and $^{00}/_{100}$ —————————— DOLLARS

FOR   *supplies*

RECEIVED BY   *Kathy Harper*

BUZZ
NEWSSTAND
Union Terminal Building, #101
Tacoma, WA 98402

RECEIPT
No. 78

January 25   20 --

RECEIVED FROM   *Lee Adkins* $ *636.00*

Six hundred thirty-six and $^{00}/_{100}$ —————————— DOLLARS

FOR   *on account*

RECEIVED BY   *Kathy Harper*

BUZZ
NEWSSTAND
Union Terminal Building, #101
Tacoma, WA 98402

DATE: *January 31, 20--*   NO. *114*

SOLD TO	*Rolling Hills Pharmacies* *16 Meadow Lane* *Tacoma, WA 98402*		
CLERK *BA*	CASH	CHARGE ✓	TERMS *2/10, n/30*

QTY.	DESCRIPTION	UNIT PRICE	AMOUNT	
30 cs	*Magazines*	*100/cs*	*3,000*	*00*

	SUBTOTAL	*3,000*	*00*
COPY	SALES TAX	*180*	*00*
	TOTAL	*3,180*	*00*

Thank You!

Problem 14-9   Recording and Posting Sales and Cash Receipts

(1)

GENERAL JOURNAL   PAGE _____

	DATE	DESCRIPTION	POST. REF.	DEBIT	CREDIT	
1						1
2						2
3						3
4						4
5						5
6						6
7						7
8						8
9						9
10						10
11						11
12						12
13						13
14						14
15						15
16						16
17						17
18						18
19						19
20						20
21						21
22						22
23						23
24						24
25						25
26						26
27						27
28						28
29						29
30						30
31						31
32						32
33						33
34						34
35						35
36						36

Problem 14-9 (continued)

(2)

GENERAL LEDGER

ACCOUNT ___Cash in Bank___ ACCOUNT NO. ___101___

DATE		DESCRIPTION	POST. REF.	DEBIT	CREDIT	BALANCE	
						DEBIT	CREDIT
20--							
Jan.	1	Balance	✓			500000	

ACCOUNT ___Accounts Receivable___ ACCOUNT NO. ___115___

DATE		DESCRIPTION	POST. REF.	DEBIT	CREDIT	BALANCE	
						DEBIT	CREDIT
20--							
Jan.	1	Balance	✓			624800	

ACCOUNT ___Supplies___ ACCOUNT NO. ___135___

DATE		DESCRIPTION	POST. REF.	DEBIT	CREDIT	BALANCE	
						DEBIT	CREDIT
20--							
Jan.	1	Balance	✓			30000	

Problem 14-9 (continued)

ACCOUNT __Sales Tax Payable__ ACCOUNT NO. __215__

DATE		DESCRIPTION	POST. REF.	DEBIT	CREDIT	BALANCE	
						DEBIT	CREDIT
20--							
Jan.	1	Balance	✓				1 3 4 8 00

ACCOUNT __Sales__ ACCOUNT NO. __401__

DATE		DESCRIPTION	POST. REF.	DEBIT	CREDIT	BALANCE	
						DEBIT	CREDIT
20--							
Jan.	1	Balance	✓				10 8 0 0 00

ACCOUNT __Sales Discounts__ ACCOUNT NO. __405__

DATE		DESCRIPTION	POST. REF.	DEBIT	CREDIT	BALANCE	
						DEBIT	CREDIT
20--							
Jan.	1	Balance	✓			2 0 0 00	

ACCOUNT __Sales Returns and Allowances__ ACCOUNT NO. __410__

DATE		DESCRIPTION	POST. REF.	DEBIT	CREDIT	BALANCE	
						DEBIT	CREDIT
20--							
Jan.	1	Balance	✓			4 0 0 00	

Problem 14-9 (concluded)

ACCOUNTS RECEIVABLE SUBSIDIARY LEDGER

Name _Adkins, Lee_

Address _720 Dogwood Lane, Seattle, WA 98101_

DATE		DESCRIPTION	POST. REF.	DEBIT	CREDIT	BALANCE
20--						
Jan.	1	Balance	✓			8 4 8 00

Name _Java Shops Inc._

Address _492 Country Place, Auburn, WA 98002_

DATE		DESCRIPTION	POST. REF.	DEBIT	CREDIT	BALANCE
20--						
Jan.	1	Balance	✓			1 5 0 0 00

Name _Nadal, Saba_

Address _943 Peachtree Drive, Spokane, WA 99210_

DATE		DESCRIPTION	POST. REF.	DEBIT	CREDIT	BALANCE
20--						
Jan.	1	Balance	✓			1 6 0 0 00

Name _Rolling Hills Pharmacies_

Address _16 Meadow Lane, Tacoma, WA 98402_

DATE		DESCRIPTION	POST. REF.	DEBIT	CREDIT	BALANCE
20--						
Jan.	1	Balance	✓			2 3 0 0 00

Analyze: _____

_____

CHAPTER  Accounting for Sales and Cash Receipts

Self-Test

Part A   True or False

Directions: *Circle the letter* T *in the Answer column if the statement is true; circle the letter* F *if the statement is false.*

Answer

T   F   **1.** Merchandising businesses sell only on a cash-and-carry basis.

T   F   **2.** Businesses are required to act as agents for local and state governments in the collection of taxes.

T   F   **3.** Customer accounts are listed in alphabetical order in the accounts receivable subsidiary ledger.

T   F   **4.** Bankcard sales are recorded like cash sales because a business expects to receive its cash within a few days of the deposit.

T   F   **5.** In most states, sales made to school districts and other governmental agencies are taxable.

T   F   **6.** Prenumbered sales slips help businesses keep track of all sales made on account.

T   F   **7.** Sales discounts increases the Sales account.

T   F   **8.** A bank card holder is a charge customer of the business from whom the purchase is being made.

T   F   **9.** "2/10, n/30" means that customers may deduct 2% of the cost of the merchandise if they pay within 10 days. Otherwise they must pay the full amount within 30 days.

T   F   **10.** Cash discounts increase the amount to be received from charge customers.

Part B   Multiple Choice

Directions: *Only one of the choices given with each of the following statements is correct. Write the letter of the correct answer in the Answer column.*

Answer

_____   **1.** Sales Returns and Allowances is classified as a
(A) revenue account.
(B) owner's equity account.
(C) contra revenue account.
(D) none of the above.

_____   **2.** If merchandise is sold for $417.80 and the sales tax rate is
4%, the amount of the sales tax is
(A) $434.5l.
(B) $16.17.
(C) $20.89.
(D) none of the above.

_____   **3.** Sales Tax Payable is classified as a(n)
(A) asset account.
(B) liability account.
(C) owner's equity account.
(D) revenue account.

_____   **4.** If merchandise is sold for $325.60 and the sales tax rate is
5%, the amount to be collected from a charge customer is
(A) $309.32.
(B) $341.88.
(C) $325.60.
(D) $331.88.

_____   **5.** If the date of an invoice is May 14, the credit terms are
2/10, n/30, and the invoice was received on May 16, the
last date the discount may be taken is
(A) May 26.
(B) June 12.
(C) May 24.
(D) May 14.

CHAPTER 15   Accounting for Purchases and Cash Payments

Study Plan

Check Your Understanding

Section 1	*Read Section 1 on pages 388–391 and complete the following exercises on page 392.*
	❏ Thinking Critically
	❏ Computing in the Business World
	❏ Problem 15-1 *Analyzing a Purchase Order*
Section 2	*Read Section 2 on pages 393–399 and complete the following exercises on page 400.*
	❏ Thinking Critically
	❏ Communicating Accounting
	❏ Problem 15-2 *Recording Purchases Transactions*
	❏ Problem 15-3 *Analyzing a Source Document*
Section 3	*Read Section 3 on pages 402–407 and complete the following exercises on page 408.*
	❏ Thinking Critically
	❏ Analyzing Accounting
	❏ Problem 15-4 *Recording Cash Payment Transactions*

Summary	*Review the Chapter 15 Summary on page 409 in your textbook.*
	❏ Key Concepts
Review and Activities	*Complete the following questions and exercises on pages 410–411 in your textbook.*
	❏ Using Key Terms
	❏ Understanding Accounting Concepts and Procedures
	❏ Case Study
	❏ Conducting an Audit with Alex
	❏ Internet Connection
	❏ Workplace Skills
Computerized Accounting	*Read the Computerized Accounting information on page 412 in your textbook.*
	❏ Making the Transition from a Manual to a Computerized System
	❏ Preparing the Payroll in Peachtree
Problems	*Complete the following end-of-chapter problems for Chapter 15.*
	❏ Problem 15-5 *Determining Due Dates and Discount Amounts*
	❏ Problem 15-6 *Analyzing Purchases and Cash Payments*
	❏ Problem 15-7 *Recording Purchases Transactions*
	❏ Problem 15-8 *Recording Cash Payment Transactions*
	❏ Problem 15-9 *Recording Purchases and Cash Payment Transactions*
Challenge Problem	❏ Problem 15-10 *Recording and Posting Purchases and Cash Payment Transactions*
Chapter Reviews and Working Papers	*Complete the following exercises for Chapter 15 in your Chapter Reviews and Working Papers.*
	❏ Chapter Review
	❏ Self-Test

CHAPTER 15 REVIEW — Accounting for Purchases and Cash Payments

Part 1 Accounting Vocabulary (19 points)

Total Points	41
Student's Score	

Directions: *Using terms from the following list, complete the sentences below. Write the letter of the term you have chosen in the space provided.*

A. accounts payable subsidiary ledger	**G.** FOB destination	**N.** purchase requisition
B. bankcard fee	**H.** FOB shipping point	**O.** Purchases account
C. cost of merchandise	**I.** invoice	**P.** purchases allowance
D. debit memorandum	**J.** packing slip	**Q.** purchases discount
E. discount period	**K.** premium	**R.** purchases return
F. due date	**L.** processing stamp	**S.** tickler file
	M. purchase order	**T.** Transportation In account

____C____ **0.** The _____ is the actual cost to the business of the merchandise to be sold to customers.

_____ **1.** A(n) _____ contains a folder for each day of the month.

_____ **2.** A(n) _____ is the period of time within which an invoice must be paid if a discount is to be taken.

_____ **3.** A(n) _____ is a written offer to a supplier to buy certain items.

_____ **4.** The _____ is a ledger that contains accounts for all creditors and the amount owed to each.

_____ **5.** A(n) _____ is a stamp placed on an invoice that outlines a set of steps to be followed in processing the invoice for payment.

_____ **6.** The _____ is the account used to record the cost of new merchandise.

_____ **7.** A written request that a certain item or items be ordered is a(n) _____.

_____ **8.** The _____ is the date by which an invoice must be paid.

_____ **9.** A bill that lists the credit terms and the quantity, description, unit price, and total cost of the items shipped to the buyer is a(n) _____.

_____ **10.** A form that lists the items included in a shipment is a(n) _____.

_____ **11.** A(n) _____ is a cash discount offered by suppliers for prompt payment.

_____ **12.** A(n) _____ is charged by a bank for handling the bankcard sales slips deposited by a business; it is usually calculated as a percentage of the total bankcard sales.

_____ **13.** _____ means that the supplier pays the shipping cost to the buyer's destination or location.

_____ **14.** The _____ is the amount paid for insurance.

_____ **15.** The cost of merchandise account that is used to record shipping charges on goods is the _____.

_____ **16.** When a business returns merchandise bought on account to the supplier for full credit, a(n) _____ occurs.

_____ **17.** _____ means that the buyer pays the shipping charge from the supplier's place of business.

_____ **18.** A price reduction received by a business for unsatisfactory merchandise kept is a(n) _____.

_____ **19.** The form a business uses to notify its supplier of a return or allowance is called a(n) _____.

Computerized Accounting Using Peachtree

Software Objectives

When you have completed this chapter, you will be able to use Peachtree to:

1. Record purchases on account transactions using the **Purchases/Receive Inventory** option.
2. Enter cash payments using the **Payments** option.
3. Record debit memorandums.
4. Print a Purchases Journal report, Cash Disbursements Journal report, and a Vendor Ledgers report.
5. Print a General Ledger report.

Problem 15-6   Analyzing Purchases and Cash Payments

Review the transactions listed in your textbook for InBeat CD Shop. As you will learn, the process for recording the purchases and cash payments transactions using the Peachtree accounting software is different than the method you learned in your text. Instead of recording these transactions in a general journal, you will record purchases on account using the **Purchases/Receive Inventory** option and process cash payments with the **Payments** option.

INSTRUCTIONS

Beginning a Session

Step 1   Select the problem set: InBeat (Prob. 15-6).
Step 2   Rename the company by adding your initials, e.g., InBeat (Prob. 15-6: XXX).
Step 3   Set the system date to March 31, 2008.

Completing the Accounting Problem

The instructions in this section explain how to enter purchases on account, debit memorandums, and cash payments using the Peachtree software. To simplify the data entry process, you will batch similar transactions by type and then enter them using the appropriate software task option.

Entering Purchases on Account

Step 4   Review the transactions for InBeat CD Shop and identify the purchases on account.
Step 5   Enter the purchase on account transaction for March 2.

March 2, Purchased merchandise on account from NightVision and Company, $2,000, Invoice NV-20, terms 2/10, n/30.

To enter the purchase on account:

- Choose **Purchases/Receive Inventory** from the *Tasks* menu to display the Purchases/Receive Inventory window.
- Type **NIG** (the account code for NightVision and Company) in the *Vendor ID* field.

    If you do not know a vendor account code, click the 🔍 button, highlight the vendor name and click the **OK** button.

DO YOU HAVE A QUESTION

Q. *Why can't you use the Peachtree **General Journal Entry** option to record purchases on account and cash payments to vendors?*

A. Unlike a manual general journal, the Peachtree software does not provide a way for you to identify the corresponding **Accounts Payable** subsidiary ledger account in the general journal. When you use the **Purchases/Receive Inventory** option, you must enter a Vendor ID. Peachtree uses this information to update both the subsidiary ledger account and the **Accounts Payable** balance.

TIP: As a shortcut, remember that you can press **SHIFT+?** when you are in a Lookup field to display a list of choices.

- Type **NV-20** for the invoice number in the *Invoice #* field.
- Type **3/2/08** to record the transaction date in the *Date* field. Make sure that Peachtree shows 2008 for the year. If not, change the system date.
- Move to the *Description* field by pressing **ENTER** or by clicking in the field.

  You can skip over the *Ship To, Ship Via, Terms,* and *A/P Account* fields. Some of these fields, such as the *Ship To* field, are optional. You do not need to complete these fields when you are entering transactions from your text, but a company using Peachtree to keep its books might need to enter this information. For other fields, such as the *Terms* field, default information is already recorded in the field. You do not need to change this field unless different terms apply.

- Type **Purchased merchandise on account** for the description.
- Skip the *GL Account* field. The **Purchases** account number already appears in this field.
- Type **2000.00** in the *Amount* field to record the purchase amount.
- Proof the information you just recorded. Check all of the information you entered. If you notice a mistake, move to that field and make the correction. Compare the information on your screen to the completed transaction shown in Figure 15-6A.

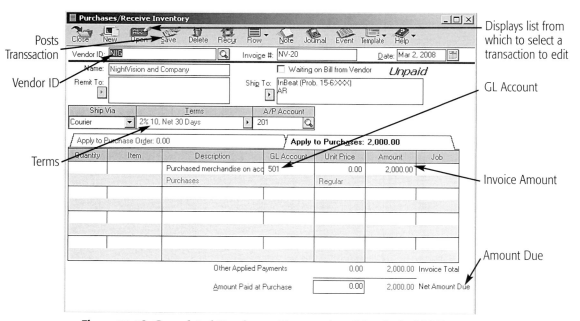

Figure 15-6A *Completed Purchases Transaction (March 2, 2008)*

- Click to display the Accounting Behind the Screens window. Use the information in this window to check your work again. (See Figure 15-6B.)

The Accounting Behind the Screens window shows you the accounts and the debit/credit amounts that Peachtree will record for this transaction. Although Peachtree does not show the **Accounts Payable** subsidiary ledger information, it will automatically update the vendor account balance, too. As you can see, Peachtree uses the data you entered to make a transaction "behind the screens" that is equivalent to the transactions you learned about in your text.

For the purchase on account transaction, the window should show the following: **Purchases** ($2,000.00 DR) and **Accounts Payable** ($2,000.00 CR).

- Click to close the Accounting Behind the Screens window.
- Click to post the transaction.

Notes

Viewing the Accounting Behind the Screens window is not required each time you enter a transaction. This information, however, is useful to understand how Peachtree records a transaction such as a purchase on account.

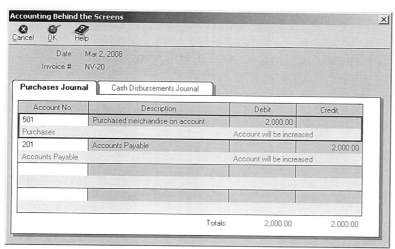

Figure 15-6B *Accounting Behind the Screens (Purchases/Receive Inventory)*

Step 6   Enter the two remaining purchases on account transactions for the month: March 7 and March 16.

Note: For the March 7 transaction, InBeat CD Shop bought supplies (not merchandise) on account. Enter **Purchased supplies on account** for the description, and make sure that the *GL Account* field shows **135 Supplies**, not **501 Purchases**. The default GL account number is set up in the vendor's record.

Recording Debit Memorandums/Purchases Returns and Allowances

Step 7   Review the transactions shown in your textbook and identify those where the company issued a debit memorandum.

Step 8   Record the debit memorandum issued on March 18.

March 18, Issued Debit Memorandum 25 for $100 to NightVision and Company for the return of merchandise.

To enter the debit memorandum:

- Choose **Purchases/Receive Inventory** from the *Tasks* menu to display the Purchases/Receive Inventory window if it is not already on your screen.
- Type **NIG** (the account code for NightVision and Company) in the *Vendor ID* field.
- Type **DM25** for the debit memo number in the *Invoice #* field.
- Type **3/18/08** to record the transaction date in the *Date* field.
- Move to the *Description* field and enter **Merchandise return** as the description.
- Type **515** (the **Purchases Returns and Allowances** account number) in the *GL Account* field.

> **Notes**
>
> *Peachtree uses "credit memo" instead of "debit memo" in reference to a transaction involving a purchase return or allowance.*

For Peachtree to apply the credit to the appropriate general ledger account, you must enter the **Purchases Returns and Allowances** account number in the *GL Account* field. If you do not enter this information, the program will apply the debit memo to purchases instead of the contra-purchases account.

TIP: If the *GL Account* field does not appear, choose **Global** from the **Options** menu. Change the setting to show the general ledger accounts for Accounts Payable. You will have to close the Purchases/Receive Inventory window and then choose this option again for the new setting to take effect.

- Type **-100.00** in the *Amount* field to record the amount.

> **Notes**
>
> *To record a debit memo, you must enter a negative amount in the Amount field.*

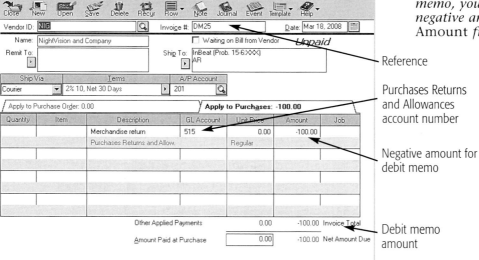

Figure 15-6C *Completed Purchases Return Transaction (March 18, Debit Memo 25)*

- Click [Save] to post the debit memorandum.
- Click [Close] to close the Purchases/Receive Inventory window.

To apply a debit memo to an invoice:

- Choose **Payments** from the *Tasks* menu.
- Click **OK** to select **Cash in Bank** as the default cash account, if prompted.
- Type **NIG** as the vendor ID for NightVision and Company.
- Enter **DM25** in the *Check Number* field.
- Type **3/18/08** in the *Date* field.
- Type **Debit Memo 25** in the *Memo* field.

- Click the **Pay** check box for the debit memo shown in the Apply to Invoices tab located in the lower portion of the window. Then, if necessary, change the discount amount to 0.00.
- Find the invoice to which you want to apply the credit memo—Invoice NV-45. Click the **Pay** check box next to this invoice and then change the discount amount to **0.00** and the amount paid to **100.00**.

TIP: If a discount amount appears when you apply a debit memo, move to the *Discount* field and delete the amount shown.

Notes

*If the debit memo is the same as the invoice , just click the **Pay** check box. You do not have to change the amount. If the debit memo is for only a portion of the total invoice, you must specify the amount to apply to the invoice.*

After you mark both the debit memo and the invoice, the *Dollars* field should show 0.00 since no cash was actually exchanged. Applying the debit memo updates the original invoice in the vendor's record.

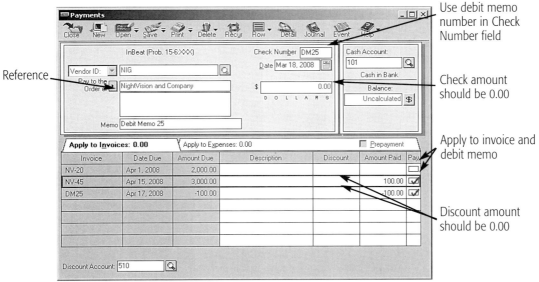

Reference — ... Check amount should be 0.00 ... Use debit memo number in Check Number field ... Apply to invoice and debit memo ... Discount amount should be 0.00

Figure 15-6D *Payments Window with Applied Debit Memo (March 18, Debit Memo 25)*

- Click [Save] to post the transaction.
- Click [Close] to close the Payments window.

Recording Cash Payments

Step 9   Review the transactions and identify the cash payment transactions for InBeat CD Shop. Notice that there are two types of payments—payments on account and other cash payments. You will learn how to enter both types.

Step 10   Enter the March 6 cash payment.

March 6, Issued Check 250 for $85 to Penn Trucking Company for delivering merchandise from NightVision and Company.

To enter a cash payment:

- Choose **Payments** from the *Tasks* menu.
- Skip the *Vendor ID* field and enter **Penn Trucking Company** in the *Pay to the Order of* field.

  A vendor account has not been set up for this company. Therefore, you must manually enter the payee information instead of providing a vendor ID.

- Type **250** in the *Check Number* field.
- Type **3/6/08** in the *Date* field.
- Type **Delivery Fee** in the *Memo* field.
- Do not change the cash account. This field should already be set to **101 Cash in Bank**.

  Peachtree lets you change the cash account if a company has more than one checking account—regular checking and payroll checking accounts, for example. For the problems in your textbook, there is only one bank account and it is setup as the default cash account.

- In the Apply to Expense tab, move to the first detail line. Type **Delivery fee** for the description, select **505 Transportation In** for the GL account, and enter **85.00** in the *Amount* field.

  For a cash payment, you do not have to include the Quantity, Item, Unit Price and Job information.

- Proof the information you just recorded. The total check amount should be $85.00. (See Figure 15-6E.) If you notice a mistake, move to that field and make the correction.
- Click [Save] to post the transaction.

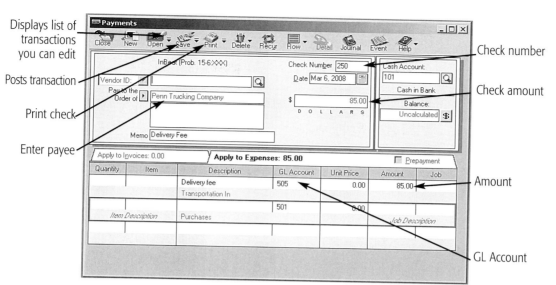

Figure 15-6E *Completed Cash Payment (March 6, Check 250)*

Step 11   Enter the cash payment on account for March 12.

> ### March 12, Issued Check 251 for $1,960 to NightVision and Company in payment of invoice NV-20 for $2,000 less a cash discount of $40.

To enter a cash payment on account:

- Choose **Payments** from the *Tasks* menu.
- Type **NIG** (the vendor code for NightVision and Company) in the *Vendor ID* field, or click 🔍 and select the code from the Lookup list.
- Type **251** in the *Check Number* field.
- Type **3/12/08** for the transaction date in the *Date* field.
- Type **Invoice NV-20** in the *Memo* field.

- In the Apply to Invoice tab, identify the invoice to which the cash payment should be applied. Click the **Pay** check box next to that invoice to indicate that you are paying it. After you mark the invoice, the Dollars field should show $1,960.00—the invoice amount ($2,000) less the discount ($40).
- Proof the information you just recorded. If you notice a mistake, move to that field and make the correction. Compare the information on your screen to the completed transaction shown in Figure 15-6F.
- Click 💾 to post the transaction.

> ❋ **Notes**
>
> *If the* Discount *and* Amount Paid *fields are not correct, you can manually change them.*

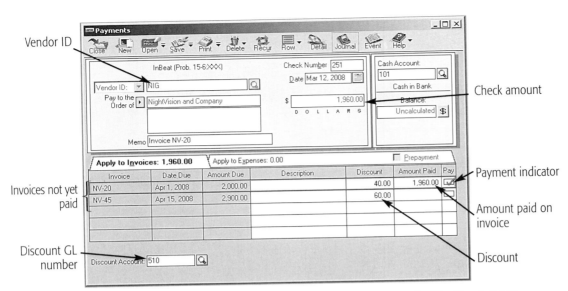

Figure 15-6F *Completed Cash Payment on Account Transaction (March 12, Check 251)*

Step 12 Record the remaining cash payment transactions: March 15, March 20, and March 22.

TIP: For payments not on account, be sure to record the correct GL account number.

Preparing Reports

Step 13 Print the Purchases Journal report.

To print a Purchases Journal report:

- Choose **Accounts Payable** from the **Reports** menu to display the Select a Report window.

- Select Purchases Journal in the report list.

- Click [Preview] and then click **OK** to display the report.

- Click [Print] to print the report.

- Click [Close] to close the report window.

Step 14 Print the Cash Disbursements Journal report.
Step 15 Print a Vendor Ledgers report.
Step 16 Proof the information shown on the reports. If there are any corrections needed, choose the corresponding task option—**Purchases/Receive Inventory** or **Payments**. Click [Open] and select the transaction to edit. Make the necessary changes and post the transaction to record your work.

Analyzing Your Work

Step 17 Choose **General Ledger** from the **Reports** menu to display the Select a Report window.

Step 18 Double-click the General Ledger report title to display the report.

Step 19 Click the **Options** button to filter the report by General Ledger Account ID 501 **(Purchases).** Print the report. What is the balance of this account?

Ending the Session

Step 20 Click the **Close Problem** button in the Glencoe Smart Guide window.

Continuing a Problem from a Previous Session

- If you were previously directed to save your work on the network, select the problem from the scrolling menu and click **OK.** The system will retrieve your files from your last session.

- If you were previously directed to save your work on a floppy disk, insert the floppy, select the corresponding problem from the scrolling menu, and click **OK.** The system will retrieve your files from the floppy disk.

Mastering Peachtree

Can you change the credit terms offered by a specific vendor? Explain your answer on a separate sheet of paper.

Problem 15-7 Recording Purchases Transactions

INSTRUCTIONS

Beginning a Session

Step 1 Select the problem set: Shutterbug Cameras (Prob. 15-7).

Step 2 Rename the company and set the system date to March 31, 2008.

Completing the Accounting Problem

Step 3 Review the transactions listed in your textbook.

> **TIP:** Refer to the instructions for Problem 15-6 if you need help entering the transactions for this problem.

Step 4 Record the purchases on accounts.

Step 5 Enter the debit memorandums.

When you apply a credit memo to a specific invoice and the vendor offers a discount, you will most likely have to change the amounts that automatically appear. Set the discount amounts to 0.00 after you mark the debit memo and invoice as "Paid." Also, if a debit memo is for an amount less than the original invoice amount, you will have to change the default amount paid to reflect the amount of the debit.

> **TIP:** Do not use the **Purchases Returns and Allowances** account when you record a debit memo for an asset account such as supplies or store equipment.

DO YOU HAVE A QUESTION

Q. *Why do you have to apply a debit memo to a specific invoice?*

A. Suppose a company purchases merchandise on account for $200 and then receives a $50 credit. When you enter the $50 credit, you record it as a "negative" invoice amount. The vendor's account correctly shows a balance of $150, but there are now two outstanding invoices—one for $200 and another for -$50. When you apply the credit, you mark the -$50 debit memo as paid and then record a $50 cash payment against the original invoice. Now there is only one outstanding invoice for $150 and the vendor's account balance remains at $150 because the net cash payment for this part of the transaction was $0.

Step 6 Print a Purchases Journal report and a Cash Disbursements Journal report.

Step 7 Proof your work. Update the transactions and print revised reports if you identify any errors.

Step 8 Print the General Ledger for account 515 **(Purchases Returns and Allowances),** and answer the Analyze question.

Checking Your Work and Ending the Session

Step 9 Click the **Close Problem** button in the Glencoe Smart Guide window.

Step 10 If your teacher has asked you to check your solution, click *Check my answer to this problem.* Review, print, and close the report.

Step 11 Click the **Close Problem** button and select the appropriate save option.

Mastering Peachtree

Peachtree displays a default General Ledger account on the Purchases/Receive Inventory data entry form when you record a new purchase. How do you set this information? Why might you want to specify a GL account other than **Purchases** for a vendor? Record your answer on a separate sheet of paper.

Problem 15-8 Recording Cash Payment Transactions

INSTRUCTIONS

Beginning a Session

Step 1 Select the problem set: Cycle Tech Bicycles (Prob. 15-8).

Step 2 Rename the company and set the system date to March 31, 2008.

Completing the Accounting Problem

Step 3 Review the transactions listed in your textbook.

> **TIP:** Refer to the instructions for Problem 15-6 if you need help entering the transactions for this problem.

DO YOU HAVE A QUESTION

Q. *Does Peachtree let you change a check amount if, for example, you forget to record a discount?*

A. Yes, you can easily change a check amount for a transaction. Choose the **Payments** option and click the **Open** button. Select a transaction, make the necessary changes, and post (save) the updated transaction.

Step 4 Record the cash payments.

Use the **Payments** option to record all of the transactions except for the March 20 transaction ($275 bankcard fees). Use the **General Journal** option to record this transaction because the fees are automatically withdrawn from the company's account.

For the March 28 and March 31 transactions, you must enter a multi-part transaction. For example, Cycle Tech purchased supplies and store equipment from Superior Store Equipment, Inc. on March 28. As you can see in Figure 15-8A, the purchase is applied to two different GL accounts—**140 Store Equipment** and **130 Supplies**.

>
>
> **TIP:** If necessary, manually enter a cash discount if Peachtree does not automatically fill in the correct amount.

Peachtree Guide

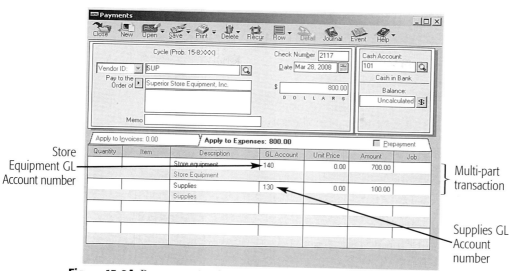

Store Equipment GL Account number

Multi-part transaction

Supplies GL Account number

Figure 15-8A *Payment Applied to Different GL Accounts (March 28)*

Step 5 Print a Cash Disbursements Journal and a General Journal report.

Step 6 Proof your work. Update the transactions and print revised reports if you identify any errors.

Step 7 Answer the Analyze question.

Ending the Session

Step 8 Click the **Close Problem** button in the Glencoe Smart Guide window.

Problem 15-9   Recording Purchases and Cash Payment Transactions

INSTRUCTIONS

Beginning a Session

Step 1 Select the problem set: River's Edge Canoe & Kayak (Prob. 15-9).

Step 2 Rename the company and set the system date to March 31, 2008.

Completing the Accounting Problem

Step 3 Review the transactions listed in your textbook and group them by type—purchases on account, debit memorandums, and cash payments.

Step 4 Record the purchases on account.

Step 5 Record the debit memorandums.

Step 6 Record the cash payments.

Step 7 Print a Purchases Journal, a General Ledger for account 501, and a Cash Disbursements Journal report.

Step 8 Proof your work. Update the transactions and print revised reports if you identify any errors.

Step 9 Answer the Analyze question.

Checking Your Work and Ending the Session

Step 10 Click the **Close Problem** button in the Glencoe Smart Guide window.

Step 11 If your teacher has asked you to check your solution, click *Check my answer to this problem*. Review, print, and close the report.

Step 12 Click the **Close Problem** button and select the appropriate save option.

> ### DO YOU HAVE A QUESTION
>
> **Q.** *Which GL account should you use to record a debit memo transaction for damaged supplies?*
>
> **A.** When you record a debit memo for damaged supplies, use the **Supplies** account, not the **Purchases Returns and Allowances** account. Only use the **Purchases Returns and Allowances** account when you record a debit memo for damaged merchandise.

Problem 15-10 Recording and Posting Purchases and Cash Payment Transactions

INSTRUCTIONS

Beginning a Session

Step 1 Select the problem set: Buzz Newsstand (Prob. 15-10).

Step 2 Rename the company and set the system date to March 31, 2008.

Completing the Accounting Problem

Step 3 Review the transactions listed in your textbook and group them by type—purchases on account, debit memorandums, and cash payments.

Step 4 Record the transactions.

Step 5 Print the following reports: Purchases Journal, Cash Disbursements Journal, Vendor Ledgers, and General Ledger.

Step 6 Proof your work. Update the transactions and print revised reports if you identify any errors.

Step 7 Answer the Analyze question.

Ending the Session

Step 8 Click the **Close Problem** button in the Glencoe Smart Guide window.

Notes

Remember that Peachtree automatically posts transactions to the general ledger and subsidiary ledger accounts.

FAQs

What steps are required to correct a purchase invoice with the wrong vendor ID code or amount?

If you enter the wrong information for a purchase invoice, you can correct it even if you have already posted the transaction. Choose the **Purchases/Receive Inventory** option and click the **Open** button. Select the invoice you want to change. When the invoice appears in the data entry window, make the necessary corrections and then post the updated transaction.

What steps are required to correct a cash payment entry?

You can change any information on a cash payment unless you have closed the current period. To make a change, choose the **Payments** option. Click the **Open** button and select the payment you want to update. Make the changes and post the corrected transaction. Peachtree automatically applies the corrections.

Why can't you use the Peachtree General Journal Entry option to record purchases on account and cash payments to vendors?

Unlike a manual general journal, the Peachtree software does not provide a way for you to identify the corresponding **Accounts Payable** subsidiary ledger account in the general journal. When you use the **Purchases/Receive Inventory** option, you must enter a Vendor ID. Peachtree uses this information to update both the subsidiary ledger account and the **Accounts Payable** balance.

Which GL account should you use to record a debit memo transaction for damaged supplies?

When you record a debit memo for damaged supplies, use the **Supplies** account, not the **Purchases Returns and Allowances** account (used when you record a debit memo for damaged merchandise).

Computerized Accounting Using Spreadsheets

Problem 15-5   Determining Due Dates and Discount Amounts

Completing the Spreadsheet

Step 1   Read the instructions for Problem 15-5 in your textbook. This problem involves determining the due date, discount amount, and amount to be paid for six invoices.

Step 2   Open the Glencoe Accounting: Electronic Learning Center software.

Step 3   From the Program Menu, click on the **Peachtree Complete® Accounting Software and Spreadsheet Applications** icon.

Step 4   Log onto the Management System by typing your user name and password.

Step 5   Under the **Problems & Tutorials** tab, select template15-5 from the Chapter 15 drop-down menu. The template should look like the one shown below.

```
PROBLEM 15-5
DETERMINING DUE DATES AND DISCOUNT AMOUNTS

(name)
(date)

   Invoice       Invoice      Credit      Invoice      Due       Discount     Amount to
   Number         Date        Terms       Amount       Date       Amount      be Paid
                                                                   $0.00        $0.00
                                                                   $0.00        $0.00
                                                                   $0.00        $0.00
                                                                   $0.00        $0.00
                                                                   $0.00        $0.00
                                                                   $0.00        $0.00

TOTAL DISCOUNTS
                                                                   $0.00
```

Step 6   Key your name and today's date in the cells containing the *(name)* and *(date)* placeholders.

Step 7   Enter the invoice number, invoice date, credit terms, and invoice amount for the first invoice in the appropriate cells of the spreadsheet template. The spreadsheet template will automatically calculate the due date, discount amount, and amount to be paid for the first invoice.

 TIP: Enter dates as month/day in the spreadsheet template. For example, enter March 5 as **3/5**. The spreadsheet will automatically convert this to a date format.

Step 8   Check your work by checking the due date, discount amount, and amount to be paid against your textbook, since the first invoice has been completed for you.

Step 9   Continue to enter the invoice number, invoice date, credit terms, and invoice amount for the remaining invoices. The due date, discount amount, and amount to be paid will be automatically calculated for each invoice.

TIP: Be careful when you enter the last invoice number: **00985**. Many spreadsheet programs will not recognize the two zeroes at the beginning of the number and will show it in the template as 985. To show the two zeroes at the beginning of the number, you must enter an apostrophe before the number: **'00985**. The apostrophe indicates that the number should be shown as a "label," meaning that all the digits will show including the beginning zeroes.

Step 10   Save the spreadsheet using the **Save** option from the *File* menu. You should accept the default location for the save as this is handled by the management system.

Step 11   Print the completed spreadsheet.

Step 12   Exit the spreadsheet program.

Step 13   In the Close Options box, select the location where you would like to save your work.

Step 14   Answer the Analyze question from your textbook for this problem.

What-If Analysis

Suppose the invoice amount for Invoice #34120 were $2,001.03. What would the discount be? What would the amount to be paid be?

_____

_____

_____

Problem 15-10 (continued)

GENERAL LEDGER

ACCOUNT ___Cash in Bank___   ACCOUNT NO. ___101___

DATE	DESCRIPTION	POST. REF.	DEBIT	CREDIT	BALANCE DEBIT	BALANCE CREDIT
20--						
Mar. 1	Balance	✓			1200000	

ACCOUNT ___Prepaid Insurance___   ACCOUNT NO. ___140___

DATE	DESCRIPTION	POST. REF.	DEBIT	CREDIT	BALANCE DEBIT	BALANCE CREDIT

ACCOUNT ___Accounts Payable___   ACCOUNT NO. ___201___

DATE	DESCRIPTION	POST. REF.	DEBIT	CREDIT	BALANCE DEBIT	BALANCE CREDIT
20--						
Mar. 1	Balance	✓				225000

Problem 15-10   (continued)

ACCOUNT ___Purchases___   ACCOUNT NO. ___501___

DATE		DESCRIPTION	POST. REF.	DEBIT	CREDIT	BALANCE	
						DEBIT	CREDIT
20--							
Mar.	1	Balance	✓			800000	

ACCOUNT ___Transportation In___   ACCOUNT NO. ___505___

DATE		DESCRIPTION	POST. REF.	DEBIT	CREDIT	BALANCE	
						DEBIT	CREDIT
20--							
Mar.	1	Balance	✓			150000	

ACCOUNT ___Purchases Discounts___   ACCOUNT NO. ___510___

DATE		DESCRIPTION	POST. REF.	DEBIT	CREDIT	BALANCE	
						DEBIT	CREDIT

ACCOUNT ___Purchases Returns and Allowances___   ACCOUNT NO. ___515___

DATE		DESCRIPTION	POST. REF.	DEBIT	CREDIT	BALANCE	
						DEBIT	CREDIT

Problem 15-10   (concluded)

ACCOUNTS PAYABLE SUBSIDIARY LEDGER

Name *ADC Publishing*

Address *5670 Mulberry Place, Seattle, WA 98101*

DATE	DESCRIPTION	POST. REF.	DEBIT	CREDIT	BALANCE

Name *American Trend Publishers*

Address *1313 Maple Drive, Seattle, WA 98148*

DATE	DESCRIPTION	POST. REF.	DEBIT	CREDIT	BALANCE
20--					
Mar. 1	Balance	✓			9 0 0 00

Name *Delta Press*

Address *One Triangle Park, Vancouver, WA 98661*

DATE	DESCRIPTION	POST. REF.	DEBIT	CREDIT	BALANCE
20--					
Mar. 1	Balance	✓			6 0 0 00

Name *Pine Forest Publications*

Address *103 Mt. Thor Road, Spokane, WA 99210*

DATE	DESCRIPTION	POST. REF.	DEBIT	CREDIT	BALANCE
20--					
Mar. 1	Balance	✓			7 5 0 00

Analyze: _____

_____

Notes

CHAPTER **15** Accounting for Purchases and Cash Payments

Self-Test

Part A   True or False

Directions: *Circle the letter* T *in the Answer column if the statement is true; circle the letter* F *if the statement is false.*

Answer

T   F   **1.** A purchase order is prepared by the buyer and sent to the supplier as an *offer* to purchase merchandise or other assets.

T   F   **2.** A tickler file contains folders for each day of the month so that invoices can be placed in the folders according to their due dates.

T   F   **3.** If the date of the invoice is October 1 and terms are 2/10, n/30, a discount may be taken if the invoice is paid on or before October 31.

T   F   **4.** Creditor accounts in the accounts payable subsidiary ledger normally have debit balances.

T   F   **5.** A cash discount is an advantage to the buyer but not the seller.

T   F   **6.** One example of a good internal control procedure requires that all cash payments be authorized.

T   F   **7.** Accounts Payable is a controlling account and it is found in the general ledger.

T   F   **8.** A business must write a check to pay the bank for service charges and bank card fees charged to its checking account.

T   F   **9.** The premium paid for insurance coverage is debited to the asset account, Prepaid Insurance.

T   F   **10.** Shipping charges are always paid by the seller.

Part B   Multiple Choice

Directions: *Only one of the choices given with each of the following statements is correct. Write the letter of the correct answer in the Answer column.*

Answer

1. If the invoice amount is $1,200.00 and the credit terms are 3/15, n/45, the amount of the discount is
   (A) $36.00.
   (B) $180.00.
   (C) $360.00.
   (D) $3.60.

2. A box of merchandise received by a business would include a(n)
   (A) invoice.
   (B) purchase requisition.
   (C) purchase order.
   (D) packing slip.

3. The document prepared by the seller is a(n)
   (A) purchase requisition.
   (B) invoice.
   (C) purchase order.
   (D) none of the above.

4. Which of the following statements is *true* about the Transportation In account?
   (A) It is classified as a cost of merchandise account.
   (B) Its balance and increase side is a debit.
   (C) Its decrease side is a credit.
   (D) Its balance shows the cost of delivery charges for merchandise.
   (E) All of the above.

5. A purchases returns and allowances transaction must be posted to the
   (A) Purchases and Accounts Payable accounts.
   (B) Accounts Payable account.
   (C) the creditor's account in the accounts payable ledger.
   (D) the Accounts Payable account and the creditor's account.

6. The Purchases account is debited when
   (A) merchandise is bought for resale.
   (B) merchandise is bought on account.
   (C) merchandise is bought for cash.
   (D) all of the above.

7. A processing stamp is placed on the
   (A) packing slip.
   (B) purchase requisition.
   (C) invoice.
   (D) purchase order.

CHAPTER 16   Special Journals: Sales and Cash Receipts

Study Plan

Check Your Understanding

Section 1	*Read Section 1 on pages 420–427 and complete the following exercises on page 428.* ❏ Thinking Critically ❏ Communication Accounting ❏ Problem 16-1 *Posting Column Totals from the Sales Journal* ❏ Problem 16-2 *Analyzing a Source Document*
Section 2	*Read Section 2 on pages 429–439 and complete the following exercises on page 440.* ❏ Thinking Critically ❏ Computing in the Business World ❏ Problem 16-3 *Completing the Cash Receipts Journal*
Summary	*Review the Chapter 16 Summary on page 441 in your textbook.* ❏ Key Concepts
Review and Activities	*Complete the following questions and exercises on pages 442–443 in your textbook.* ❏ Using Key Terms ❏ Understanding Accounting Concepts and Procedures ❏ Case Study ❏ Conducting an Audit with Alex ❏ Internet Connection ❏ Workplace Skills
Computerized Accounting	*Read the Computerized Accounting information on page 444 in your textbook.* ❏ *Making the Transition from a Manual to a Computerized System* ❏ *Mastering Sales and Cash Receipts in Peachtree*
Problems	*Complete the following end-of-chapter problems for Chapter 16 in your textbook.* ❏ Problem 16-4 *Recording and Posting Sales and Cash Receipts* ❏ Problem 16-5 *Recording and Posting Cash Receipts*
Challenge Problem	❏ Problem 16-6 *Recording and Posting Sales and Cash Receipts*
Chapter Reviews and Working Papers	*Complete the following exercises for Chapter 16 in your Chapter Reviews and Working Papers.* ❏ Chapter Review ❏ Self-Test

CHAPTER  REVIEW   Special Journals: Sales and Cash Receipts

Part 1   Accounting Vocabulary (4 points)

Total Points	38
Student's Score	

Directions: *Using terms from the following list, complete the sentences below. Write the letter of the term you have chosen in the space provided.*

A. cash receipts journal	**C.** sales journal	**E.** special journal
B. footing	**D.** schedule of accounts receivable	

    ___E___   **0.** A(n) _____, which simplifies the journalizing and posting process, has special columns that are used for recording specific types of business transactions.

    _____   **1.** The sale of merchandise on account is recorded in the _____.

    _____   **2.** A column total written in small pencil figures is called a(n) _____.

    _____   **3.** All transactions in which cash is received are recorded in the _____.

    _____   **4.** A(n) _____ is a report listing each charge customer's name and account balance and the total amount due from all charge customers.

Part 2   Recording Transactions in Special Journals (11 points)

Directions: *Record the following entries in the sales journal. Use page 6 in the sales journal.*

20--
May   3   Sold $800 in merchandise, plus $48 sales taxes, on account to Molly Brian, Sales Slip 120.
      10   Sold $1,200 in merchandise to Spring Branch School District on account, Sales Slip 121. They are a tax exempt organization.

SALES JOURNAL

PAGE _____

	DATE	SALES SLIP NO.	CUSTOMER'S ACCOUNT DEBITED	POST. REF.	SALES CREDIT	SALES TAX PAYABLE CREDIT	ACCOUNTS RECEIVABLE DEBIT	
1								1
2								2
3								3
4								4

Part 3   Posting to the Accounts Receivable Subsidiary Ledger (8 points)

Directions: *Post the first journal entry completed in Part 2 to Molly Brian's account in the accounts receivable subsidiary ledger. Indicate on the sales journal in Part 2 that the transaction has been posted.*

ACCOUNTS RECEIVABLE SUBSIDIARY LEDGER

Name **Molly Brian**

Address **865 Elmwood Place, Cincinnati, OH 45202**

DATE	DESCRIPTION	POST. REF.	DEBIT	CREDIT	BALANCE

Part 4   Analyzing Sales and Cash Receipts Transactions (15 points)

Directions: *Read each of the following statements to determine whether the statement is true or false. Write your answer in the space provided.*

    True   **0.** Two advantages of special journals are that they save time in recording and posting.

              1. Only the sale of merchandise on account is recorded in the sales journal.

              2. Every transaction recorded in the cash receipts journal results in a debit to cash.

              3. The sales journal is used to record any sale of merchandise, whether on account or for cash.

              4. Sales taxes are not usually charged on sales of merchandise to government agencies.

              5. The individual amounts in the Accounts Receivable Debit column of the sales journal are to be posted on a monthly basis.

              6. The column totals from the sales journal are posted to three general ledger accounts.

              7. When posting from a special journal, post information moving from top to bottom rather than left to right.

              8. When first totaling amount columns in a special journal always use a pencil.

              9. The total of an amount column in a special journal is posted to the general ledger account named in the column heading.

              10. At the end of the month, a check mark is placed below the double ruling in the General Credit column of the cash receipts journal to indicate the column total is not posted.

              11. Amounts recorded in the General Credit column of the cash receipts journal are to be posted when the transaction is journalized.

              12. When preparing a schedule of accounts receivable, enter only accounts that have a balance.

              13. Cash discounts are computed on the total amount of merchandise sold and not the sales tax.

              14. When journalizing cash sales and bankcard sales transactions, a dash is placed in the Posting Reference column.

              15. Most transactions recorded in the cash receipts journal result in a debit to Accounts Receivable and a credit to Cash in Bank.

Working Papers *for Section Problems*

Problem 16-1   Posting Column Totals from the Sales Journal

SALES JOURNAL   PAGE ____4____

	DATE	SALES SLIP NO.	CUSTOMER'S ACCOUNT DEBITED	POST. REF.	SALES CREDIT	SALES TAX PAYABLE CREDIT	ACCOUNTS RECEIVABLE DEBIT	
1	20--							1
2	Apr. 1	47	Amy Anderson	✓	8 0 0 00	4 8 00	8 4 8 00	2
31	30		Totals		1 2 0 0 0 00	7 2 0 00	1 2 7 2 0 00	31
32								32
33								33

GENERAL LEDGER

ACCOUNT __Accounts Receivable_____   ACCOUNT NO. ____115____

DATE	DESCRIPTION	POST. REF.	DEBIT	CREDIT	BALANCE DEBIT	BALANCE CREDIT
20--						
Apr. 1	Balance	✓			1 5 0 0 0 00	

ACCOUNT __Sales Tax Payable_____   ACCOUNT NO. ____220____

DATE	DESCRIPTION	POST. REF.	DEBIT	CREDIT	BALANCE DEBIT	BALANCE CREDIT
20--						
Apr. 1	Balance	✓				1 3 0 0 00

ACCOUNT __Sales_____   ACCOUNT NO. ____401____

DATE	DESCRIPTION	POST. REF.	DEBIT	CREDIT	BALANCE DEBIT	BALANCE CREDIT
20--						
Apr. 1	Balance	✓				2 5 0 0 0 00

Problem 16-2   Analyzing a Source Document

SALES JOURNAL

PAGE _____

	DATE	SALES SLIP NO.	CUSTOMER'S ACCOUNT DEBITED	POST. REF.	SALES CREDIT	SALES TAX PAYABLE CREDIT	ACCOUNTS RECEIVABLE DEBIT	
1								1
2								2
3								3
4								4
5								5
6								6
7								7

Name *M&M Consultants*

Address *2816 Mt. Odin Drive, Williamsburg, VA 23185*

DATE		DESCRIPTION	POST. REF.	DEBIT	CREDIT	BALANCE	
20--							
June	*1*	*Balance*	✓			*30000*	

Problem 16-3 Completing the Cash Receipts Journal

Name _____ Date _____ Class _____

CASH RECEIPTS JOURNAL PAGE 10

	DATE	DOC. NO.	ACCOUNT NAME	POST. REF.	GENERAL CREDIT	SALES CREDIT	SALES TAX PAYABLE CREDIT	ACCOUNTS RECEIVABLE CREDIT	SALES DISCOUNTS DEBIT	CASH IN BANK DEBIT
1	20--									
2	Jan. 3	R502	Jennifer Smith	✓				8000		8000
3	5	R503	Wilton High School	✓				310000	6200	303800
4	8	R504	Store Equipment	155	7500					7500
5	15	T42	Cash Sales	—		500000	30000			530000
6	15	T42	Bankcard Sales	—		120000	7200			127200
7	20	R505	Norwin High School	✓				240000	4800	235200
8	30	R506	Supplies	115	3000					3000

Computerized Accounting Using Peachtree

Software Objectives

When you have completed this chapter, you will be able to use Peachtree to:

1. Record sales on account using the **Sales/Invoicing** option.
2. Record cash receipts using the **Receipts** option.
3. Record partial payments by customers.

Problem 16-4   Recording and Posting Sales and Cash Receipts

INSTRUCTIONS

Beginning a Session

Step 1   Select the problem set: Shutterbug Cameras (Prob. 16-4).
Step 2   Rename the company and set the system date to May 31, 2008.

Completing the Accounting Problem

Step 3   Review the transactions in your textbook.

TIP: It is often faster to group and then enter transactions in batches. For example, enter all of the sales on account and then enter the cash sales.

Step 4   Record the sales on account using the **Sales/Invoicing** option.

TIP: Refer to Problem 14-5 if you need help entering the transactions for this problem.

Step 5   Record the cash sales using the **Receipts** option.
Step 6   Print a Sales Journal report and a Cash Receipts Journal report.
Step 7   Proof your work.
Step 8   Print a Customer Ledgers report.
Step 9   Print a General Ledger report.
Step 10   Answer the Analyze question.

Ending the Session

Step 11   Click the **Close Problem** button in the Glencoe Smart Guide window.

Mastering Peachtree

While entering a sales on account transaction, can you enter a new customer account on the fly? Explain your answer on a separate sheet of paper.

DO YOU HAVE A QUESTION

Q. *Does Peachtree include a Schedule of Accounts Receivable report?*

A. No, Peachtree does not have a report called *Schedule of Accounts Receivable.* The Customer Ledgers report is a combination of an Accounts Receivable Subsidiary Ledger report and a Schedule of Accounts Receivable. The Customer Ledgers report shows the transaction detail and a running balance. You can also print the Aged Receivables report to display the A/R balance.

Notes

Peachtree uses the term invoice, *not* sales slip, *on the sales data entry form.*

Problem 16-5   Recording and Posting Cash Receipts

INSTRUCTIONS

Beginning a Session

Step 1   Select the problem set: River's Edge Canoe & Kayak (Prob. 16-5).
Step 2   Rename the company and set the system date to May 31, 2008.

Completing the Accounting Problem

Step 3   Review the transactions in your textbook.
Step 4   Record the cash receipts.

TIP: Remember to change the *GL Account* field for cash receipts transactions not involving the **Sales** account.

Step 5   Print a Cash Receipts Journal report.
Step 6   Proof your work.
Step 7   Print a Customer Ledgers report.
Step 8   Print a General Ledger report.
Step 9   Answer the Analyze question.

Notes

To record a partial customer payment on an invoice, enter the amount received in the Amount Paid *field.*

Checking Your Work and Ending the Session

Step 10   Click the **Close Problem** button in the Glencoe Smart Guide window.
Step 11   If your teacher has asked you to check your solution, click *Check my answer to this problem*. Review, print, and close the report.
Step 12   Click the **Close Problem** button and select the appropriate save option.

Problem 16-6   Recording and Posting Sales and Cash Receipts

INSTRUCTIONS

Beginning a Session

Step 1   Select the problem set: Buzz Newsstand (Prob. 16-6).
Step 2   Rename the company and set the system date to May 31, 2008.

Completing the Accounting Problem

Step 3   Review the transactions in your textbook.
Step 4   Record the sales on account.
Step 5   Record the cash receipts.
Step 6   Print a Sales Journal report and a Cash Receipts Journal report.
Step 7   Proof your work.
Step 8   Print a Customer Ledgers report.
Step 9   Print a General Ledger report.
Step 10   Answer the Analyze question.

Ending the Session

Step 11   Click the **Close Problem** button in the Glencoe Smart Guide window.

Working Papers *for End-of-Chapter Problems*

Problem 16-4   Recording and Posting Sales and Cash Receipts
(1), (3)

SALES JOURNAL

PAGE _____

	DATE	SALES SLIP NO.	CUSTOMER'S ACCOUNT DEBITED	POST. REF.	SALES CREDIT	SALES TAX PAYABLE CREDIT	ACCOUNTS RECEIVABLE DEBIT	
1								1
2								2
3								3
4								4
5								5
6								6
7								7
8								8
9								9
10								10
11								11
12								12
13								13
14								14
15								15
16								16
17								17
18								18
19								19
20								20
21								21
22								22
23								23
24								24
25								25
26								26
27								27
28								28
29								29
30								30
31								31
32								32
33								33

Problem 16-4 (continued)

(1)

PAGE _____

CASH RECEIPTS JOURNAL

DATE	DOC. NO.	ACCOUNT NAME	POST. REF.	GENERAL CREDIT	SALES CREDIT	SALES TAX PAYABLE CREDIT	ACCOUNTS RECEIVABLE CREDIT	SALES DISCOUNTS DEBIT	CASH IN BANK DEBIT	
										1
										2
										3
										4
										5
										6
										7
										8
										9
										10
										11
										12
										13
										14
										15
										16
										17
										18
										19
										20
										21
										22
										23
										24
										25

Problem 16-4   (continued)

(2)

ACCOUNTS RECEIVABLE SUBSIDIARY LEDGER

Name **FastForward Productions**

Address **3 Oakhill Mall, Decatur, AL 35601**

DATE	DESCRIPTION	POST. REF.	DEBIT	CREDIT	BALANCE

Name **Yoko Nakata**

Address **19 Hawthorne Street, Tuscaloosa, AL 35401**

DATE		DESCRIPTION	POST. REF.	DEBIT	CREDIT	BALANCE
20--						
May	1	Balance	✓			600 00

Name **Heather Sullivan**

Address **835 Aspen Lane, Huntsville, AL 35801**

DATE		DESCRIPTION	POST. REF.	DEBIT	CREDIT	BALANCE
20--						
May	1	Balance	✓			50 00

Problem 16-4 (concluded)

(4)

GENERAL LEDGER

ACCOUNT _Cash in Bank_ ACCOUNT NO. ___101___

	DATE	DESCRIPTION	POST. REF.	DEBIT	CREDIT	BALANCE DEBIT	BALANCE CREDIT
	20--						
	May 1	Balance	✓			500000	

ACCOUNT _Accounts Receivable_ ACCOUNT NO. ___115___

	DATE	DESCRIPTION	POST. REF.	DEBIT	CREDIT	BALANCE DEBIT	BALANCE CREDIT
	20--						
	May 1	Balance	✓			65000	

ACCOUNT _Sales Tax Payable_ ACCOUNT NO. ___215___

	DATE	DESCRIPTION	POST. REF.	DEBIT	CREDIT	BALANCE DEBIT	BALANCE CREDIT
	20--						
	May 1	Balance	✓				120000

ACCOUNT _Sales_ ACCOUNT NO. ___401___

	DATE	DESCRIPTION	POST. REF.	DEBIT	CREDIT	BALANCE DEBIT	BALANCE CREDIT
	20--						
	May 1	Balance	✓				3000000

(5)

_____

Analyze: _____

Problem 16-5   Recording and Posting Cash Receipts

CASH RECEIPTS JOURNAL

DATE	DOC. NO.	ACCOUNT NAME	POST. REF.	GENERAL CREDIT	SALES CREDIT	SALES TAX PAYABLE CREDIT	ACCOUNTS RECEIVABLE CREDIT	SALES DISCOUNTS DEBIT	CASH IN BANK DEBIT	
										1
										2
										3
										4
										5
										6
										7
										8
										9
										10
										11
										12
										13
										14
										15
										16
										17
										18
										19
										20
										21
										22
										23
										24
										25
										26

Problem 16-5 (continued)

GENERAL LEDGER

ACCOUNT ___Cash in Bank___ ACCOUNT NO. ___101___

DATE		DESCRIPTION	POST. REF.	DEBIT	CREDIT	BALANCE	
						DEBIT	CREDIT
20--							
May	1	Balance	✓			7 5 0 0 00	

ACCOUNT ___Accounts Receivable___ ACCOUNT NO. ___115___

DATE		DESCRIPTION	POST. REF.	DEBIT	CREDIT	BALANCE	
						DEBIT	CREDIT
20--							
May	1	Balance	✓			6 9 0 0 00	

ACCOUNT ___Store Equipment___ ACCOUNT NO. ___150___

DATE		DESCRIPTION	POST. REF.	DEBIT	CREDIT	BALANCE	
						DEBIT	CREDIT
20--							
May	1	Balance	✓			3 0 0 0 00	

ACCOUNT ___Sales Tax Payable___ ACCOUNT NO. ___215___

DATE		DESCRIPTION	POST. REF.	DEBIT	CREDIT	BALANCE	
						DEBIT	CREDIT
20--							
May	1	Balance	✓				2 5 0 00

ACCOUNT ___Sales___ ACCOUNT NO. ___401___

DATE		DESCRIPTION	POST. REF.	DEBIT	CREDIT	BALANCE	
						DEBIT	CREDIT
20--							
May	1	Balance	✓				4 0 0 0 00

ACCOUNT ___Sales Discounts___ ACCOUNT NO. ___405___

DATE		DESCRIPTION	POST. REF.	DEBIT	CREDIT	BALANCE	
						DEBIT	CREDIT
20--							
May	1	Balance	✓			1 2 0 0 00	

Problem 16-5 (continued)

ACCOUNTS RECEIVABLE SUBSIDIARY LEDGER

Name **Adventure River Tours**

Address **Box 101, Jackson, WY 83001**

DATE		DESCRIPTION	POST. REF.	DEBIT	CREDIT	BALANCE
20--						
May	1	Balance	✓			3 0 0 0 00

Name **Paul Drake**

Address **125 Rodeo Road, Cody, WY 82414**

DATE		DESCRIPTION	POST. REF.	DEBIT	CREDIT	BALANCE
20--						
May	1	Balance	✓			8 0 0 00

Name **Celeste Everett**

Address **1824 Grays Gable, Laramie, WY 82070**

DATE		DESCRIPTION	POST. REF.	DEBIT	CREDIT	BALANCE
20--						
May	1	Balance	✓			4 0 0 00

Name **Isabel Rodriguez**

Address **626 Buffalo Road, Cheyenne, WY 82001**

DATE		DESCRIPTION	POST. REF.	DEBIT	CREDIT	BALANCE
20--						
May	1	Balance	✓			2 0 0 00

Problem 16-5 (concluded)

Name **Wildwood Resorts**

Address **601 Ponderosa Trail, Moose, WY 83012**

DATE		DESCRIPTION	POST. REF.	DEBIT	CREDIT	BALANCE
20--						
May	1	Balance	✓			2 50 00 00

Analyze: _____

_____

Problem 16-6   Source Documents

Instructions: *Use the following source documents to record the transactions for this problem.*

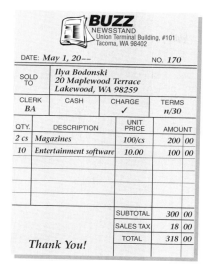

BUZZ NEWSSTAND — Union Terminal Building, #101, Tacoma, WA 98402

DATE: May 1, 20--   NO. 170

SOLD TO: Ilya Bodonski, 20 Maplewood Terrace, Lakewood, WA 98259

CLERK BA   CASH   CHARGE ✓   TERMS n/30

QTY.	DESCRIPTION	UNIT PRICE	AMOUNT	
2 cs	Magazines	100/cs	200	00
10	Entertainment software	10.00	100	00
		SUBTOTAL	300	00
		SALES TAX	18	00
		TOTAL	318	00

Thank You!

BUZZ NEWSSTAND — Union Terminal Building, #101, Tacoma, WA 98402

RECEIPT   No. 147

May 7   20 --

RECEIVED FROM Rothwell Management Inc. $ 294.00

Two hundred ninety-four and 00/100 ———— DOLLARS

FOR on account (300 less 2% discount)

RECEIVED BY Kathy Harper

BUZZ NEWSSTAND — Union Terminal Building, #101, Tacoma, WA 98402

DATE: April 30, 20--   NO. 162

SOLD TO: Rothwell Management Inc., 16 University Place, Vancouver, WA 98661

CLERK BA   CASH   CHARGE ✓   TERMS 2/10, n/30

QTY.	DESCRIPTION	UNIT PRICE	AMOUNT	
2 cs	Magazines	100/cs	200	00
75	Daily newspapers	1.00	75	00
1 bx	Pocket combs	8.02	8	02
		SUBTOTAL	283	02
		SALES TAX	16	98
		TOTAL	300	00

COPY   *Thank You!*

BUZZ NEWSSTAND — Union Terminal Building, #101, Tacoma, WA 98402

RECEIPT   No. 145

May 3   20 --

RECEIVED FROM Katz Properties $ 490.00

Four hundred ninety and 00/100 ———— DOLLARS

FOR on account (500.00 less 2% discount)

RECEIVED BY Kathy Harper

BUZZ NEWSSTAND — Union Terminal Building, #101, Tacoma, WA 98402

DATE: April 26, 20--   NO. 159

SOLD TO: Katz Properties, 103 Prospect Point, Bellevue, WA 98009

CLERK BA   CASH   CHARGE ✓   TERMS 2/10, n/30

QTY.	DESCRIPTION	UNIT PRICE	AMOUNT	
4 cs	Magazines	100/cs	400	00
2	Desk lamps	35.85	71	70
		SUBTOTAL	471	70
		SALES TAX	28	30
		TOTAL	500	00

COPY   *Thank You!*

BUZZ NEWSSTAND — Union Terminal Building, #101, Tacoma, WA 98402

DATE: May 9, 20--   NO. 171

SOLD TO: Saba Nadal, 943 Peachtree Drive, Spokane, WA 99210

CLERK BA   CASH   CHARGE ✓   TERMS n/30

QTY.	DESCRIPTION	UNIT PRICE	AMOUNT	
25	Paperback books	4.00	100	00
		SUBTOTAL	100	00
		SALES TAX	6	00
		TOTAL	106	00

Thank You!

BUZZ NEWSSTAND — Union Terminal Building, #101, Tacoma, WA 98402

DATE: May 10, 20--   NO. 172

SOLD TO: Java Shops, Inc., 449 Country Place, Auburn, WA 98002

CLERK BA   CASH   CHARGE ✓   TERMS 2/10, n/30

QTY.	DESCRIPTION	UNIT PRICE	AMOUNT	
3 cs	Magazines	100/cs	300	00
12	Hardcover books	25.00	300	00
		SUBTOTAL	600	00
		SALES TAX	36	00
		TOTAL	636	00

Thank You!

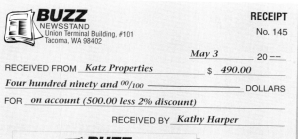

BUZZ NEWSSTAND — Union Terminal Building, #101, Tacoma, WA 98402

RECEIPT   No. 146

May 5   20 --

RECEIVED FROM Straka Stores $ 60.00

Sixty and 00/100 ———— DOLLARS

FOR supplies

RECEIVED BY Kathy Harper

Problem 16-6 (continued)

BUZZ NEWSSTAND
Union Terminal Building, #101
Tacoma, WA 98402

DATE: *May 12, 20--*   NO. *173*

SOLD TO	Lee Adkins 720 Dogwood Lane Seattle, WA 98101		
CLERK BA	CASH	CHARGE ✓	TERMS n/30

QTY.	DESCRIPTION	UNIT PRICE	AMOUNT	
½ cs	Magazines	100/cs	50	00
		SUBTOTAL	50	00
		SALES TAX	3	00
		TOTAL	53	00

Thank You!

BUZZ NEWSSTAND
Union Terminal Building, #101
Tacoma, WA 98402

DATE: *May 18, 20--*   NO. *174*

SOLD TO	Katz Properties 103 Prospect Point Bellevue, WA 98009		
CLERK BA	CASH	CHARGE ✓	TERMS 2/10, n/30

QTY.	DESCRIPTION	UNIT PRICE	AMOUNT	
32	Hardcover books	25.00	800	00
200	Daily newspapers	1.00	200	00
		SUBTOTAL	1,000	00
		SALES TAX	60	00
		TOTAL	1,060	00

Thank You!

BUZZ NEWSSTAND
Union Terminal Building, #101
Tacoma, WA 98402

RECEIPT
No. 148

May 15   20 --

RECEIVED FROM *Rolling Hills Pharmacies* $ *196.00*

One hundred ninety-six and ⁰⁰/₁₀₀ ——————— DOLLARS

FOR *on account ($200.00 less 2% discount)*

RECEIVED BY *Kathy Harper*

BUZZ NEWSSTAND
Union Terminal Building, #101
Tacoma, WA 98402

RECEIPT
No. 149

May 20   20 --

RECEIVED FROM *Ilya Bodonski* $ *100.00*

One hundred and ⁰⁰/₁₀₀ ——————— DOLLARS

FOR *on account*

RECEIVED BY *Kathy Harper*

```
May  15
Tape 33


        2,400.00    CA
          144.00    ST
```

```
May  15
Tape 33


        2,000.00    BCS
          120.00    ST
```

BUZZ NEWSSTAND
Union Terminal Building, #101
Tacoma, WA 98402

DATE: *May 22, 20--*   NO. *175*

SOLD TO	Rothwell Management Inc. 16 University Place Vancouver, WA 98661		
CLERK BA	CASH	CHARGE ✓	TERMS 2/10, n/30

QTY.	DESCRIPTION	UNIT PRICE	AMOUNT	
4 cs	Magazines	100/cs	400	00
200	Daily newspapers	1.00	200	00
5	Travel planning software	40.00	200	00
		SUBTOTAL	800	00
		SALES TAX	48	00
		TOTAL	848	00

Thank You!

Problem 16-6 (continued)

BUZZ NEWSSTAND
Union Terminal Building, #101
Tacoma, WA 98402

RECEIPT
No. 150

May 23 20 --

RECEIVED FROM _Lee Adkins_ $ _53.00_

Fifty-three and ⁰⁰/₁₀₀ ——————— DOLLARS

FOR _on account_

RECEIVED BY _Kathy Harper_

BUZZ NEWSSTAND
Union Terminal Building, #101
Tacoma, WA 98402

RECEIPT
No. 153

May 28 20 --

RECEIVED FROM _Brown's Books and More_ $ _75.00_

Seventy-five and ⁰⁰/₁₀₀ ——————— DOLLARS

FOR _used store equipment_

RECEIVED BY _Kathy Harper_

BUZZ NEWSSTAND
Union Terminal Building, #101
Tacoma, WA 98402

RECEIPT
No. 151

May 24 20 --

RECEIVED FROM _Saba Nadal_ $ _106.00_

One hundred six and ⁰⁰/₁₀₀ ——————— DOLLARS

FOR _on account_

RECEIVED BY _Kathy Harper_

May 30
Tape 34

2,600.00   CA
156.00   ST

BUZZ NEWSSTAND
Union Terminal Building, #101
Tacoma, WA 98402

RECEIPT
No. 152

May 26 20 --

RECEIVED FROM _Java Shops, Inc._ $ _200.00_

Two hundred and ⁰⁰/₁₀₀ ——————— DOLLARS

FOR _on account_

RECEIVED BY _Kathy Harper_

May 30
Tape 34

2,200.00   BCS
132.00   ST

BUZZ NEWSSTAND
Union Terminal Building, #101
Tacoma, WA 98402

DATE: _May 27, 20--_   NO. _176_

SOLD TO	_Lee Adkins_ _720 Dogwood Lane_ _Seattle, WA 98101_		

CLERK _BA_	CASH	CHARGE ✓	TERMS _n/30_

QTY.	DESCRIPTION	UNIT PRICE	AMOUNT	
2 cs	Magazines	100/cs	200	00
	SUBTOTAL		200	00
	SALES TAX		12	00
	TOTAL		212	00

Thank You!

Problem 16-6   Recording and Posting Sales and Cash Receipts
(1), (4)

SALES JOURNAL

PAGE _____

	DATE	SALES SLIP NO.	CUSTOMER'S ACCOUNT DEBITED	POST. REF.	SALES CREDIT	SALES TAX PAYABLE CREDIT	ACCOUNTS RECEIVABLE DEBIT	
1								1
2								2
3								3
4								4
5								5
6								6
7								7
8								8
9								9
10								10
11								11
12								12
13								13
14								14
15								15
16								16
17								17
18								18
19								19
20								20
21								21
22								22
23								23
24								24
25								25
26								26
27								27
28								28
29								29
30								30
31								31
32								32
33								33
34								34
35								35

Problem 16-6 (continued)

CASH RECEIPTS JOURNAL

PAGE _____

DATE	DOC. NO.	ACCOUNT NAME	POST. REF.	GENERAL CREDIT	SALES CREDIT	SALES TAX PAYABLE CREDIT	ACCOUNTS RECEIVABLE CREDIT	SALES DISCOUNTS DEBIT	CASH IN BANK DEBIT
1									
2									
3									
4									
5									
6									
7									
8									
9									
10									
11									
12									
13									
14									
15									
16									
17									
18									
19									
20									
21									
22									
23									
24									
25									
26									

Problem 16-6   (continued)

(2)

ACCOUNTS RECEIVABLE SUBSIDIARY LEDGER

Name _Lee Adkins_

Address _720 Dogwood Lane, Seattle, WA 98101_

DATE	DESCRIPTION	POST. REF.	DEBIT	CREDIT	BALANCE

Name _Ilya Bodonski_

Address _20 Maplewood Terrace, Lakewood, WA 98259_

DATE	DESCRIPTION	POST. REF.	DEBIT	CREDIT	BALANCE

Name _Java Shops Inc._

Address _449 Country Place, Auburn, WA 98002_

DATE	DESCRIPTION	POST. REF.	DEBIT	CREDIT	BALANCE

Name _Katz Properties_

Address _103 Prospect Point, Bellevue, WA 98009_

DATE		DESCRIPTION	POST. REF.	DEBIT	CREDIT	BALANCE
20--						
May	1	Balance	✓			500 00

Problem 16-6 (continued)

Name *Saba Nadal*

Address *943 Peachtree Drive, Spokane, WA 99210*

DATE	DESCRIPTION	POST. REF.	DEBIT	CREDIT	BALANCE

Name *Rolling Hills Pharmacies*

Address *16 Meadow Lane, Tacoma, WA 98402*

DATE		DESCRIPTION	POST. REF.	DEBIT	CREDIT	BALANCE
20--						
May	5	Balance	✓			20000

Name *Rothwell Management Inc.*

Address *16 University Place, Vancouver, WA 98661*

DATE		DESCRIPTION	POST. REF.	DEBIT	CREDIT	BALANCE
20--						
May	1	Balance	✓			30000

Problem 16-6 (continued)

GENERAL LEDGER (PARTIAL)

ACCOUNT __*Cash in Bank*__ ACCOUNT NO. __101__

DATE		DESCRIPTION	POST. REF.	DEBIT	CREDIT	BALANCE DEBIT	BALANCE CREDIT
20--							
May	1	Balance	✓			500000	

ACCOUNT __*Accounts Receivable*__ ACCOUNT NO. __115__

DATE		DESCRIPTION	POST. REF.	DEBIT	CREDIT	BALANCE DEBIT	BALANCE CREDIT
20--							
May	1	Balance	✓			100000	

ACCOUNT __*Supplies*__ ACCOUNT NO. __135__

DATE		DESCRIPTION	POST. REF.	DEBIT	CREDIT	BALANCE DEBIT	BALANCE CREDIT
20--							
May	1	Balance	✓			30000	

ACCOUNT __*Store Equipment*__ ACCOUNT NO. __150__

DATE		DESCRIPTION	POST. REF.	DEBIT	CREDIT	BALANCE DEBIT	BALANCE CREDIT
20--							
May	1	Balance	✓			400000	

ACCOUNT __*Sales Tax Payable*__ ACCOUNT NO. __215__

DATE		DESCRIPTION	POST. REF.	DEBIT	CREDIT	BALANCE DEBIT	BALANCE CREDIT
20--							
May	1	Balance	✓				40000

Problem 16-6   (concluded)

ACCOUNT _Sales_   ACCOUNT NO. _401_

DATE		DESCRIPTION	POST. REF.	DEBIT	CREDIT	BALANCE DEBIT	BALANCE CREDIT
20--							
May	1	Balance	✓				2000000

ACCOUNT _Sales Discounts_   ACCOUNT NO. _405_

DATE		DESCRIPTION	POST. REF.	DEBIT	CREDIT	BALANCE DEBIT	BALANCE CREDIT
20--							
May	1	Balance	✓			50000	

(7)

Analyze: _____

CHAPTER **16** Special Journals: Sales and Cash Receipts

Self-Test

Part A   True or False

Directions: *Circle the letter* T *in the Answer column if the statement is true; circle the letter* F *if the statement is false.*

Answer

T   F   **1.** Two major advantages of a special journal are that it saves time in recording and posting business transactions.

T   F   **2.** The cash register tape is the source document for recording cash and bank card sales.

T   F   **3.** Only customer accounts with balances are entered on the schedule of accounts receivable.

T   F   **4.** A business handles bank card sales like cash sales because it receives cash from the bank within several days after depositing its bank card sales slips.

T   F   **5.** Cash sales are recorded in the sales journal.

T   F   **6.** Footings are always completed in pencil, and column totals are written in ink.

T   F   **7.** Merchandising businesses sell strictly on the cash basis.

T   F   **8.** Customer accounts are listed in alphabetical order in the accounts receivable subsidiary ledger.

T   F   **9.** After a posting has been made from the sales or cash receipts journal to a customer's account, the customer account number is entered in the posting reference column of the journal.

T   F   **10.** Postings are made to customer accounts usually at the end of the week.

Part B   Multiple Choice

Directions: *Only one of the choices given with each of the following questions is correct. Write the letter of the correct answer in the Answer column.*

Answer

_____   **1.** The Accounts Receivable Debit column of the sales journal is used to record
(A) the amount of the sale.
(B) the amount of the sales tax.
(C) the total amount to be received from a customer.
(D) the total amount to be paid to a customer.

_____   **2.** The value of the merchandise sold is recorded in the sales journal in the
(A) Sales Tax Payable Credit column.
(B) Sales Credit column.
(C) Customer's Account Debited column.
(D) Accounts Receivable Debit column.

_____   **3.** The totals of the Sales Credit column and the Sales Tax Payable Credit column in the sales journal should be posted
(A) at the end of the month.
(B) at the end of the week.
(C) once every two weeks.
(D) on a daily basis.

_____   **4.** Special journals do the following:
(A) simplify the posting process.
(B) simplify the recording process.
(C) organize the transactions of a business.
(D) all of the above.

_____   **5.** The amounts recorded in the General Credit column of the cash receipts journal are posted
(A) on the day the transaction occurred.
(B) at the end of the week.
(C) at the end of the month.
(D) once every two weeks.

_____   **6.** The amounts recorded in the Accounts Receivable Credit column of the cash receipts journal are posted
(A) once a week.   (C) once a month.
(B) once every two weeks.   (D) on a daily basis.

_____   **7.** The following kinds of transactions are recorded in the cash receipts journal except
(A) sale of merchandise on account.
(B) cash sale of merchandise.
(C) bankcard sales.
(D) sale of other assets for cash.

CHAPTER 17

Special Journals: Purchases and Cash Payments

Study Plan

Check Your Understanding

Section 1	*Read Section 1 on pages 450–456 and complete the following exercises on page 457.*
	❏ Thinking Critically
	❏ Communication Accounting
	❏ Problem 17-1 *Recording Transactions in the Purchases Journal*

Section 2	*Read Section 2 on pages 458–470 and complete the following exercises on page 471.*
	❏ Thinking Critically
	❏ Computing in the Business World
	❏ Problem 17-2 *Preparing a Cash Proof*
	❏ Problem 17-3 *Analyzing a Source Document*

Summary   *Review the Chapter 17 Summary on page 473 in your textbook.*
❏ Key Concepts

Review and Activities   *Complete the following questions and exercises on pages 474–475 in your textbook.*
❏ Using Key Terms
❏ Understanding Accounting Concepts and Procedures
❏ Case Study
❏ Conducting an Audit with Alex
❏ Internet Connection
❏ Workplace Skills

Computerized Accounting   *Read the Computerized Accounting information on page 476 in your textbook.*
❏ *Mastering Purchases and Cash Payments in Peachtree*
❏ *Making the Transition from a Manual to a Computerized System*

Problems   *Complete the following end-of-chapter problems for Chapter 17 in your textbook.*
❏ Problem 17-4   *Recording Payment of the Payroll*
❏ Problem 17-5   *Recording Transactions in the Purchases Journal*
❏ Problem 17-6   *Recording and Posting Purchases*
❏ Problem 17-7   *Recording and Posting Cash Payments*

Challenge Problem   ❏ Problem 17-8   *Recording and Posting Purchases and Cash Payments*

Chapter Reviews and Working Papers   *Complete the following exercises for Chapter 17 in your Chapter Reviews and Working Papers.*
❏ Chapter Review
❏ Self-Test

CHAPTER 17 REVIEW

Special Journals: Purchases and Cash Payments

Part 1   Accounting Vocabulary (12 points)

Total Points	36
Student's Score	

Directions: *Using terms from the following list, complete the sentences below. Write the letter of the term you have chosen in the space provided.*

A. Accounts Payable	**D.** accounts receivable subsidiary ledger	**G.** controlling account	**K.** purchases journal
B. accounts payable subsidiary ledger	**E.** cash payments journal	**H.** due date	**L.** sales journal
C. Accounts Receivable	**F.** cash receipts journal	**I.** general journal **J.** proving cash	**M.** schedule of accounts payable

_H_   **0.** The _____ is the date by which the invoice must be paid if a discount is to be taken.

_____   **1.** A(n) _____ is a special journal used for recording the sale of merchandise on account.

_____   **2.** A(n) _____ is a separate ledger that contains accounts for all creditors.

_____   **3.** _____ is an account in the general ledger that controls the accounts receivable subsidiary ledger.

_____   **4.** The _____ is a special journal used for recording all cash received by the business.

_____   **5.** _____ is an account in the general ledger that controls the accounts payable subsidiary ledger.

_____   **6.** The _____ is an all-purpose journal used for recording transactions that do not fit into a special journal.

_____   **7.** The process of determining whether the amount of cash recorded in the accounting records of a business agrees with the amount recorded in its checkbook is called _____.

_____   **8.** A special journal used for recording all cash paid out of the business is the _____.

_____   **9.** The _____ is a separate ledger that contains all charge customer accounts.

_____   **10.** The _____ is a special journal used for recording all purchases of assets on account.

_____   **11.** The _____ is a list of all creditors in the accounts payable subsidiary ledger, the balance of each account, and the total owed to all creditors.

_____   **12.** A(n) _____ is an account in the general ledger that acts as a control on the accuracy of the accounts in the subsidiary ledger.

Part 2   Recording a Transaction in the Purchases Journal (4 points)

Directions: *The following transaction would be recorded in a purchases journal. Indicate where the transaction information would be recorded in the purchases journal below by writing the correct identifying letter in the space provided.*

Received Invoice 4208 on August 9 from Bosco Enterprises for merchandise purchased on account.

_E_   **0.** Amount of the credit

_____   **1.** Name of the creditor

_____   **2.** Amount of the debit

_____   **3.** The date

_____   **4.** Invoice number

PURCHASES JOURNAL   PAGE _____

	DATE	INVOICE NO.	CREDITOR'S ACCOUNT CREDITED	POST. REF.	ACCOUNTS PAYABLE CREDIT	PURCHASES DEBIT	GENERAL			
							ACCOUNT DEBITED	POST. REF.	DEBIT	
1	**Ⓐ**	**Ⓑ**	**Ⓒ**	**Ⓓ**	**Ⓔ**	**Ⓕ**	**Ⓖ**	**Ⓗ**	**Ⓘ**	1
2										2
3										3
4										4
5										5
6										6

Part 3   Analyzing Purchases and Cash Payment Transactions (10 points)

Directions: *Read each of the following statements to determine whether the statement is true or false. Write your answer in the space provided.*

True	**0.** The account Purchases is a cost of merchandise account.
_____	**1.** Purchases of items for cash are recorded in the purchases journal.
_____	**2.** The General Debit column of the purchases journal is used to record purchases of all items other than merchandise on account.
_____	**3.** The source document for recording information in the purchases journal is an invoice.
_____	**4.** Amounts entered in the Accounts Payable Credit column of the purchases journal are posted at the end of the month.
_____	**5.** A check mark is entered in parentheses below the double rule of the General Debit column of the purchases journal to indicate the column total is not posted.
_____	**6.** The source document for recording a bank card fee is the check stub.
_____	**7.** Every transaction recorded in the cash payments journal results in a credit to cash.
_____	**8.** The schedule of accounts payable lists only creditor accounts showing a balance.
_____	**9.** The totals of the special amount columns of the cash payments journal are posted at the end of the month.
_____	**10.** Amounts entered in the Accounts Payable Debit column of the cash payments journal are posted at the end of the week.

Part 4   Recording Transactions in the Cash Payments Journal (10 points)

Directions: *The following transactions would be recorded in a cash payments journal. Indicate where the information for each transaction would be recorded in the cash payments journal below by writing the correct identifying letter(s) in the space provided.*

1. June 8, issued Check 148 for $882 to Franklin Brothers Distributors for merchandise purchased on account, Invoice 337 for $900 less a cash discount of $18

A	**a.** Date
_____	**b.** Check number
_____	**c.** Name of the account to be debited
_____	**d.** Amount of the debit
_____	**e.** Amount of the credit

2. May 3, issued Check 601 to KM-Realty for the May rent, $500

_____	**a.** Check number
_____	**b.** Amount of the credit
_____	**c.** Amount of the debit
_____	**d.** Date
_____	**e.** Name of the account to be debited

CASH PAYMENTS JOURNAL   PAGE _____

	DATE	DOC. NO.	ACCOUNT NAME	POST. REF.	GENERAL DEBIT	GENERAL CREDIT	ACCOUNTS PAYABLE DEBIT	PURCHASES DISCOUNTS CREDIT	CASH IN BANK CREDIT	
1	**A**	**B**	**C**	**D**	**E**	**F**	**G**	**H**	**I**	1
2										2
3										3
4										4
5										5
6										6

Working Papers *for Section Problems*

Problem 17-1   Recording Transactions in the Purchases Journal

PURCHASES JOURNAL

PAGE _____

DATE	INVOICE NO.	CREDITOR'S ACCOUNT CREDITED	POST. REF.	ACCOUNTS PAYABLE CREDIT	PURCHASES DEBIT	GENERAL ACCOUNT DEBITED	POST. REF.	DEBIT
1								
2								
3								
4								
5								
6								
7								
8								
9								
10								
11								
12								
13								
14								
15								
16								
17								
18								
19								
20								
21								
22								
23								
24								
25								
26								

Problem 17-2   Preparing a Cash Proof

Problem 17-3   **Analyzing a Source Document**

Source document check No. 104:

$ 873.00			No. 104
Date November 2			20 —
To Colonial Products Inc.			
For Inv. 323—$900 less 3% disc., $27.00			
	Dollars	Cents	
Balance brought forward	3,468	29	
Add deposits			
Total	3,468	29	
Less this check	873	00	
Balance carried forward	2,595	29	

CASH PAYMENTS JOURNAL   PAGE _____

DATE	DOC. NO.	ACCOUNT NAME	POST. REF.	GENERAL DEBIT	GENERAL CREDIT	ACCOUNTS PAYABLE DEBIT	PURCHASES DISCOUNTS CREDIT	CASH IN BANK CREDIT	
									1
									2
									3
									4

Computerized Accounting Using Peachtree

Software Objectives

When you have completed this chapter, you will be able to use Peachtree to:

1. Record purchases on account using the **Purchases/Receive Inventory** option.
2. Record cash payments using the **Payments** option.
3. Record partial payments on account to vendors.

Problem 17-4   Recording Payment of the Payroll

INSTRUCTIONS

Beginning a Session

Step 1   Select the problem set: Denardo's Country Store (Prob. 17-4).
Step 2   Rename the company and set the system date to July 15, 2008.

Completing the Accounting Problem

Step 3   Review the payroll information in your textbook.
Step 4   Record the payment of the payroll using **Payments** from the **Tasks** menu. Refer to Problem 15-6 if you need help entering a cash payment transaction.

TIP: Record the debit to **Salaries Expense** and enter the credits (as negative amounts) to the payroll liability accounts as a multi-part entry.

Step 5   Print a Cash Disbursements Journal report.
Step 6   Proof your work.
Step 7   Answer the Analyze question.

Ending the Session

Step 8   Click the **Close Problem** button in the Glencoe Smart Guide window.

Problem 17-5   Recording Transactions in the Purchases Journal

INSTRUCTIONS

Beginning a Session

Step 1   Select the problem set: Sunset Surfwear (Prob. 17-5).
Step 2   Rename the company and set the system date to July 31, 2008.

DO YOU HAVE A QUESTION

Q. *Do you have to use the* **Payments** *option to record all payments?*

A. For payments on account to vendors, you must use the **Payments** option so that Peachtree can update the subsidiary ledger account. Remember that the general journal does not provide a way to identify the vendor. For payments not involving vendors (e.g., payment of an insurance premium), you could use the general journal. However, the **Payments** option is designed for these kinds of transactions just like special journals are more efficient in a manual system.

Peachtree Guide

Completing the Accounting Problem

Step 3 Review the transactions in your textbook.

TIP: Refer to Problem 15-6 if you need help entering purchases on account transactions.

Step 4 Record the purchases on account transactions using the **Purchases/Receive Inventory** option.
Step 5 Print a Purchases Journal report.
Step 6 Proof your work.
Step 7 Answer the Analyze question.

Checking Your Work and Ending the Session

Step 8 Click the **Close Problem** button in the Glencoe Smart Guide window.
Step 9 If your teacher has asked you to check your solution, click *Check my answer to this problem*. Review, print, and close the report.
Step 10 Click the **Close Problem** button and select the appropriate save option.

Problem 17-6 Recording and Posting Purchases

INSTRUCTIONS

Beginning a Session

Step 1 Select the problem set: Shutterbug Cameras (Prob. 17-6).
Step 2 Rename the company and set the system date to July 31, 2008.

Completing the Accounting Problem

Step 3 Review the transactions in your textbook.
Step 4 Record the purchases on account.

TIP: Verify the *GL Account* field each time you record an entry.

Step 5 Print a Purchases Journal report.
Step 6 Proof your work.
Step 7 Print a Vendor Ledgers report and a General Ledger report.
Step 8 Answer the Analyze question.

Ending the Session

Step 9 Click the **Close Problem** button in the Glencoe Smart Guide window and select the appropriate save option.

Notes

Peachtree automatically updates the general ledger accounts and subsidiary ledger accounts when you post a purchase on account entry.

Problem 17-7   Recording and Posting Cash Payments

INSTRUCTIONS

Beginning a Session

Step 1   Select the problem set: River's Edge Canoe & Kayak (Prob. 17-7).
Step 2   Rename the company and set the system date to July 31, 2008.

Completing the Accounting Problem

Step 3   Review the transactions in your textbook.
Step 4   Record the cash payments using the **Payments** option.

TIP: Record a payment on account using the **Apply to Invoices** tab. For cash payments other than those on account, use the **Apply to Expenses** tab. You may have to manually enter the cash discount for some payment transactions.

Step 5   Print a Cash Disbursements Journal report.
Step 6   Proof your work.
Step 7   Print a Vendor Ledgers report and a General Ledger report.
Step 8   Answer the Analyze question.

Checking Your Work and Ending the Session

Step 9   Click the **Close Problem** button in the Glencoe Smart Guide window.
Step 10   If your teacher has asked you to check your solution, click *Check my answer to this problem*. Review, print, and close the report.
Step 11   Click the **Close Problem** button and select the appropriate save option.

Problem 17-8   Recording and Posting Purchases and Cash Payments

INSTRUCTIONS

In Peachtree there are different ways to record cash disbursements. In this example, record bank service charges and bankcard fees using the **Payments** option.

Beginning a Session

Step 1   Select the problem set: Buzz Newsstand (Prob. 17-8).
Step 2   Rename the company and set the system date to July 31, 2008.

Completing the Accounting Problem

Step 3   Review the transactions in your textbook.
Step 4   Enter the purchases on account transactions.
Step 5   Record the cash payments.
Step 6   Print the following reports: Purchases Journal, Cash Disbursements Journal, and Vendor Ledgers.
Step 7   Proof your work.
Step 8   Print a General Ledger report.
Step 9   Answer the Analyze question.

Ending the Session

Step 10   Click the **Close Problem** button in the Glencoe Smart Guide window.

Peachtree Guide

FAQs

What steps are required to correct a purchase invoice with the wrong vendor ID code or amount?

If you enter the wrong information for a purchase invoice, you can correct it even if you have already posted the transaction. Choose the **Purchases/Receive Inventory** option and click the **Open** button. Select the invoice you want to change. When the invoice appears in the data entry window, make the necessary corrections and then post the updated transaction.

What steps are required to correct a cash payment entry?

You can change any information on a cash payment unless you have closed the current period. To make a change, choose the **Payments** option. Click the **Open** button and select the payment you want to update. Make the changes and post the corrected transaction. Peachtree automatically applies the corrections.

The GL Account fields do not appear in the Purchases/Receive Inventory or Payments data entry windows.

Peachtree lets you hide the general ledger fields in the data entry windows. However, you will need to enter information in these fields to record certain kinds of transactions. To display the general ledger account fields, choose **Global** from the *Options* menu. Change the setting to show the general ledger accounts by deselecting the Accounts Payable check box. **Note:** If the Purchases/Receive Inventory or Payments window is open, you must close the window and then choose the task option again before the new setting will take effect.

Working Papers *for End-of-Chapter Problems*
Problem 17-4   Recording Payment of the Payroll

CASH PAYMENTS JOURNAL

PAGE _____

DATE	DOC. NO.	ACCOUNT NAME	POST. REF.	GENERAL DEBIT	GENERAL CREDIT	ACCOUNTS PAYABLE DEBIT	PURCHASES DISCOUNTS CREDIT	CASH IN BANK CREDIT	
									1
									2
									3
									4
									5
									6
									7

Analyze:

Problem 17-5   Recording Transactions in the Purchases Journal

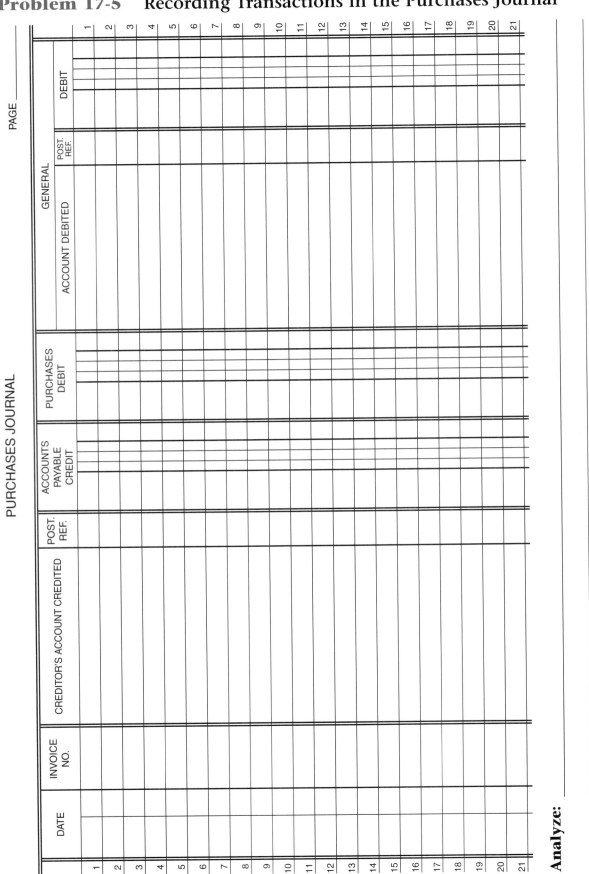

PURCHASES JOURNAL   PAGE _____

Analyze:

Problem 17-6   Recording and Posting Purchases

(1)

PURCHASES JOURNAL

PAGE _____

DATE	INVOICE NO.	CREDITOR'S ACCOUNT CREDITED	POST. REF.	ACCOUNTS PAYABLE CREDIT	PURCHASES DEBIT	GENERAL ACCOUNT DEBITED	POST. REF.	DEBIT
1								
2								
3								
4								
5								
6								
7								
8								
9								
10								
11								
12								
13								
14								
15								
16								
17								
18								
19								
20								
21								
22								
23								
24								
25								

Problem 17-6 (continued)

(2), (3), (4), (5)

ACCOUNTS PAYABLE SUBSIDIARY LEDGER

Name *Allen's Repair*

Address *Two Deauville Place, Birmingham, AL 35203*

DATE	DESCRIPTION	POST. REF.	DEBIT	CREDIT	BALANCE

Name *Digital Precision Equipment*

Address *16 Military Complex, Huntsville, AL 35801*

DATE		DESCRIPTION	POST. REF.	DEBIT	CREDIT	BALANCE
20--						
July	1	Balance	✓			1 000 00

Name *Photo Emporium*

Address *Center Mall, Mobile, AL 36601*

DATE	DESCRIPTION	POST. REF.	DEBIT	CREDIT	BALANCE

Name *ProStudio Supply*

Address *Penn Center Blvd., Montgomery, AL 36104*

DATE		DESCRIPTION	POST. REF.	DEBIT	CREDIT	BALANCE
20--						
July	1	Balance	✓			2 000 00

Problem 17-6 (continued)

Name **State Street Office Supply**

Address **16 Garden Drive, Tuscaloosa, AL 35401**

DATE		DESCRIPTION	POST. REF.	DEBIT	CREDIT	BALANCE

Name **U-Tech Products**

Address **42 Ridgeway Drive, Decatur, AL 35601**

DATE		DESCRIPTION	POST. REF.	DEBIT	CREDIT	BALANCE
20--						
July	1	Balance	✓			1 5 0 0 00

Name **Video Optics Inc.**

Address **Three Oxford Place, Auburn, AL 36830**

DATE		DESCRIPTION	POST. REF.	DEBIT	CREDIT	BALANCE

Problem 17-6 (continued)

GENERAL LEDGER (PARTIAL)

ACCOUNT _Supplies_ ACCOUNT NO. __130__

DATE		DESCRIPTION	POST. REF.	DEBIT	CREDIT	BALANCE	
						DEBIT	CREDIT
20--							
July	1	Balance	✓			30000	

ACCOUNT _Store Equipment_ ACCOUNT NO. __140__

DATE		DESCRIPTION	POST. REF.	DEBIT	CREDIT	BALANCE	
						DEBIT	CREDIT
20--							
July	1	Balance	✓			250000	

ACCOUNT _Accounts Payable_ ACCOUNT NO. __201__

DATE		DESCRIPTION	POST. REF.	DEBIT	CREDIT	BALANCE	
						DEBIT	CREDIT
20--							
July	1	Balance	✓				450000

ACCOUNT _Purchases_ ACCOUNT NO. __501__

DATE		DESCRIPTION	POST. REF.	DEBIT	CREDIT	BALANCE	
						DEBIT	CREDIT
20--							
July	1	Balance	✓			1500000	

ACCOUNT _Maintenance Expense_ ACCOUNT NO. __640__

DATE		DESCRIPTION	POST. REF.	DEBIT	CREDIT	BALANCE	
						DEBIT	CREDIT
20--							
July	1	Balance	✓			20000	

Problem 17-6 (concluded)

(6)

Analyze: _____

_____

_____

_____

Problem 17-7 Recording and Posting Cash Payments

(1), (4), (5) _____

CASH PAYMENTS JOURNAL

PAGE _____

DATE	DOC. NO.	ACCOUNT NAME	POST. REF.	GENERAL DEBIT	GENERAL CREDIT	ACCOUNTS PAYABLE DEBIT	PURCHASES DISCOUNTS CREDIT	CASH IN BANK CREDIT	
									1
									2
									3
									4
									5
									6
									7
									8
									9
									10
									11
									12
									13
									14
									15
									16
									17
									18
									19
									20
									21
									22

Problem 17-7   (continued)

(2)

ACCOUNTS PAYABLE SUBSIDIARY LEDGER

Name **Mohican Falls Kayak Wholesalers**

Address **Box 17, Buffalo Road, Jackson, WY 83001**

DATE		DESCRIPTION	POST. REF.	DEBIT	CREDIT	BALANCE
20--						
July	1	Balance	✓			500 00

Name **North American Waterways Suppliers**

Address **Horse Creek Road, Casper, WY 82601**

DATE		DESCRIPTION	POST. REF.	DEBIT	CREDIT	BALANCE
20--						
July	1	Balance	✓			1 400 00

Name **Office Max**

Address **142 Park Plaza, Cody, WY 82414**

DATE		DESCRIPTION	POST. REF.	DEBIT	CREDIT	BALANCE
20--						
July	1	Balance	✓			150 00

Name **Pacific Wholesalers**

Address **497 State Street, Laramie, WY 82070**

DATE		DESCRIPTION	POST. REF.	DEBIT	CREDIT	BALANCE
20--						
July	1	Balance	✓			1 300 00

Problem 17-7 (continued)

Name _Rollins Plumbing Service_

Address _14 Ponderosa Road, Gillette, WY 82716_

DATE		DESCRIPTION	POST. REF.	DEBIT	CREDIT	BALANCE
20--						
July	1	Balance	✓			200 00

Name _StoreMart Supply_

Address _Box 182 Yellowstone Creek, Sheridan, WY 82801_

DATE		DESCRIPTION	POST. REF.	DEBIT	CREDIT	BALANCE
20--						
July	1	Balance	✓			900 00

Name _Trailhead Canoes_

Address _800 Trail Road, Cheyenne, WY 82001_

DATE		DESCRIPTION	POST. REF.	DEBIT	CREDIT	BALANCE
20--						
July	1	Balance	✓			700 00

Problem 17-7   (continued)

(3)

GENERAL LEDGER (PARTIAL)

ACCOUNT **Cash in Bank**   ACCOUNT NO. **101**

DATE		DESCRIPTION	POST. REF.	DEBIT	CREDIT	BALANCE DEBIT	BALANCE CREDIT
20--							
July	1	Balance	✓			800000	
	31		CR18	700000		1500000	

ACCOUNT **Supplies**   ACCOUNT NO. **135**

DATE		DESCRIPTION	POST. REF.	DEBIT	CREDIT	BALANCE DEBIT	BALANCE CREDIT
20--							
July	1	Balance	✓			15000	

ACCOUNT **Prepaid Insurance**   ACCOUNT NO. **140**

DATE		DESCRIPTION	POST. REF.	DEBIT	CREDIT	BALANCE DEBIT	BALANCE CREDIT

ACCOUNT **Store Equipment**   ACCOUNT NO. **150**

DATE		DESCRIPTION	POST. REF.	DEBIT	CREDIT	BALANCE DEBIT	BALANCE CREDIT
20--							
July	1	Balance	✓			300000	

ACCOUNT **Accounts Payable**   ACCOUNT NO. **201**

DATE		DESCRIPTION	POST. REF.	DEBIT	CREDIT	BALANCE DEBIT	BALANCE CREDIT
20--							
July	1	Balance	✓				515000

Problem 17-7   (continued)

ACCOUNT **_Transportation In_**   ACCOUNT NO. **505**

DATE		DESCRIPTION	POST. REF.	DEBIT	CREDIT	BALANCE DEBIT	BALANCE CREDIT
20--							
July	1	Balance	✓			65000	

ACCOUNT **_Purchases Discounts_**   ACCOUNT NO. **510**

DATE		DESCRIPTION	POST. REF.	DEBIT	CREDIT	BALANCE DEBIT	BALANCE CREDIT
20--							
July	1	Balance	✓				150000

ACCOUNT **_Advertising Expense_**   ACCOUNT NO. **601**

DATE		DESCRIPTION	POST. REF.	DEBIT	CREDIT	BALANCE DEBIT	BALANCE CREDIT
20--							
July	1	Balance	✓			180000	

ACCOUNT **_Miscellaneous Expense_**   ACCOUNT NO. **655**

DATE	DESCRIPTION	POST. REF.	DEBIT	CREDIT	BALANCE DEBIT	BALANCE CREDIT

Problem 17-7 (concluded)

(6), (7)

Analyze: _____

_____

_____

Problem 17-8 Source Documents

Instructions: *Use the following source documents to record the transactions for this problem.*

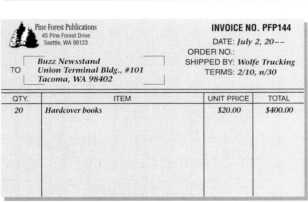

$ 1,552.00		**No. 2455**
Date *July 1*		20--
To *ADC Publishing*		
For *on account*		
	Dollars	Cents
Balance brought forward	9,000	00
Add deposits 7/1	800	00
Total	9,800	00
Less this check	1,552	00
Balance carried forward	8,248	00

$ 1,358.00		**No. 2456**
Date *July 2*		20--
To *Candlelight Software*		
For *on account*		
	Dollars	Cents
Balance brought forward	8,248	00
Add deposits		
Total	8,248	00
Less this check	1,358	00
Balance carried forward	6,890	00

ADC PUBLISHING
14003 Chen Street
San Francisco, CA 94122

INVOICE NO. ADC63
DATE: *June 22, 20--*
ORDER NO.:
SHIPPED BY: *Express Shipping*
TERMS: *3/10, n/30*

TO *Buzz Newsstand*
Union Terminal Bldg., #101
Tacoma, WA 98402

QTY.	ITEM	UNIT PRICE	TOTAL
2,400	*Newspapers*	$ .50	$1,200.00
5/cs	*Magazines*	60.00/cs	300.00
50	*Paperback books*	2.00	100.00
			$1,600.00

Date to be paid: 7/1
Discount: $48.00
Amount to be paid: $1,552.00
Check No.: 2455
REC'D JUN 24

Candlelight Software
1466 San Diego Avenue
Tacoma, WA 98407

INVOICE NO. CS92
DATE: *June 18, 20--*
ORDER NO.:
SHIPPED BY: *Picked up*
TERMS: *3/15, n/30*

TO *Buzz Newsstand*
Union Terminal Bldg., #101
Tacoma, WA 98402

QTY.	ITEM	UNIT PRICE	TOTAL
40	*Travel planning software*	$25.00	$1,000.00
80	*Entertainment software*	5.00	400.00
			$1,400.00

Date to be paid: 7/2
Discount: $42.00
Amount to be paid: $1,358.00
Check No.: 2456
REC'D JUN 20

Pine Forest Publications
45 Pine Forest Drive
Seattle, WA 98123

INVOICE NO. PFP144
DATE: *July 2, 20--*
ORDER NO.:
SHIPPED BY: *Wolfe Trucking*
TERMS: *2/10, n/30*

TO *Buzz Newsstand*
Union Terminal Bldg., #101
Tacoma, WA 98402

QTY.	ITEM	UNIT PRICE	TOTAL
20	*Hardcover books*	$20.00	$400.00

$ 350.00		**No. 2457**
Date *July 4*		20--
To *Nomad Computer Sales*		
For *on account*		
	Dollars	Cents
Balance brought forward	6,890	00
Add deposits		
Total	6,890	00
Less this check	350	00
Balance carried forward	6,540	00

CorpTech Office Supply
818 McCain Street
Tacoma, WA 98402

INVOICE NO. CT67
DATE: *July 5, 20--*
ORDER NO.:
SHIPPED BY: *Picked up*
TERMS: *n/30*

TO *Buzz Newsstand*
Union Terminal Bldg., #101
Tacoma, WA 98402

QTY.	ITEM	UNIT PRICE	TOTAL
5	*Bookshelves*	$300.00	$1,500.00
2	*Book racks*	193.00	386.00
			$1,886.00
		TX	114.00
			$2,000.00

$ 125.00		**No. 2458**
Date *July 7*		20--
To *Wolfe Trucking*		
For *transportation charges*		
	Dollars	Cents
Balance brought forward	6,540	00
Add deposits 7/5	1,000	00
Total	7,540	00
Less this check	125	00
Balance carried forward	7,415	00

Problem 17-8 (continued)

American Trend Publishers
766 Goldrush Way
Denver, CO 80207

INVOICE NO. ATP98

TO
Buzz Newsstand
Union Terminal Bldg., #101
Tacoma, WA 98402

DATE: *July 9, 20--*
ORDER NO.:
SHIPPED BY: *Wolfe Trucking*
TERMS: *2/10, n/30*

QTY.	ITEM	UNIT PRICE	TOTAL
250	Paperback books	$2.00	$500.00
800	Newspapers	.50	400.00
			$900.00

$ 882.00		No. 2461
Date July 16		20--
To American Trend Publishers		
For on account (900 less 2% disc.)		

	Dollars	Cents
Balance brought forward	7,265	00
Add deposits		
Total	7,265	00
Less this check	882	00
Balance carried forward	6,383	00

CorpTech Office Supply
818 McCain Street
Tacoma, WA 98402

INVOICE NO. CT72

TO
Buzz Newsstand
Union Terminal Bldg., #101
Tacoma, WA 98402

DATE: *July 12, 20--*
ORDER NO.:
SHIPPED BY: *Picked up*
TERMS: *n/30*

QTY.	ITEM	UNIT PRICE	TOTAL
2 cs	Office paper	$66.50/cs	$133.00
3	Inkjet ink cartridges	50.00/ea	150.00
			$283.00
		TX	17.00
			$300.00

Candlelight Software
1466 San Diego Avenue
Tacoma, WA 98407

INVOICE NO. CS101

TO
Buzz Newsstand
Union Terminal Bldg., #101
Tacoma, WA 98402

DATE: *July 18, 20--*
ORDER NO.:
SHIPPED BY: *Picked up*
TERMS: *n/30*

QTY.	ITEM	UNIT PRICE	TOTAL
100	Entertainment software	$5.00	$500.00

$ 750.00		No. 2459
Date July 14		20--
To Delta Press		
For on account		

	Dollars	Cents
Balance brought forward	7,415	00
Add deposits		
Total	7,415	00
Less this check	750	00
Balance carried forward	6,665	00

Nomad COMPUTER SALES
1601 San Diego Avenue
Tacoma, WA 98407

INVOICE NO. NC56

TO
Buzz Newsstand
Union Terminal Bldg., #101
Tacoma, WA 98402

DATE: *July 20, 20--*
ORDER NO.:
SHIPPED BY: *Picked up*
TERMS: *2/10, n/30*

QTY.	ITEM	UNIT PRICE	TOTAL
20	Pocket electronic organizers	$10.00	$200.00

$ 1,600.00		No. 2460
Date July 15		20--
To SeaTac Insurance Co.		
For prepaid insurance		

	Dollars	Cents
Balance brought forward	6,665	00
Add deposits 7/15	2,200	00
Total	8,865	00
Less this check	1,600	00
Balance carried forward	7,265	00

$ 100.00		No. 2462
Date July 22		20--
To Pine Forest Publications		
For on account		

	Dollars	Cents
Balance brought forward	6,383	00
Add deposits 7/18	1,500	00
7/20	2,000	00
Total	9,883	00
Less this check	100	00
Balance carried forward	9,783	00

Problem 17-8 (continued)

$ 2,000.00		No. 2463
Date *July 23*		20—
To *CorpTech Office Supply*		
For *on account*		
	Dollars	Cents
Balance brought forward	9,783	00
Add deposits		
Total	9,783	00
Less this check	2,000	00
Balance carried forward	7,783	00

INTEROFFICE MEMORANDUM

BUZZ
NEWSSTAND
Union Terminal Building, #101
Tacoma, WA 98402

TO: *Accounting Clerk*
FROM: *Kathy Harper, Manager*
DATE: *7/30/20--*
SUBJECT: *Bank Statement Fees*

The July bank statement had the following fees:
Bank Service Charge $ 25.00
Bankcard Fees 130.00

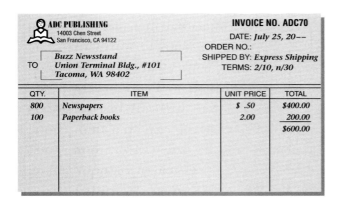

ADC PUBLISHING
14003 Chen Street
San Francisco, CA 94122

INVOICE NO. ADC70

TO *Buzz Newsstand*
Union Terminal Bldg., #101
Tacoma, WA 98402

DATE: *July 25, 20--*
ORDER NO.:
SHIPPED BY: *Express Shipping*
TERMS: *2/10, n/30*

QTY.	ITEM	UNIT PRICE	TOTAL
800	*Newspapers*	$ .50	$400.00
100	*Paperback books*	2.00	200.00
			$600.00

$		No. 2465
Date		20—
To		
For		
	Dollars	Cents
Balance brought forward	7,333	00
Add deposits 7/30	3,000	00
Bk. Stmt.	– 155	00
Total	10,178	00
Less this check		
Balance carried forward		

$ 450.00		No. 2464
Date *July 28*		20—
To *Nomad Computer Sales*		
For *on account*		
	Dollars	Cents
Balance brought forward	7,783	00
Add deposits		
Total	7,783	00
Less this check	450	00
Balance carried forward	7,333	00

Problem 17-8   Recording and Posting Purchases and Cash Payments

(1), (3), (4)

PAGE _____

CASH PAYMENTS JOURNAL

DATE	DOC. NO.	ACCOUNT NAME	POST. REF.	GENERAL DEBIT	GENERAL CREDIT	ACCOUNTS PAYABLE DEBIT	PURCHASES DISCOUNTS CREDIT	CASH IN BANK CREDIT
								1
								2
								3
								4
								5
								6
								7
								8
								9
								10
								11
								12
								13
								14
								15
								16
								17
								18
								19
								20
								21
								22

Problem 17-8   (continued)

(1), (3), (4)

PURCHASES JOURNAL   PAGE _____

DATE	INVOICE NO.	CREDITOR'S ACCOUNT CREDITED	POST. REF.	ACCOUNTS PAYABLE CREDIT	PURCHASES DEBIT	GENERAL — ACCOUNT DEBITED	POST. REF.	DEBIT
1								
2								
3								
4								
5								
6								
7								
8								
9								
10								
11								
12								
13								
14								
15								
16								
17								
18								
19								
20								
21								
22								

Problem 17-8 (continued)

(5), (6)

GENERAL LEDGER (PARTIAL)

ACCOUNT __Cash in Bank__   ACCOUNT NO. __101__

DATE		DESCRIPTION	POST. REF.	DEBIT	CREDIT	BALANCE DEBIT	BALANCE CREDIT
20--							
July	1	Balance	✓			9 00 000	
	30		CR12	10 50 000		19 50 000	

ACCOUNT __Supplies__   ACCOUNT NO. __135__

DATE		DESCRIPTION	POST. REF.	DEBIT	CREDIT	BALANCE DEBIT	BALANCE CREDIT
20--							
July	1	Balance	✓			1 50 00	

ACCOUNT __Prepaid Insurance__   ACCOUNT NO. __140__

DATE		DESCRIPTION	POST. REF.	DEBIT	CREDIT	BALANCE DEBIT	BALANCE CREDIT

ACCOUNT __Store Equipment__   ACCOUNT NO. __150__

DATE		DESCRIPTION	POST. REF.	DEBIT	CREDIT	BALANCE DEBIT	BALANCE CREDIT
20--							
July	1	Balance	✓			6 00 000	

ACCOUNT __Accounts Payable__   ACCOUNT NO. __201__

DATE		DESCRIPTION	POST. REF.	DEBIT	CREDIT	BALANCE DEBIT	BALANCE CREDIT
20--							
July	1	Balance	✓				5 70 000

Problem 17-8   (continued)

ACCOUNT ___Purchases_____ ACCOUNT NO. ___501___

DATE		DESCRIPTION	POST. REF.	DEBIT	CREDIT	BALANCE	
						DEBIT	CREDIT
20--							
July	1	Balance	✓			2500000	

ACCOUNT ___Transportation In_____ ACCOUNT NO. ___505___

DATE		DESCRIPTION	POST. REF.	DEBIT	CREDIT	BALANCE	
						DEBIT	CREDIT
20--							
July	1	Balance	✓			80000	

ACCOUNT ___Purchases Discounts_____ ACCOUNT NO. ___510___

DATE		DESCRIPTION	POST. REF.	DEBIT	CREDIT	BALANCE	
						DEBIT	CREDIT
20--							
July	1	Balance	✓				160000

ACCOUNT ___Bankcard Fees Expense_____ ACCOUNT NO. ___605___

DATE		DESCRIPTION	POST. REF.	DEBIT	CREDIT	BALANCE	
						DEBIT	CREDIT
20--							
July	1	Balance	✓			110000	

ACCOUNT ___Miscellaneous Expense_____ ACCOUNT NO. ___650___

DATE		DESCRIPTION	POST. REF.	DEBIT	CREDIT	BALANCE	
						DEBIT	CREDIT
20--							
July	1	Balance	✓			20000	

Problem 17-8   (continued)

(2)

ACCOUNTS PAYABLE SUBSIDIARY LEDGER

Name *ADC Publishing*

Address *5670 Mulberry Place, Seattle, WA 98101*

DATE		DESCRIPTION	POST. REF.	DEBIT	CREDIT	BALANCE
20--						
July	1	Balance	✓			1 6 0 0 00

Name *American Trend Publishers*

Address *1313 Maple Drive, Seattle, WA 98148*

DATE	DESCRIPTION	POST. REF.	DEBIT	CREDIT	BALANCE

Name *Candlelight Software*

Address *Six Evergreen Park, Tacoma, WA 98402*

DATE		DESCRIPTION	POST. REF.	DEBIT	CREDIT	BALANCE
20--						
July	1	Balance	✓			1 4 0 0 00

Name *CorpTech Office Supply*

Address *601 Cascade Park, Bellevue, WA 98009*

DATE	DESCRIPTION	POST. REF.	DEBIT	CREDIT	BALANCE

Problem 17-8 (continued)

Name **Delta Press** _____

Address **One Triangle Park, Vancouver, WA 98661** _____

DATE		DESCRIPTION	POST. REF.	DEBIT	CREDIT	BALANCE
20--						
July	1	Balance	✓			1 50 0 00

Name **Nomad Computer Sales** _____

Address **16 Point Drive, Ft. Lewis, WA 98433** _____

DATE		DESCRIPTION	POST. REF.	DEBIT	CREDIT	BALANCE
20--						
July	1	Balance	✓			1 20 0 00

Name **Pine Forest Publications** _____

Address **103 Mt. Thor Road, Spokane, WA 99210** _____

DATE		DESCRIPTION	POST. REF.	DEBIT	CREDIT	BALANCE

Problem 17-8 (concluded)

(7)

(8)

Analyze: _____

_____

Notes

CHAPTER 17 Special Journals: Purchases and Cash Payments

Self-Test

Part A   True or False

Directions: *Circle the letter* T *in the Answer column if the statement is true; circle the letter* F *if the statement is false.*

Answer

T   F   **1.** Accounts in the accounts payable subsidiary ledger have normal debit balances.

T   F   **2.** An invoice dated June 17 with terms 2/10, n/30 must be paid by July 17 to take advantage of the cash discount.

T   F   **3.** The Accounts Payable controlling account is found in the general ledger.

T   F   **4.** Because a bank service charge is automatically deducted from the checking account of a business, it is not necessary to record an entry in the cash payments journal.

T   F   **5.** Special journals should be footed before column totals are proved.

T   F   **6.** The purchases journal is used only for recording the purchase of merchandise on account.

T   F   **7.** Transactions recorded in either the purchases or cash payments journals which affect accounts payable subsidiary ledger accounts should be posted on a daily basis, that is, as soon as the transaction is recorded.

T   F   **8.** Cash should always be proved at the end of the month.

T   F   **9.** Check stubs and the bank statement are the only two source documents used for recording transactions in the cash payments journal.

T   F   **10.** The column totals of special journals are always posted at the end of the month.

Part B   Matching

Directions: *The general journal and special journals listed below are identified by a letter. Following the list is a series of business transactions. For each transaction, indicate in which journal the transaction would be recorded by writing the identifying letter in the Answer column.*

A. cash payments journal	**C.** general journal	**E.** sales journal
B. cash receipts journal	**D.** purchases journal	

Answer

_____ **1.** Recorded bank card fees for the month.

_____ **2.** Purchased store equipment on account.

_____ **3.** Issued a debit memorandum for the return of merchandise purchased on account.

_____ **4.** Recorded bank card sales for the day.

_____ **5.** Sold merchandise on account.

_____ **6.** Purchased office supplies for cash.

_____ **7.** Issued a credit memorandum to a charge customer for the return of merchandise purchased on account.

_____ **8.** Recorded cash sales for the day.

_____ **9.** Discovered that a payment made by Port Co., a charge customer, was incorrectly credited to Porter Co.

_____ **10.** Purchased merchandise for cash.

_____ **11.** Paid the monthly utility bill.

_____ **12.** Purchased merchandise on account.

_____ **13.** Recorded the bank service charge.

CHAPTER 18 Adjustments and the Ten-Column Work Sheet

Study Plan

Check Your Understanding

Section 1	*Read Section 1 on pages 484–489 and complete the following exercises on page 490.* ❑ Thinking Critically ❑ Analyzing Accounting ❑ Problem 18-1 *Analyzing the Adjustment for Merchandise Inventory*
Section 2	*Read Section 2 on pages 491–494 and complete the following exercises on page 495.* ❑ Thinking Critically ❑ Computing in the Business World ❑ Problem 18-2 *Analyzing Adjustments*
Section 3	*Read Section 3 on pages 496–503 and complete the following exercises on page 504.* ❑ Thinking Critically ❑ Communicating Accounting ❑ Problem 18-3 *Analyzing the Work Sheet* ❑ Problem 18-4 *Analyzing a Source Document*
Summary	*Review the Chapter 18 Summary on page 505 in your textbook.* ❑ Key Concepts
Review and Activities	*Complete the following questions and exercises on pages 506–507 in your textbook.* ❑ Using Key Terms ❑ Understanding Accounting Concepts and Procedures ❑ Case Study ❑ Conducting an Audit with Alex ❑ Internet Connection ❑ Workplace Skills
Computerized Accounting	*Read the Computerized Accounting information on page 508 in your textbook.* ❑ *Making the Transition from a Manual to a Computerized System* ❑ *Recording Adjusting Entries in Peachtree*
Problems	*Complete the following end-of-chapter problems for Chapter 18 in your textbook.* ❑ Problem 18-5 *Completing a Ten-Column Work Sheet* ❑ Problem 18-6 *Completing a Ten-Column Work Sheet* ❑ Problem 18-7 *Completing a Ten-Column Work Sheet* ❑ Problem 18-8 *Completing a Ten-Column Work Sheet*
Challenge Problem	❑ Problem 18-9 *Locating Errors on the Work Sheet*
Chapter Reviews and Working Papers	*Complete the following exercises for Chapter 18 in your Chapter Reviews and Working Papers.* ❑ Chapter Review ❑ Self-Test

CHAPTER 18 REVIEW

Adjustments and the Ten-Column Work Sheet

Total Points	45
Student's Score	

Part 1   Accounting Vocabulary (3 points)

Directions: *Using terms from the following list, complete the sentences below. Write the letter of the term you have chosen in the space provided.*

A. adjustment	B. beginning inventory	C. ending inventory	D. physical inventory

_B_   **0.** The merchandise a business has on hand at the beginning of a fiscal period is the _____.

_____   **1.** An amount that is added to or subtracted from an account balance to bring that balance up to date is known as a(n) _____.

_____   **2.** The merchandise on hand at the end of a fiscal period is the _____.

_____   **3.** An actual count of all the merchandise on hand and available for sale is called a(n) _____.

Part 2   Examining End-of-Period Adjustments (12 points)

Directions: *Read each of the following statements to determine whether the statement is true or false. Write your answer in the space provided.*

True   **0.** The amount of merchandise on hand at the end of a fiscal period is determined by a physical count.

_____   **1.** The value of the ending inventory for merchadise is always less than the value of the beginning inventory.

_____   **2.** A corporation is not required to estimate in advance its federal income taxes for the year.

_____   **3.** The amount entered in the Adjustments Credit column on the line for Supplies is the amount of supplies used during the fiscal period.

_____   **4.** If the ending inventory is less than the beginning inventory, the Income Summary account is credited for the difference.

_____   **5.** The Federal Corporate Income Tax Payable account normally has a debit balance.

_____   **6.** The work sheet is the source of information for preparing the end-of-period financial statements and journal entries.

_____   **7.** Changes in account balances are not always the result of daily business transactions.

_____   **8.** Supplies used in the operation of a business are initially recorded as assets and eventually become expenses as they are consumed.

_____   **9.** A trial balance is prepared only at the end of the fiscal period.

_____   **10.** Prepaid Insurance will normally be credited in the Adjustments section of the work sheet.

_____   **11.** The Trial Balance section of the work sheet does not include general ledger accounts with zero balances.

_____   **12.** Because there are no source documents that show changes in account balances caused by internal operations, these changes must be shown through adjusting entries.

Part 3   Analyzing Adjustments (8 points)

Directions: *Using account names from the following list, determine the accounts to be debited and credited for the adjustments below. Write your answers in the space provided.*

A. Federal Corporate Income Tax Expense	**C.** Income Summary	**F.** Prepaid Insurance
B. Federal Corp. Income Tax Payable	**D.** Insurance Expense	**G.** Supplies
	E. Merchandise Inventory	**H.** Supplies Expense

Debit	**Credit**	
D	*F*	**0.** Adjustment for the insurance premiums expired.
_____	_____	**1.** The beginning inventory is $34,946; and the ending inventory is $36,496.
_____	_____	**2.** Adjustment for additional federal income taxes owed.
_____	_____	**3.** Adjustment for the supplies on hand.
_____	_____	**4.** The beginning inventory is $73,937; and the ending inventory is $72,094.

Part 4   Extending Account Balances (22 points)

Directions: *For each of the following account names, indicate whether the account will have a normal debit or credit balance in the Adjusted Trial Balance section of the work sheet. Place a check mark in the appropriate column.*

Account Name	Debit Balance	Credit Balance
0. Cash in Bank	✓	____
1. Medicare Tax Payable	____	____
2. Capital Stock	____	____
3. Sales Returns and Allowances	____	____
4. Transportation In	____	____
5. Rent Expense	____	____
6. Federal Corporate Income Tax Payable	____	____
7. Sales Discounts	____	____
8. Purchases	____	____
9. Accounts Payable	____	____
10. Supplies Expense	____	____
11. Prepaid Insurance	____	____
12. Accounts Receivable	____	____
13. Office Equipment	____	____
14. Retained Earnings	____	____
15. Payroll Tax Expense	____	____
16. Purchases Discounts	____	____
17. Sales	____	____
18. Supplies	____	____
19. Merchandise Inventory	____	____
20. Sales Tax Payable	____	____
21. Employees' State Income Tax Payable	____	____
22. Federal Unemployment Tax Payable	____	____

Working Papers *for Section Problems*

Problem 18-1   Analyzing the Adjustment for Merchandise Inventory

1. _____

2. _____

3. _____

4. _____

Problem 18-2   Analyzing Adjustments

1. Amount of Adjustment _____

Account Debited _____

Account Credited _____

2. Amount of Adjustment _____

Account Debited _____

Account Credited _____

3. Amount of Adjustment _____

Account Debited _____

Account Credited _____

Problem 18-3   Analyzing the Work Sheet

1. Amount? _____

2. Section? _____

3. Amount? _____

4. Amount? _____

Problem 18-4   Analyzing a Source Document

1. _____

2. _____

3. _____

4. _____

5. _____

Computerized Accounting Using Peachtree

Software Objectives

When you have completed this chapter, you will be able to use Peachtree to:

1. Print a Working Trial Balance report.
2. Record adjusting entries using the **General Journal Entry** option.
3. Print an adjusted trial balance.

Problem 18-6   Completing a Ten-Column Work Sheet

INSTRUCTIONS

Beginning a Session

Step 1   Select the problem set: Shutterbug Cameras (Prob. 18-6).
Step 2   Rename the company and set the system date to August 31, 2008.

Completing the Accounting Problem

Step 3   Review the information in your textbook.
Step 4   Print a Working Trial Balance report and use it to record the adjustments.

> **DO YOU HAVE A QUESTION**
>
> **Q.** *Does Peachtree include an option to print a ten-column work sheet?*
>
> **A.** The ten-column work sheet is a convenient way to manually prepare an adjusted trial balance, an income statement, and a balance sheet. When you use Peachtree, the program automatically prepares the financial statements so a work sheet is not needed. Peachtree does, however, include a Working Trial Balance report that you can use as an aid when recording the adjustments.

The procedures to record the adjustments using Peachtree are the same as a manual system except for the adjustment to **Merchandise Inventory**. In a manual system, you make an adjustment to the **Income Summary** and **Merchandise Inventory** accounts. For a computerized accounting system such as Peachtree you must use the **Inventory Adjustment** account instead of the **Income Summary** account. The inventory adjustment for Shutterbug Cameras is shown below.

Account	Debit	Credit
Inventory Adjustment	4,309	
Merchandise Inventory Adjusting Entry		4,309

The adjustment shown above achieves the same result as the adjustment you learned how to make in your textbook. After the adjustment, the **Merchandise Inventory** account balance reflects the value of the physical inventory at the end of the period. This change in value is posted to the **Inventory Adjustment** account. As you will see in the next chapter, this account appears in the Cost of Merchandise Sold section on an Income Statement.

Step 5   Record the adjustments using the **General Journal Entry** option.

TIP: You can record the adjustments as one multi-part general journal entry to save time.

> **Notes**
>
> *Use **ADJ. ENT.** for the reference and **Adjusting Entry** as the description when you record the adjustments.*

Step 6   Print a General Journal and an Adjusted Trial Balance report. Set the Transaction Reference filter options to include only the adjusting entries on the report.
Step 7   Proof your work.
Step 8   Answer the Analyze question.

Ending the Session

    Step 9   Click the **Close Problem** button in the Glencoe Smart Guide window.

Mastering Peachtree

Change the report title on the General Ledger Trial balance to *Adjusted Trial Balance*. Also, widen the Account Description column so that it accommodates the longest description. Print your resulting report.

Problem 18-7   Completing a Ten-Column Work Sheet

INSTRUCTIONS

Beginning a Session

    Step 1   Select the problem set: Cycle Tech Bicycles (Prob. 18-7).
    Step 2   Rename the company and set the system date to August 31, 2008.

Completing the Accounting Problem

    Step 3   Review the information in your textbook.
    Step 4   Print a Working Trial Balance report from the General Ledger Report List and use it to help you prepare the adjustments.
    Step 5   Record the adjustments using the **General Journal Entry** option.
    Step 6   Print a General Journal report.
    Step 7   Proof your work.
    Step 8   Answer the Analyze question.

Checking Your Work and Ending the Session

    Step 9   Click the **Close Problem** button in the Glencoe Smart Guide window.
    Step 10   If your teacher has asked you to check your solution, click *Check my answer to this problem*. Review and print the report.
    Step 11   Click the **Close Problem** button and save your work.

Problem 18-8   Completing a Ten-Column Work Sheet

INSTRUCTIONS

Beginning a Session

    Step 1   Select the problem set: River's Edge Canoe & Kayak (Prob. 18-8).
    Step 2   Rename the company and set the system date to August 31, 2008.

Completing the Accounting Problem

    Step 3   Review the information in your textbook.
    Step 4   Print a Working Trial Balance report and use it to help you prepare the adjustments.
    Step 5   Record the adjustments.
    Step 6   Print a General Journal report and a General Ledger report.
    Step 7   Proof your work.
    Step 8   Answer the Analyze question.

Ending the Session

    Step 9   Click the **Close Problem** button in the Glencoe Smart Guide window.

DO YOU HAVE A QUESTION

Q. *Why do you have to use the Inventory Adjustment account instead of the Income Summary account when you adjust the Merchandise Inventory?*

A. A computerized system such as Peachtree needs to calculate the cost of merchandise sold so that it can determine the net income. The **Inventory Adjustment** account is a cost of sales account as opposed to the **Income Summary** account, which is a temporary capital account. As you will learn in the next chapter, the **Inventory Adjustment** account appears in the Cost of Merchandise Sold section on an income statement.

Computerized Accounting Using Spreadsheets

Problem 18-5   Completing a Ten-Column Work Sheet

Completing the Spreadsheet

Step 1   Read the instructions for Problem 18-5 in your textbook. This problem involves completing a ten-column work sheet for InBeat CD Shop.

Step 2   Open the Glencoe Accounting: Electronic Learning Center software.

Step 3   From the Program Menu, click on the **Peachtree Complete® Accounting Software and Spreadsheet Applications** icon.

Step 4   Log onto the Management System by typing your user name and password.

Step 5   Under the **Problems & Tutorials** tab, select template 18-5 from the Chapter 18 drop-down menu. The template should look like the one shown below.

```
PROBLEM 18-5
COMPLETING A TEN-COLUMN WORK SHEET

(name)
(date)


INBEAT CD SHOP WORK SHEET
FOR THE YEAR ENDED AUGUST 31, 20--
```

ACCOUNT NUMBER	ACCOUNT NAME	TRIAL BALANCE DEBIT	CREDIT	ADJUSTMENTS DEBIT	CREDIT	> <	BALANCE SHEET DEBIT	CREDIT
101	Cash in Bank	14,974.00				> <	14,974.00	
115	Accounts Receivable	3,774.00				> <	3,774.00	
130	Merchandise Inventory	86,897.00			AMOUNT	> <	86,897.00	
135	Supplies	2,940.00			AMOUNT	> <	2,940.00	
140	Prepaid Insurance	1,975.00			AMOUNT	> <	1,975.00	
150	Office Equipment	10,819.00				> <	10,819.00	
201	Accounts Payable		7,740.00			> <		7,740.00
207	Federal Corporate Income Tax Payable				AMOUNT	> <		0.00
210	Employees' Federal Income Tax Payable		291.00			> <		291.00
211	Employees' State Income Tax Payable		86.00			> <		86.00
212	Social Security Tax Payable		106.00			> <		106.00
213	Medicare Tax Payable		21.00			> <		21.00
215	Fed. Unemployment Tax Payable		32.00			> <		32.00
216	State Unemployment Tax Payable		106.00			> <		106.00
217	Sales Tax Payable		1,370.00			> <		1,370.00
301	Capital Stock		55,000.00			> <		55,000.00
305	Retained Earnings		30,928.00			> <		30,928.00
310	Income Summary	----	----	AMOUNT		> <		
401	Sales		149,136.00			> <		
501	Purchases	93,874.00				> <		
625	Federal Corporate Income Tax	2,200.00		AMOUNT		> <		
630	Insurance Expense			AMOUNT		> <		
647	Payroll Tax Expense	2,170.00				> <		
650	Miscellaneous Expense	3,662.00				> <		
655	Rent Expense	9,225.00				> <		
660	Salaries Expense	12,306.00				> <		
665	Supplies Expense			AMOUNT		> <		
		244,816.00	244,816.00	0.00	0.00	> <	121,379.00	95,680.00
	Net Income					> <		25,699.00
						> <	121,379.00	121,379.00

Step 6 Key your name and today's date in the cells containing the *(name)* and *(date)* placeholders.

Step 7 The trial balance amounts are given for you. The first adjustment that must be made is to adjust beginning merchandise inventory of $86,897 to an ending balance of $77,872. To make this adjustment, you must debit Income Summary and credit Merchandise Inventory for the difference between the beginning and ending merchandise inventory amounts. Enter the Income Summary adjustment in cell E29 and the Merchandise Inventory adjustment in cell F14.

Notice that, as you enter the adjustments, the balances for the affected accounts in the adjusted trial balance change accordingly.

Step 8 Enter the remaining adjustments into the Adjustments section of the spreadsheet template. When you have entered all of the adjustments, move the cell pointer into the Adjusted Trial Balance, Income Statement, and Balance Sheet sections of the spreadsheet template. Notice that the amounts for the Adjusted Trial Balance, Income Statement, and Balance Sheet are automatically entered. The program also calculates the column totals and the net income for InBeat CD Shop.

Step 9 Save the spreadsheet using the **Save** option from the *File* menu. You should accept the default location for the save as this is handled by the management system.

Step 10 Print the completed spreadsheet.

 TIP: If your spreadsheet is too wide to fit on an 8.5-inch wide piece of paper, you can change your print settings to print the worksheet *landscape.* Landscape means that the worksheet will be printed broadside on the page. Some spreadsheet applications also allow you to choose a "fit to page" option. This function will reduce the width and/or depth of the worksheet to fit on one page.

Step 11 Exit the spreadsheet program.

Step 12 In the Close Options box, select the location where you would like to save your work.

Step 13 Answer the Analyze question from your textbook for this problem.

What-If Analysis

If Merchandise Inventory on August 31 were $80,123, what adjustments would be made? What would be the effect on net income?

_____

_____

Cameras

Sheet

August 31, 20--

	ADJUSTED TRIAL BALANCE		INCOME STATEMENT		BALANCE SHEET		
	DEBIT	CREDIT	DEBIT	CREDIT	DEBIT	CREDIT	
							1
							2
							3
							4
							5
							6
							7
							8
							9
							10
							11
							12
							13
							14
							15
							16
							17
							18
							19
							20
							21
							22
							23
							24
							25
							26
							27
							28
							29
							30
							31
							32

Problem 18-6   (concluded)

Shutterbug

Work Sheet

For the Month Ended

	ACCT. NO.	ACCOUNT NAME	TRIAL BALANCE		ADJUSTMENTS	
			DEBIT	CREDIT	DEBIT	CREDIT
1		*Brought Forward*				
2						
3	630	*Insurance Expense*				
4	640	*Maintenance Expense*				
5	645	*Miscellaneous Expense*				
6	647	*Payroll Tax Expense*				
7	650	*Rent Expense*				
8	655	*Salaries Expense*				
9	660	*Supplies Expense*				
10	670	*Utilities Expense*				
11						
12						
13						
14						
15						
16						
17						
18						
19						
20						
21						
22						
23						
24						
25						
26						
27						
28						
29						
30						
31						
32						

Cameras

(continued)

August 31, 20––

ADJUSTED TRIAL BALANCE		INCOME STATEMENT		BALANCE SHEET		
DEBIT	CREDIT	DEBIT	CREDIT	DEBIT	CREDIT	
						1
						2
						3
						4
						5
						6
						7
						8
						9
						10
						11
						12
						13
						14
						15
						16
						17
						18
						19
						20
						21
						22
						23
						24
						25
						26
						27
						28
						29
						30
						31
						32

Analyze: _____

_____

Problem 18-7   Completing a Ten-Column Work Sheet

Cycle Tech

Work

For the Month Ended

	ACCT. NO.	ACCOUNT NAME	TRIAL BALANCE		ADJUSTMENTS	
			DEBIT	CREDIT	DEBIT	CREDIT
1	101	Cash in Bank				
2	115	Accounts Receivable				
3	125	Merchandise Inventory				
4	130	Supplies				
5	135	Prepaid Insurance				
6	140	Store Equipment				
7	145	Office Equipment				
8	201	Accounts Payable				
9	210	Fed. Corporate Income Tax Pay.				
10	211	Employees' Fed. Inc. Tax Pay.				
11	212	Employees' State Inc. Tax Pay.				
12	213	Social Security Tax Payable				
13	214	Medicare Tax Payable				
14	215	Sales Tax Payable				
15	216	Fed. Unemployment Tax Pay.				
16	217	State Unemployment Tax Pay.				
17	301	Capital Stock				
18	305	Retained Earnings				
19	310	Income Summary				
20	401	Sales				
21	405	Sales Discounts				
22	410	Sales Returns and Allowances				
23	501	Purchases				
24	505	Transportation In				
25	510	Purchases Discounts				
26	515	Purchases Returns and Allow.				
27	601	Advertising Expense				
28	605	Bankcard Fees Expense				
29		Carried Forward				
30						
31						
32						

Bicycles

Sheet

August 31, 20– –

ADJUSTED TRIAL BALANCE		INCOME STATEMENT		BALANCE SHEET		
DEBIT	CREDIT	DEBIT	CREDIT	DEBIT	CREDIT	
						1
						2
						3
						4
						5
						6
						7
						8
						9
						10
						11
						12
						13
						14
						15
						16
						17
						18
						19
						20
						21
						22
						23
						24
						25
						26
						27
						28
						29
						30
						31
						32

Problem 18-7   (concluded)

	ACCT. NO.	ACCOUNT NAME	TRIAL BALANCE		ADJUSTMENTS	
			DEBIT	CREDIT	DEBIT	CREDIT
1		*Brought Forward*				
2						
3	625	*Fed. Corporate Income Tax Exp.*				
4	630	*Insurance Expense*				
5	645	*Maintenance Expense*				
6	650	*Miscellaneous Expense*				
7	655	*Payroll Tax Expense*				
8	657	*Rent Expense*				
9	660	*Salaries Expense*				
10	665	*Supplies Expense*				
11	675	*Utilities Expense*				
12						
13						
14						
15						
16						
17						
18						
19						
20						
21						
22						
23						
24						
25						
26						
27						
28						
29						
30						
31						
32						

**Bicycles**

**(continued)**

**August 31, 20--**

	ADJUSTED TRIAL BALANCE		INCOME STATEMENT		BALANCE SHEET		
	DEBIT	CREDIT	DEBIT	CREDIT	DEBIT	CREDIT	
							1
							2
							3
							4
							5
							6
							7
							8
							9
							10
							11
							12
							13
							14
							15
							16
							17
							18
							19
							20
							21
							22
							23
							24
							25
							26
							27
							28
							29
							30
							31
							32

Analyze: _____

_____

Problem 18-8 Completing a Ten-Column Work Sheet
(1)

River's Edge

Work

For the Month Ended

	ACCT. NO.	ACCOUNT NAME	TRIAL BALANCE		ADJUSTMENTS	
			DEBIT	CREDIT	DEBIT	CREDIT
1	101	Cash in Bank				
2	115	Accounts Receivable				
3	130	Merchandise Inventory				
4	135	Supplies				
5	140	Prepaid Insurance				
6	145	Delivery Equipment				
7	150	Store Equipment				
8	201	Accounts Payable				
9	204	Fed. Corporate Income Tax Pay.				
10	210	Employees' Fed. Inc. Tax Pay.				
11	211	Employees' State Inc. Tax Pay.				
12	212	Social Security Tax Payable				
13	213	Medicare Tax Payable				
14	215	Sales Tax Payable				
15	216	Fed. Unemployment Tax Pay.				
16	217	State Unemployment Tax Pay.				
17	219	U.S. Savings Bonds Payable				
18	301	Capital Stock				
19	305	Retained Earnings				
20	310	Income Summary				
21	401	Sales				
22	405	Sales Discounts				
23	410	Sales Returns and Allowances				
24	501	Purchases				
25	505	Transportation In				
26	510	Purchases Discounts				
27	515	Purchases Returns and Allow.				
28	601	Advertising Expense				
29		Carried Forward				
30						
31						
32						

Canoe & Kayak

Sheet

August 31, 20– –

	ADJUSTED TRIAL BALANCE		INCOME STATEMENT		BALANCE SHEET		
	DEBIT	CREDIT	DEBIT	CREDIT	DEBIT	CREDIT	
							1
							2
							3
							4
							5
							6
							7
							8
							9
							10
							11
							12
							13
							14
							15
							16
							17
							18
							19
							20
							21
							22
							23
							24
							25
							26
							27
							28
							29
							30
							31
							32

Problem 18-8   (continued)

River's Edge

Work Sheet

For the Month Ended

	ACCT. NO.	ACCOUNT NAME	TRIAL BALANCE		ADJUSTMENTS	
			DEBIT	CREDIT	DEBIT	CREDIT
1		*Brought Forward*				
2						
3	605	*Bankcard Fees Expense*				
4	625	*Fed. Corporate Income Tax Exp.*				
5	635	*Insurance Expense*				
6	650	*Maintenance Expense*				
7	655	*Miscellaneous Expense*				
8	658	*Payroll Tax Expense*				
9	660	*Rent Expense*				
10	665	*Salaries Expense*				
11	670	*Supplies Expense*				
12	680	*Utilities Expense*				
13						
14						
15						
16						
17						
18						
19						
20						
21						
22						
23						
24						
25						
26						
27						
28						
29						
30						
31						
32						

Canoe & Kayak

(continued)

August 31, 20--

	ADJUSTED TRIAL BALANCE		INCOME STATEMENT		BALANCE SHEET		
	DEBIT	CREDIT	DEBIT	CREDIT	DEBIT	CREDIT	
							1
							2
							3
							4
							5
							6
							7
							8
							9
							10
							11
							12
							13
							14
							15
							16
							17
							18
							19
							20
							21
							22
							23
							24
							25
							26
							27
							28
							29
							30
							31
							32

Problem 18-8   (continued)

(2)

GENERAL JOURNAL   PAGE _____

	DATE	DESCRIPTION	POST. REF.	DEBIT	CREDIT	
1						1
2						2
3						3
4						4
5						5
6						6
7						7
8						8
9						9
10						10
11						11
12						12

(3)

GENERAL LEDGER

ACCOUNT _Merchandise Inventory_ _____ ACCOUNT NO. ____130____

DATE		DESCRIPTION	POST. REF.	DEBIT	CREDIT	BALANCE DEBIT	BALANCE CREDIT
20--							
Aug.	1	Balance	✓			4920500	

ACCOUNT _Supplies_ _____ ACCOUNT NO. ____135____

DATE		DESCRIPTION	POST. REF.	DEBIT	CREDIT	BALANCE DEBIT	BALANCE CREDIT
20--							
Aug.	1	Balance	✓			302700	

ACCOUNT _Prepaid Insurance_ _____ ACCOUNT NO. ____140____

DATE		DESCRIPTION	POST. REF.	DEBIT	CREDIT	BALANCE DEBIT	BALANCE CREDIT
20--							
Aug.	1	Balance	✓			168000	

Problem 18-8 (concluded)

ACCOUNT _Federal Corporate Income Tax Payable_   ACCOUNT NO. _204_

DATE	DESCRIPTION	POST. REF.	DEBIT	CREDIT	BALANCE DEBIT	BALANCE CREDIT

ACCOUNT _Income Summary_   ACCOUNT NO. _310_

DATE	DESCRIPTION	POST. REF.	DEBIT	CREDIT	BALANCE DEBIT	BALANCE CREDIT

ACCOUNT _Federal Corporate Income Tax Expense_   ACCOUNT NO. _625_

DATE	DESCRIPTION	POST. REF.	DEBIT	CREDIT	BALANCE DEBIT	BALANCE CREDIT
20--						
Aug. 1	Balance	✓			2 4 8 0 00	

ACCOUNT _Insurance Expense_   ACCOUNT NO. _635_

DATE	DESCRIPTION	POST. REF.	DEBIT	CREDIT	BALANCE DEBIT	BALANCE CREDIT

ACCOUNT _Supplies Expense_   ACCOUNT NO. _670_

DATE	DESCRIPTION	POST. REF.	DEBIT	CREDIT	BALANCE DEBIT	BALANCE CREDIT

Analyze: _____

_____

Problem 18-9   Locating Errors on a Work Sheet

Buzz Newsstand

Work Sheet

For the Month Ended August 31, 20--

	ACCT. NO.	ACCOUNT NAME	TRIAL BALANCE DEBIT	TRIAL BALANCE CREDIT	ADJUSTMENTS DEBIT	ADJUSTMENTS CREDIT
1	101	Cash in Bank	8131 00			
2	115	Accounts Receivable	363 00			
3	130	Merchandise Inventory	5120 00			(a) 12950 00
4	135	Supplies	974 00			(b) 454 00
5	140	Prepaid Insurance	980 00			(c) 245 00
6	145	Delivery Equipment	7600 00			
7	150	Store Equipment	2854 00			
8	201	Accounts Payable		4515 00		
9	204	Fed. Corporate Income Tax Pay.		—		(d) 249 00
10	210	Employees' Fed. Inc. Tax Pay.		149 00		
11	211	Employees' State Inc. Tax Pay.	26 00			
12	215	Sales Tax Payable		421 00		
13	216	Social Security Tax Payable		79 00		
14	217	Medicare Tax Payable		10 00		
15	301	Capital Stock		25000 00		
16	305	Retained Earnings		5120 00		
17	310	Income Summary		—	(a) 12950 00	
18	401	Sales		11034 00		
19	410	Sales Returns and Allowances		126 00		
20	501	Purchases	16819 00			
21	510	Purchases Discounts		—		
22	515	Purchases Returns and Allow.		246 00		
23	601	Advertising Expense	125 00			
24	625	Fed. Corporate Income Tax Exp.	—		(d) 249 00	
25	635	Insurance Expense	—		(c) 245 00	
26	650	Miscellaneous Expense	45 00			
27	655	Rent Expense	1700 00			
28	657	Payroll Tax Expense	156 00			
29	660	Salaries Expense	1265 00			
30	665	Supplies Expense	—			
31	675	Utilities Expense	342 00			
32			41380 00	41580 00		
33		Corrected TOTALS				

Analyze: _____

_____

CHAPTER 18

Adjustments and the Ten-Column Work Sheet

Self-Test

Part A   True or False

Directions: *Circle the letter* T *in the Answer column if the statement is true; circle the letter* F *if the statement is false.*

Answer

T   F   **1.** The purpose of the end-of-period reports is to provide essential information about the financial position of a business organization.

T   F   **2.** The five amount sections of the ten-column work sheet are: Trial Balance, Adjustments, Adjusted Balance Sheet, Income Statement, and Ending Balance.

T   F   **3.** When preparing a work sheet, every general ledger account should be listed, even if it has a zero balance.

T   F   **4.** At the end of a period, adjustments are made to transfer the costs originally recorded in the expense accounts (temporary accounts) to the asset accounts (permanent accounts).

T   F   **5.** An account balance must be adjusted if the balance shown in the account is not up-to-date as of the last day of the fiscal period.

T   F   **6.** The ending inventory for one period is not the beginning inventory for the next period.

T   F   **7.** A physical inventory is always taken at the end of a period.

T   F   **8.** The totals of the Adjustments Debit and Credit columns do not have to be the same.

T   F   **9.** The source of information for journalizing adjusting entries at the end of a period is the Adjustments section of the work sheet.

T   F   **10.** The completed work sheet only lists the general ledger accounts and their updated balances. It does not show net income or net loss.

Part B   Fill in the Missing Term

Directions: *Using the terms in the following list, complete the sentences below. Write the letter of the term you have chosen in the space provided.*

| **A.** adjustment | **C.** ending inventory | **E.** ten-column work sheet |
| **B.** beginning inventory | **D.** physical inventory | |

Answer

_____   **1.** The _____ has ten amount columns, including columns titled Adjustments and Adjusted Trial Balance.

_____   **2.** A(n) _____ is an amount that is added to or subtracted from an account balance to bring the balance up-to-date.

_____   **3.** A(n) _____ is an actual count of all the merchandise on hand and available for sale.

_____   **4.** The _____ is the merchandise a business has on hand and available for sale at the beginning of a period.

_____   **5.** The _____ is the merchandise a business has on hand at the end of a period.

CHAPTER 19 Financial Statements for a Corporation

Study Plan

Check Your Understanding

Section 1	*Read Section 1 on pages 516–520 and complete the following exercises on page 521.*
	❑ Thinking Critically
	❑ Communicating Accounting
	❑ Problem 19-1 *Analyzing Stockholders' Equity Accounts*
	❑ Problem 19-2 *Analyzing a Source Document*
Section 2	*Read Section 2 on pages 522–528 and complete the following exercises on page 529.*
	❑ Thinking Critically
	❑ Analyzing Accounting
	❑ Problem 19-3 *Calculating Amounts on the Income Statement*
Section 3	*Read Section 3 on pages 531–535 and complete the following exercises on page 536.*
	❑ Thinking Critically
	❑ Computing in the Business World
	❑ Problem 19-4 *Analyzing a Balance Sheet*
Summary	*Review the Chapter 19 Summary on page 537 in your textbook.*
	❑ Key Concepts
Review and Activities	*Complete the following questions and exercises on pages 538–539 in your textbook.*
	❑ Using Key Terms
	❑ Understanding Accounting Concepts and Procedures
	❑ Case Study
	❑ Conducting an Audit with Alex
	❑ Internet Connection
	❑ Workplace Skills
Computerized Accounting	*Read the Computerized Accounting information on page 540 in your textbook.*
	❑ Making the Transition from a Manual to a Computerized System
	❑ Preparing Financial Statements in Peachtree
Problems	*Complete the following end-of-chapter problems for Chapter 19 in your textbook.*
	❑ Problem 19-5 *Preparing an Income Statement*
	❑ Problem 19-6 *Preparing a Statement of Retained Earnings and a Balance Sheet*
	❑ Problem 19-7 *Preparing Financial Statements*
	❑ Problem 19-8 *Completing a Work Sheet and Financial Statements*
Challenge Problem	❑ Problem 19-9 *Evaluating the Effect of an Error on the Income Statement*
Chapter Reviews and Working Papers	*Complete the following exercises for Chapter 19 in your Chapter Reviews and Working Papers.*
	❑ Chapter Review
	❑ Self-Test

CHAPTER 19 REVIEW — Financial Statements for a Corporation

Part 1 Accounting Vocabulary (19 points)

Total Points	57
Student's Score	

Directions: *Using terms from the following list, complete the sentences below. Write the letter of the term you have chosen in the space provided.*

A. administrative expenses	**H.** materiality	**O.** retained earnings
B. base year	**I.** net purchases	**P.** selling expenses
C. capital stock	**J.** net sales	**Q.** statement of retained earnings
D. comparability	**K.** operating expenses	**R.** stockholders' equity
E. full disclosure	**L.** operating income	**S.** vertical analysis
F. gross profit on sales	**M.** relevance	**T.** working capital
G. horizontal analysis	**N.** reliability	

_____Q_____ **0.** The changes that have taken place in the Retained Earnings account during the period are reported on the _____.

_____ **1.** _____ represents the increase in stockholders' equity from net income held by a corporation and not distributed to the stockholders as a return on their investment.

_____ **2.** _____ is the taxable income of a corporation, or the amount of income before federal income taxes.

_____ **3.** _____ is the amount of profit made during the period before expenses are deducted.

_____ **4.** The amount of sales for the period less any sales discounts, returns, or allowances is _____.

_____ **5.** The cash spent or the assets consumed to earn revenue for a business are _____.

_____ **6.** _____ represents the total investment in the corporation by its stockholders.

_____ **7.** The _____ is the amount of all costs related to merchandise purchased during the period.

_____ **8.** _____ is the value of the stockholders' claims to the assets of the corporation.

_____ **9.** _____ allows accounting information to be compared from one fiscal period to another.

_____ **10.** _____ guarantees that financial reports include enough information to be complete.

_____ **11.** _____ relates to the confidence that users have that the financial information is reasonably free from bias and error.

_____ **12.** In accounting, _____ means that the information "makes a difference" to a user in reaching a business decision.

_____ **13.** _____ in financial reporting means that information deemed relevant should be included in the reports.

_____ **14.** _____ are incurred to sell or market the merchandise sold.

_____ **15.** Costs related to the management of the business are called _____.

_____ **16.** With _____, each dollar amount reported on a financial statement is also reported as a percentage of another amount.

_____ **17.** The comparison of the same items on financial reports for two or more accounting periods is called _____.

_____ **18.** A _____ is a year or period used for comparison purposes.

_____ **19.** The amount by which current assets exceed current liabilities is known as _____.

Part 2 Listing Accounts on the Financial Statements (20 points)

Directions: *Using the following codes, indicate the financial statement(s) on which each account title would appear. Write your answer in the space provided.*

B Balance Sheet	**I** Income Statement	**S** Statement of Retained Earnings

_**I**_	**0.** Rent Expense		**10.** Federal Corporate Income Tax Payable
	1. Accounts Payable		**11.** Computer Equipment
	2. Supplies Expense		**12.** Prepaid Insurance
	3. Merchandise Inventory		**13.** Salaries Expense
	4. Sales		**14.** Insurance Expense
	5. Supplies		**15.** Purchases
	6. Capital Stock		**16.** Retained Earnings
	7. Transportation In		**17.** Purchases Discounts
	8. Purchases Returns and Allowances		**18.** Federal Income Tax Expense
	9. Cash in Bank		**19.** Sales Returns and Allowances
			20. State Unemployment Tax Payable

Part 3 Reporting Information on Financial Statements (18 points)

Directions: *Read each of the following statements to determine whether the statement is true or false. Write your answer in the space provided.*

**True** **0.** The balance sheet reports the balances of all permanent accounts as of a specific date.

1. The balance of the Capital Stock account should change every period.

2. The total sales amount of a merchandising business includes the cost of merchandise sold and the profit made from selling that merchandise.

3. Transportation charges increase the cost of merchandise purchased during the period.

4. The cost of merchandise sold is obtained by subtracting the beginning inventory from the cost of merchandise available for sale.

5. The three financial statements prepared by a merchandising corporation are the income statement, the statement of retained earnings, and the balance sheet.

6. Federal Corporate Income Tax Expense is not an operating expense because it represents cash paid out as a result of the revenue earned rather than cash spent to earn revenue.

7. Financial reports are prepared so that managers can evaluate past decisions and make future decisions.

8. The stockholders' equity section of the balance sheet lists the accounts Capital Stock and Cash in Bank.

9. The general ledger is the source of information for preparing the financial statements of a business.

10. The federal income tax amount is listed separately on the income statement so that the operating income can be more easily seen.

11. The statement of retained earnings is prepared before the income statement.

12. The balance of the Retained Earnings account will always increase at the end of a period.

13. The income statement reports the balances of the contra revenue accounts Sales Discounts and Sales Returns and Allowances.

14. A net loss has no direct effect on the Retained Earnings account.

15. The amount of gross profit for the period is the difference between net sales and the cost of merchandise sold.

16. Purchases Returns and Allowances and Purchases Discounts are both contra cost of merchandise accounts.

17. The balance sheet is prepared from the information in the Balance Sheet section of the work sheet and from the income statement.

18. The statement of retained earnings consists of only the balance of the Retained Earnings account at the beginning of the period plus net income before taxes.

Working Papers *for Section Problems*

Problem 19-1   Analyzing Stockholders' Equity Accounts

1. _____

2. _____

3. _____

Problem 19-2   Analyzing a Source Document

Cindy's Curtains
432 Meadowbrook Street
Wilcoxson, GA 30345-8417
404-555-2488

DATE: June 26, 20——   NO. 1441

SOLD TO:
Rachel C. Washington
59 Priscilla Drive
Park Ridge, IL 60068

CLERK K.C.	CASH ✓	CHARGE	TERMS

QTY.	DESCRIPTION	UNIT PRICE	AMOUNT	
2	Curtain Rods #21847	$ 14.95	$ 29	09
4	Anchor Pieces #23104	6.75	27	00
15	Feet of ribbon/per ft.	.89	13	00
		SUBTOTAL	$ 69	09
		SALES TAX	2	76
		TOTAL	$ 71	85

Thank You!

Problem 19-3   Calculating Amounts on the Income Statement

1. Cost of merchandise available for sale   _____

2. Gross profit on sales   _____

3. Cost of delivered merchandise   _____

4. Cost of merchandise sold   _____

Problem 19-4   Analyzing a Balance Sheet

1. _____

2. _____

3. _____

4. _____

5. _____

   _____

   _____

   _____

   _____

   _____

   _____

6. _____

   _____

Problem 19-5   Preparing an Income Statement

Sunset

Work

For the Year Ended

	ACCT. NO.	ACCOUNT NAME	TRIAL BALANCE DEBIT	TRIAL BALANCE CREDIT	ADJUSTMENTS DEBIT	ADJUSTMENTS CREDIT
1	101	Cash in Bank	15 274 00			
2	115	Accounts Receivable	4 124 00			
3	130	Merchandise Inventory	84 097 00			(a) 9 025 00
4	135	Supplies	3 740 00			(b) 2 722 00
5	140	Prepaid Insurance	1 584 00			(c) 528 00
6	145	Store Equipment	7 231 00			
7	150	Office Equipment	4 619 00			
8	201	Accounts Payable		9 340 00		
9	204	Fed. Corporate Income Tax Pay.				(d) 122 00
10	205	Employees' Fed. Inc. Tax Pay.		311 00		
11	208	Employees' State Inc. Tax Pay.		89 00		
12	210	Social Security Tax Payable		132 00		
13	211	Medicare Tax Payable		21 00		
14	212	Fed. Unemployment Tax Pay.		37 00		
15	213	State Unemployment Tax Pay.		134 00		
16	215	Sales Tax Payable		2 670 00		
17	301	Capital Stock		60 000 00		
18	305	Retained Earnings		14 920 00		
19	310	Income Summary			(a) 9 025 00	
20	401	Sales		137 711 00		
21	405	Sales Discounts	2 336 00			
22	410	Sales Returns and Allowances	4 188 00			
23	501	Purchases	71 097 00			
24	505	Transportation In	928 00			
25	510	Purchases Discounts		1 823 00		
26	515	Purchases Returns and Allow.		2 108 00		
27	601	Advertising Expense	840 00			
28	605	Bankcard Fees Expense	374 00			
29	630	Fed. Corporate Income Tax Exp.	2 600 00		(d) 122 00	
30		*Carried Forward*	203 032 00	229 296 00	9 147 00	12 397 00
31						
32						
33						

Surfwear

Sheet

December 31, 20--

	ADJUSTED TRIAL BALANCE		INCOME STATEMENT		BALANCE SHEET		
	DEBIT	CREDIT	DEBIT	CREDIT	DEBIT	CREDIT	
	15274 00				15274 00		1
	4124 00				4124 00		2
	75072 00				75072 00		3
	1018 00				1018 00		4
	1056 00				1056 00		5
	7231 00				7231 00		6
	4619 00				4619 00		7
		9340 00				9340 00	8
		122 00				122 00	9
		311 00				311 00	10
		89 00				89 00	11
		132 00				132 00	12
		21 00				21 00	13
		37 00				37 00	14
		134 00				134 00	15
		2670 00				2670 00	16
		60000 00				60000 00	17
		14920 00				14920 00	18
	9025 00		9025 00				19
		137711 00		137711 00			20
	2336 00		2336 00				21
	4188 00		4188 00				22
	71097 00		71097 00				23
	928 00		928 00				24
		1823 00		1823 00			25
		2108 00		2108 00			26
	840 00		840 00				27
	374 00		374 00				28
	2722 00		2722 00				29
	199904 00	229418 00	91510 00	141642 00	108394 00	87776 00	30
							31
							32
							33

Problem 19-5   (continued)

Sunset

Work Sheet

For the Year Ended

	ACCT. NO.	ACCOUNT NAME	TRIAL BALANCE		ADJUSTMENTS	
			DEBIT	CREDIT	DEBIT	CREDIT
1		**Brought Forward**	20303200	22929600	914700	1239700
2						
3	635	**Insurance Expense**			(c) 52800	
4	645	**Maintenance Expense**	123100			
5	650	**Miscellaneous Expense**	286000			
6	652	**Payroll Tax Expense**	217000			
7	655	**Rent Expense**	927000			
8	660	**Salaries Expense**	1073300			
9	685	**Supplies Expense**			(b) 272200	
10			22929600	22929600	1239700	1239700
11		**Net Income**				
12						
13						
14						
15						
16						
17						
18						
19						
20						
21						
22						
23						
24						
25						
26						
27						
28						
29						
30						
31						
32						
33						

Surfwear

(continued)

December 31, 20--

ADJUSTED TRIAL BALANCE		INCOME STATEMENT		BALANCE SHEET		
DEBIT	CREDIT	DEBIT	CREDIT	DEBIT	CREDIT	
199904 00	229418 00	91510 00	141642 00	108394 00	87776 00	1
						2
	528 00	528 00				3
1231 00		1231 00				4
2860 00		2860 00				5
2170 00		2170 00				6
9270 00		9270 00				7
10733 00		10733 00				8
2722 00		2722 00				9
229418 00	229418 00	121024 00	141642 00	108394 00	87776 00	10
			20618 00		20618 00	11
		141642 00	141642 00	108394 00	108394 00	12
						13
						14
						15
						16
						17
						18
						19
						20
						21
						22
						23
						24
						25
						26
						27
						28
						29
						30
						31
						32
						33

Computerized Accounting Using Peachtree

Software Objectives

When you have completed this chapter, you will be able to use Peachtree to:

1. Print a Balance Sheet for a corporation.
2. Print an Income Statement and a Statement of Retained Earnings for a corporation.
3. Print an adjusted trial balance.

Problem 19-6   Preparing a Statement of Retained Earnings and a Balance Sheet

INSTRUCTIONS

Beginning a Session

Step 1   Select the problem set: Sunset Surfwear (Prob. 19-6).
Step 2   Rename the company and set the system date to December 31, 2008.

Completing the Accounting Problem

Step 3   Review the information in your textbook.
Step 4   Print a Statement of Retained Earnings and Balance Sheet. The Statement of Retained Earnings and Balance Sheet reports are available in the Financial Statements report area.

       Choose the <Standard> Balance Sheet report whenever you are instructed to print a Balance Sheet. This report option is the default layout provided by Peachtree. As you can see, the Peachtree Balance Sheet is similar to the Balance Sheet report shown in your text. However, the Peachtree report is divided into more segments including the following parts: Current Assets, Property and Equipment, Other Assets, Current Liabilities, Long-Term Liabilities, and Capital.

Step 5   Answer the Analyze question.

Ending the Session

Step 6   Click the **Close Problem** button in the Glencoe Smart Guide window and select a save option.

> **DO YOU HAVE A QUESTION**
>
> **Q.** *Where is the Statement of Retained Earnings report option located?*
>
> **A.** Select **Financial Statements** from the **Reports** menu. Select **<Standard> Retained Earnings**.

Problem 19-7   Preparing Financial Statements

INSTRUCTIONS

Beginning a Session

Step 1   Select the problem set: Shutterbug Cameras (Prob. 19-7).
Step 2   Rename the company and set the system date to December 31, 2008.

Completing the Accounting Problem

Step 3   Review the information in your textbook.
Step 4   Print the following reports: Adjusted Trial Balance, Balance Sheet, Statement of Retained Earnings, and Income Statement.

       Choose the Peachtree <Standard> Income Statement report when you are instructed to print an Income Statement. As you will notice, the report includes more information than the Income Statements shown in your textbook. The default Peachtree report shows the results for the current month and the current year. For the problems in this text, these two columns will always be the same.

The Cost of Sales (or Cost of Merchandise Sold) on a Peachtree Income Statement is very different from the style used in the textbook. Where a manual report shows both the beginning and ending **Merchandise Inventory** account balance, the Peachtree report does not show this information. Instead, it includes the **Inventory Adjustment** account balance. The net effect is the same, in that, this account reflects the difference in the cost of the physical inventory.

Step 5   Answer the Analyze question.

Ending the Session

Step 6   Click the **Close Problem** button in the Glencoe Smart Guide window and select a save option.

Problem 19-8   Completing a Work Sheet and Financial Statements

INSTRUCTIONS

Beginning a Session

Step 1   Select the problem set: Cycle Tech Bicycles (Prob. 19-8).
Step 2   Rename the company and set the system date to December 31, 2008.

Completing the Accounting Problem

Step 3   Review the information in your textbook.
Step 4   Print a Working Trial Balance report and use it to prepare the adjustments.
Step 5   Record the adjusting entries using the **General Journal Entry** option.
Step 6   Print a General Journal report and proof your work.
Step 7   Print the following reports: Adjusted Trial Balance, Income Statement, Statement of Retained Earnings, and Balance Sheet.
Step 8   Answer the Analyze question.

Ending the Session

Step 9   Click the **Close Problem** button in the Glencoe Smart Guide window and select a save option.

DO YOU HAVE A QUESTION

Q. *Why is the Cost of Merchandise Sold section on Peachtree's standard Income Statement different from a manual report?*

A. In Chapter 18, you learned how to make the adjusting entry for **Merchandise Inventory** using a computerized system such as the Peachtree Complete® Accounting software. The adjustment was made to an account called **Inventory Adjustment**, which is the difference between the beginning and ending inventory value. Peachtree uses the **Inventory Adjustment** account balance along with the other cost of sales accounts (**Purchases, Transportation In, Purchases Discounts,** and **Purchases Returns and Allowances**) to calculate the total cost of sales. The net effect is the same as the result obtained using the manual method.

Notes

Whenever you are instructed to print an Adjusted Trial Balance, choose the General Ledger Trial Balance report. You can change the report title if you want.

FAQs

Where is the Statement of Retained Earnings report option located?

Select **Financial Statements** from the *Reports* menu. Select **<Standard> Retained Earnings**.

Computerized Accounting Using Spreadsheets

Problem 19-5 Preparing an Income Statement

Completing the Spreadsheet

Step 1 Read the instructions for Problem 19-5 in your textbook. This problem involves preparing an income statement for Sunset Surfwear.

Step 2 Open the Glencoe Accounting: Electronic Learning Center software.

Step 3 From the Program Menu, click on the **Peachtree Complete®** **Accounting Software and Spreadsheet Applications** icon.

Step 4 Log onto the Management System by typing your user name and password.

Step 5 Under the **Problems & Tutorials** tab, select template 19-5 from the Chapter 19 drop-down menu. The template should look like the one shown below.

```
PROBLEM 19-5
PREPARING AN INCOME STATEMENT

(name)
(date)


SUNSET SURFWEAR
INCOME STATEMENT
FOR THE YEAR ENDED DECEMBER 31, 20--

Revenue:
    Sales                                                       AMOUNT
    Less: Sales Discounts                          AMOUNT
          Sales Returns & Allowances              AMOUNT      0.00
          Net Sales                                                        0.00
Cost of Merchandise Sold:
    Merchandise Inventory January 1                             AMOUNT
    Purchases                              AMOUNT
    Plus: Transportation In                AMOUNT
    Cost of Delivered Merchandise                     0.00
    Less: Purchases Discounts              AMOUNT
          Purchases Returns & Allowances   AMOUNT    0.00
    Net Purchases                                               0.00
    Cost of Merchandise Available                               0.00
    Merchandise Inventory December 31                          AMOUNT
        Cost of Merchandise Sold                                         0.00
Gross Profit on Sales                                                     0.00
Operating Expenses:
    Advertising Expense                                         AMOUNT
    Bankcard Fees Expense                                       AMOUNT
    Insurance Expense                                           AMOUNT
    Maintenance Expense                                         AMOUNT
    Miscellaneous Expense                                       AMOUNT
    Payroll Tax Expense                                         AMOUNT
    Rent Expense                                                AMOUNT
    Salaries Expense                                            AMOUNT
    Supplies Expense                                            AMOUNT
    Total Operating Expenses                                             0.00
Operating Income                                                         0.00
    Less: Federal Corporate Income Tax Expense                          AMOUNT
Net Income                                                               0.00
```

Step 6 Key your name and today's date in the cells containing the *(name)* and *(date)* placeholders.

Step 7 Using the data provided in the work sheet for Sunset Surfwear given in your working papers, enter the income statement data into the spreadsheet template in the cells containing the AMOUNT placeholders. The spreadsheet template will automatically calculate the net sales, cost of merchandise sold, gross profit on sales, total operating expenses, operating income, and net income. Remember, it is not necessary to enter a comma or the decimal point and ending zeroes when entering the amounts.

Step 8 Save the spreadsheet using the **Save** option from the *File* menu. You should accept the default location for the save as this is handled by the management system.

Step 9 Print the completed spreadsheet.

Step 10 Exit the spreadsheet program.

Step 11 In the Close Options box, select the location where you would like to save your work.

Step 12 Answer the Analyze question from your textbook for this problem.

What-If Analysis

If Sunset Surfwear's merchandise inventory on January 1 were $100,000, what would the cost of merchandise sold be? What would net income be?

_____

_____

Working Papers *for End-of-Chapter Problems*

Problem 19-5   (concluded)

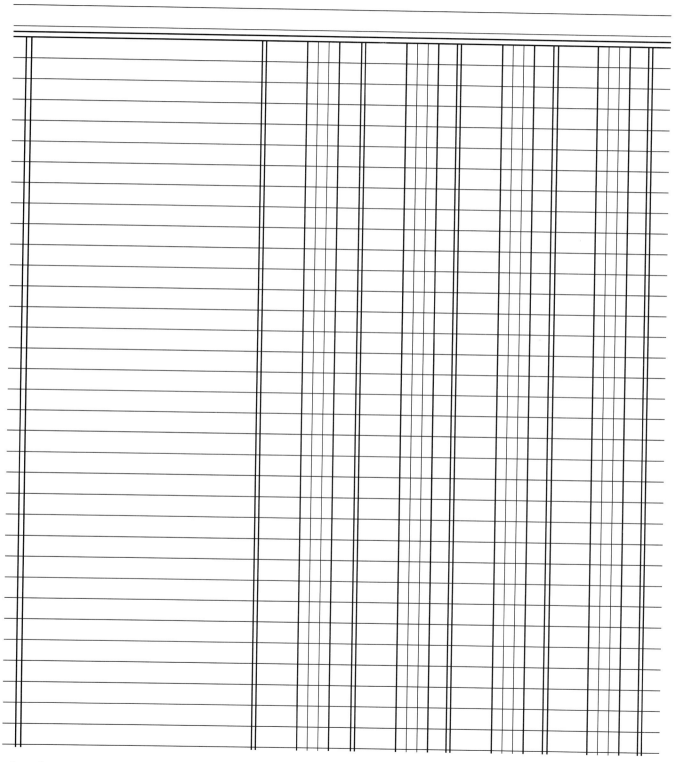

Analyze: _____

Problem 19-6 Preparing a Statement of Retained Earnings and a Balance Sheet

Problem 19-6   (concluded)

Analyze: _____

_____

Problem 19-7 Preparing Financial Statements
(1)

Shutterbug

Work

For the Year Ended

	ACCT. NO.	ACCOUNT NAME	TRIAL BALANCE DEBIT	TRIAL BALANCE CREDIT	ADJUSTMENTS DEBIT	ADJUSTMENTS CREDIT
1	101	Cash in Bank	1360300			
2	115	Accounts Receivable	541800			(a) 445100
3	125	Merchandise Inventory	8276300			(b) 203900
4	130	Supplies	252200			(c) 47500
5	135	Prepaid Insurance	135000			
6	140	Store Equipment	2676900			
7	201	Accounts Payable		1448100		
8	207	Fed. Corporate Income Tax Pay.				(d) 26100
9	210	Employees' Fed. Inc. Tax Pay.		18900		
10	211	Employees' State Inc. Tax Pay.		5200		
11	212	Social Security Tax Payable		13800		
12	213	Medicare Tax Payable		2800		
13	215	Sales Tax Payable		89100		
14	216	Fed. Unemployment Tax Pay.		1900		
15	217	State Unemployment Tax Pay.		9600		
16	301	Capital Stock		8000000		
17	305	Retained Earnings		1919200		
18	310	Income Summary			(a) 445100	
19	401	Sales		9286700		
20	405	Sales Discounts	10500			
21	410	Sales Returns and Allowances	88500			
22	501	Purchases	3749100			
23	505	Transportation In	180500			
24	510	Purchases Discounts		64400		
25	515	Purchases Returns and Allow.		23100		
26	601	Advertising Expense	65000			
27	605	Bankcard Fees Expense	21300			
28	620	Fed. Corporate Income Tax Exp.	172000		(d) 26100	
29		Carried Forward	17529400	20882800	471200	722600
30						
31						
32						
33						

Cameras

Sheet

December 31, 20--

| ADJUSTED TRIAL BALANCE | | INCOME STATEMENT | | BALANCE SHEET | | |
DEBIT	CREDIT	DEBIT	CREDIT	DEBIT	CREDIT	
1360300						1
541800						2
7831200						3
48300						4
87500						5
2676900						6
	1448100					7
	26100					8
	18900					9
	5200					10
	13800					11
	2800					12
	89100					13
	1900					14
	9600					15
	8000000					16
	1919200					17
445100						18
	9286700					19
10500						20
88500						21
3749100						22
180500						23
	64400					24
	23100					25
65000						26
21300						27
198100						28
17304100	20908900					29
						30
						31
						32
						33

Problem 19-7   (continued)

Shutterbug

Work Sheet

For the Year Ended

	ACCT. NO.	ACCOUNT NAME	TRIAL BALANCE		ADJUSTMENTS	
			DEBIT	CREDIT	DEBIT	CREDIT
1		**Brought Forward**	17 5 29 4 00	20 8 82 8 00	4 7 1 2 00	7 2 2 6 00
2						
3	630	**Insurance Expense**			(c) 4 7 5 00	
4	640	**Maintenance Expense**	2 5 5 2 00			
5	645	**Miscellaneous Expense**	2 8 5 00			
6	647	**Payroll Tax Expense**	1 9 2 0 00			
7	650	**Rent Expense**	9 7 0 0 00			
8	655	**Salaries Expense**	18 7 2 0 00			
9	660	**Supplies Expense**			(b) 2 0 3 9 00	
10	670	**Utilities Expense**	3 5 7 00			
11						
12			20 8 82 8 00	20 8 82 8 00	7 2 2 6 00	7 2 2 6 00
13						
14						
15						
16						
17						
18						
19						
20						
21						
22						
23						
24						
25						
26						
27						
28						
29						
30						
31						
32						

Cameras
(continued)
December 31, 20--

| ADJUSTED TRIAL BALANCE | | INCOME STATEMENT | | BALANCE SHEET | | |
DEBIT	CREDIT	DEBIT	CREDIT	DEBIT	CREDIT	
17304100	20908900					1
						2
47500						3
255200						4
28500						5
192000						6
970000						7
1872000						8
203900						9
35700						10
						11
20908900	20908900					12
						13
						14
						15
						16
						17
						18
						19
						20
						21
						22
						23
						24
						25
						26
						27
						28
						29
						30
						31
						32

Problem 19-7   (continued)

(2)

Problem 19-7   (concluded)

(3)

(4)

Analyze: _____

Problem 19-8 Completing a Work Sheet and Financial Statements

(1)

Cycle Tech

Work

For the Year Ended

	ACCT. NO.	ACCOUNT NAME	TRIAL BALANCE DEBIT	TRIAL BALANCE CREDIT	ADJUSTMENTS DEBIT	ADJUSTMENTS CREDIT
1	101	Cash in Bank	2193100			
2	115	Accounts Receivable	178200			
3	125	Merchandise Inventory	2402800			
4	130	Supplies	415900			
5	135	Prepaid Insurance	180000			
6	140	Store Equipment	2489500			
7	145	Office Equipment	1611300			
8	201	Accounts Payable		1122400		
9	210	Fed. Corporate Income Tax Pay.				
10	211	Employees' Fed. Inc. Tax Pay.		52200		
11	212	Employees' State Inc. Tax Pay.		14400		
12	213	Social Security Tax Payable		41300		
13	214	Medicare Tax Payable		13400		
14	215	Sales Tax Payable		191500		
15	216	Fed. Unemployment Tax Pay.		5400		
16	217	State Unemployment Tax Pay.		27100		
17	301	Capital Stock		4000000		
18	305	Retained Earnings		1109100		
19	310	Income Summary				
20	401	Sales		12715100		
21	405	Sales Discounts	24600			
22	410	Sales Returns and Allowances	132800			
23	501	Purchases	6610700			
24	505	Transportation In	98300			
25	510	Purchases Discounts		82200		
26	515	Purchases Returns and Allow.		37600		
27	601	Advertising Expense	238000			
28	605	Bankcard Fees Expense	18100			
29	625	Fed. Corporate Income Tax Exp.	334000			
30		Carried Forward	16927300	19411700		
31						
32						

Bicycles _____

Sheet _____

December 31, 20– –

ADJUSTED TRIAL BALANCE		INCOME STATEMENT		BALANCE SHEET		
DEBIT	CREDIT	DEBIT	CREDIT	DEBIT	CREDIT	
						1
						2
						3
						4
						5
						6
						7
						8
						9
						10
						11
						12
						13
						14
						15
						16
						17
						18
						19
						20
						21
						22
						23
						24
						25
						26
						27
						28
						29
						30
						31
						32

Problem 19-8 (continued)

Cycle Tech

Work Sheet

For the Year Ended

	ACCT. NO.	ACCOUNT NAME	TRIAL BALANCE		ADJUSTMENTS	
			DEBIT	CREDIT	DEBIT	CREDIT
1		**Brought Forward**	16927300	19411700		
2						
3	630	**Insurance Expense**				
4	645	**Maintenance Expense**	195000			
5	650	**Miscellaneous Expense**	183100			
6	655	**Payroll Tax Expense**	83400			
7	657	**Rent Expense**	1080000			
8	660	**Salaries Expense**	473400			
9	665	**Supplies Expense**				
10	675	**Utilities Expense**	469500			
11						
12			19411700	19411700		
13						
14						
15						
16						
17						
18						
19						
20						
21						
22						
23						
24						
25						
26						
27						
28						
29						
30						
31						
32						

Bicycles

(continued)

December 31, 20--

ADJUSTED TRIAL BALANCE		INCOME STATEMENT		BALANCE SHEET		
DEBIT	CREDIT	DEBIT	CREDIT	DEBIT	CREDIT	
						1
						2
						3
						4
						5
						6
						7
						8
						9
						10
						11
						12
						13
						14
						15
						16
						17
						18
						19
						20
						21
						22
						23
						24
						25
						26
						27
						28
						29
						30
						31
						32

Problem 19-8   (continued)
(2)

Problem 19-8 (concluded)

(3)

(4)

Analyze: _____

Problem 19-9   Evaluating the Effect of an Error on the Income Statement

River's Edge Canoe & Kayak
Income Statement
For the Year Ended December 31, 20--

Revenue:					
Sales				32478400	
Less: Sales Discounts			383900		
Sales Returns and Allowances			120900	504800	
Net Sales					31973600
Cost of Merchandise Sold:					
Merchandise Inv., Jan. 1, 20--				8492100	
Purchases	20841600				
Cost of Delivered Merchandise			20841600		
Less: Purchases Discounts	962300				
Purchases Returns and Allow.	472100		1434400		
Net Purchases				19407200	
Cost of Merchandise Available				27899300	
Merchandise Inv., Dec. 31, 20--				8138500	
Cost of Merchandise Sold					19760800
Gross Profit on Sales					12212800
Operating Expenses:					
Advertising Expense				257000	
Bankcard Fees Expense				418200	
Insurance Expense				27500	
Maintenance Expense				355200	
Miscellaneous Expense				34400	
Payroll Tax Expense				382400	
Rent Expense				1500000	
Salaries Expense				2938100	
Supplies Expense				371000	
Utilities Expense				237800	
Total Operating Expenses					6521600
Operating Income					5691200
Less: Fed. Corporate Inc. Tax Exp.					943600
Net Income					4747600

Problem 19-9   (concluded)

1. _____

2. _____

3. _____

4. _____

5. _____

Analyze: _____

_____

CHAPTER 19   Financial Statements for a Corporation

Self-Test

Part A   True or False

Directions: *Circle the letter* T *in the Answer column if the statement is true; circle the letter* F *if the statement is false.*

Answer

T	F	**1.**	The balances of all permanent accounts as of a specific date are reported on the balance sheet.
T	F	**2.**	The statement of retained earnings is prepared before the balance sheet.
T	F	**3.**	Purchases Discounts is a contra account of Purchases.
T	F	**4.**	The account Transportation In reduces the amount of merchandise available for sale.
T	F	**5.**	The Retained Earnings account always increases at the end of the period.
T	F	**6.**	The source of information for preparing the income statement is the general ledger.
T	F	**7.**	Capital Stock is the only general ledger account classified as stockholders' equity.
T	F	**8.**	The cost of merchandise sold is determined by subtracting the ending inventory from the cost of merchandise available for sale.
T	F	**9.**	The amount of profit earned before expenses are subtracted is the gross profit on sales.
T	F	**10.**	Changes in the Cash in Bank account are reported in the statement of retained earnings.
T	F	**11.**	Net purchases is added to the beginning inventory to get the merchandise available for sale for the period.
T	F	**12.**	Corporate Federal Income Tax Expense is not included in the total operating expenses for the period.
T	F	**13.**	The income statement reports the financial position of the business as of a specific date.
T	F	**14.**	The balance in the Capital Stock account never changes unless stock in the business is purchased or sold.

Part B   Fill in the Missing Term

Directions: *In the Answer column, write the letter of the word or phrase that best completes the sentence. Some answers may be used more than once.*

A. capital stock	**C.** gross profit on sales	**F.** statement of retained
B. cost of merchandise	**D.** net purchases	earnings
available for sale	**E.** net sales	**G.** operating expenses

Answer

1. _____ is the amount of beginning inventory plus net purchases.

2. The _____ reports any changes that have taken place in the Retained Earnings account during the period.

3. _____ is the total cost of merchandise bought during the period, plus transportation charges, less returns, allowances and discounts.

4. _____ is the account used to record any investments by stockholders.

5. _____ are the assets consumed or the cash spent to earn revenue for a business.

6. On the income statement, net sales less the cost of merchandise sold is the _____.

7. The amount of sales revenue remaining after sales returns and allowances and sales discounts have been subtracted is _____.

MINI PRACTICE SET 4

In-Touch Electronics

CHART OF ACCOUNTS

ASSETS
101   Cash in Bank
105   Accounts Receivable
110   Merchandise Inventory
115   Supplies
120   Prepaid Insurance
150   Store Equipment
155   Office Equipment

LIABILITIES
201   Accounts Payable
205   Sales Tax Payable
210   Employees' Federal Income Tax Payable
211   Employees' State Income Tax Payable
212   Social Security Tax Payable
213   Medicare Tax Payable
214   Federal Unemployment Tax Payable
215   State Unemployment Tax Payable

STOCKHOLDERS' EQUITY
301   Capital Stock
302   Retained Earnings
303   Income Summary

REVENUE
401   Sales
405   Sales Discounts
410   Sales Returns and Allowances

COST OF MERCHANDISE
501   Purchases
505   Transportation In
510   Purchases Discounts
515   Purchases Returns
        and Allowances

EXPENSES
605   Advertising Expense
610   Bankcard Fees Expense
615   Miscellaneous Expense
620   Payroll Tax Expense
625   Rent Expense
630   Salaries Expense
635   Utilities Expense

Accounts Receivable Subsidiary Ledger
LOR   Sam Lorenzo
MAR   Marianne Martino
MCC   Mark McCormick
SCO   Sue Ellen Scott
TRO   Tom Trout

Accounts Payable Subsidiary Ledger
COM   Computer Systems, Inc.
DES   Desktop Wholesalers
HIT   Hi-Tech Electronics Outlet
LAS   Laser & Ink Jet Products
OFF   Office Suppliers, Inc.

Mini Practice Set 4   Source Documents

Instructions: *Use the following source documents to record the transactions for this practice set.*

In-Touch Electronics
612 Kent Avenue
Brooklyn, NY 11205

DATE: *May 16, 20--*   NO. *607*

SOLD TO: *Sam Lorenzo*

CLERK	CASH	CHARGE	TERMS

QTY.	DESCRIPTION	UNIT PRICE	AMOUNT	
2	stereo speakers	$60.00	$120	00
		SUBTOTAL	$120	00
		SALES TAX	6	00
		TOTAL	$126	00

Thank You!

In-Touch Electronics   **893**
612 Kent Avenue   74-103
Brooklyn, NY 11205   720

DATE *May 19*   20--

PAY TO THE ORDER OF *Computer Systems, Inc.* $ *1,200.00*

One thousand two hundred and 00/100 ———————— DOLLARS

UB Union Bank

MEMO *on account*   *Pedro Cordova*

⑆0720 01033⑆ 6171 5222⑈ 0893

In-Touch Electronics   **894**
612 Kent Avenue   74-103
Brooklyn, NY 11205   720

DATE *May 19*   20--

PAY TO THE ORDER OF *Hi-Tech Electronics Outlet* $ *1,750.00*

One thousand seven hundred fifty and 00/100 ———————— DOLLARS

UB Union Bank

MEMO *on account*   *Tina Cordova*

⑆0720 01033⑆ 6171 5222⑈ 0894

In-Touch Electronics   **RECEIPT**
612 Kent Avenue   No. 356
Brooklyn, NY 11205

May 17   20--

RECEIVED FROM *Tom Trout* $ *126.00*

One hundred twenty-six and 00/100 ———————— DOLLARS

FOR *Payment on account*

RECEIVED BY *Tina Cordova*

In-Touch Electronics   **895**
612 Kent Avenue   74-103
Brooklyn, NY 11205   720

DATE *May 19*   20--

PAY TO THE ORDER OF *Office Suppliers, Inc.* $ *770.00*

Seven hundred seventy and 00/100 ———————— DOLLARS

UB Union Bank

MEMO *on account*   *Pedro Cordova*

⑆0720 01033⑆ 6171 5222⑈ 0895

In-Touch Electronics   **892**
612 Kent Avenue   74-103
Brooklyn, NY 11205   720

DATE *May 17*   20--

PAY TO THE ORDER OF *Desktop Wholesalers* $ *800.00*

Eight hundred and 00/100 ———————— DOLLARS

UB Union Bank

MEMO ————— *Pedro Cordova*

⑆0720 01033⑆ 6171 5222⑈ 0892

In-Touch Electronics   **RECEIPT**
612 Kent Avenue   No. 357
Brooklyn, NY 11205

May 20   20--

RECEIVED FROM *Bob Bell* $ *90.00*

Ninety and 00/100 ———————— DOLLARS

FOR *employee purchase of office equipment*

RECEIVED BY *Pedro Cordova*

DEBIT MEMORANDUM   No. 38

Date: *May 18, 20--*
Invoice No.: *N/A*

In-Touch Electronics
612 Kent Avenue
Brooklyn, NY 11205

To: *Laser & Ink Jet Products*
1412 Abrams Avenue
Brooklyn, NY 11205

This day we have debited your account as follows:

Quantity	Item	Unit Price	Total
1	Ink Jet Cartridge	$75.00	$75.00

Desktop Wholesalers
6190 Grand Street
Bronx, NY 10451

INVOICE NO. DW87

DATE: *May 20, 20--*
ORDER NO.:
SHIPPED BY:
TERMS: *n/30*

TO   *In-Touch Electronics*
612 Kent Avenue
Brooklyn, NY 11205

QTY.	ITEM	UNIT PRICE	TOTAL
3	Car Stereo System	$ 300.00	$ 900.00
1	Television	300.00	300.00
	Total		$ 1,200.00

In-Touch Electronics
612 Kent Avenue
Brooklyn, NY 11205

RECEIPT
No. 358

May 20   20 --

RECEIVED FROM *Mark McCormick* $ *210.00*

Two hundred ten and ⁰⁰/₁₀₀ —— DOLLARS

FOR *Payment on account*

RECEIVED BY *Kelly Briggs*

In-Touch Electronics
612 Kent Avenue
Brooklyn, NY 11205

RECEIPT
No. 359

May 20   20 --

RECEIVED FROM *Sue Ellen Scott* $ *308.70*

Three hundred eight and ⁷⁰/₁₀₀ —— DOLLARS

FOR *Paid $315 less 2% on account*

RECEIVED BY *Tina Cordova*

Hi-Tech Electronics Outlet
265 Pixie Drive
New York, NY 10006

INVOICE NO. HT99

DATE: *May 21, 20--*
ORDER NO.:
SHIPPED BY:
TERMS: *2/10, n/30*

TO   *In-Touch Electronics*
612 Kent Avenue
Brooklyn, NY 11205

QTY.	ITEM	UNIT PRICE	TOTAL
2	Sony VCR Systems	$ 250.00	$ 500.00
5	Intercom Systems	200.00	1,000.00
			$ 1,500.00

In-Touch Electronics
612 Kent Avenue
Brooklyn, NY 11205

RECEIPT
No. 360

May 21   20 --

RECEIVED FROM *Marianne Martino* $ *94.50*

Ninety-four and ⁵⁰/₁₀₀ —— DOLLARS

FOR *Payment on account*

RECEIVED BY *Pedro Cordova*

In-Touch Electronics
612 Kent Avenue
Brooklyn, NY 11205

DATE: *May 21, 20--*   NO. *608*

SOLD TO   *Mark McCormick*

CLERK	CASH	CHARGE	TERMS 2/10, n/30

QTY.	DESCRIPTION	UNIT PRICE	AMOUNT
5	answering machines	$80.00	$400 00
		SUBTOTAL	$400 00
		SALES TAX	20 00
		TOTAL	$420 00

Thank You!

In-Touch Electronics
612 Kent Avenue
Brooklyn, NY 11205

896
74-103
720

DATE *May 22*   20 --

PAY TO THE ORDER OF *Desktop Wholesalers* $ *1,200.00*

One thousand two hundred and ⁰⁰/₁₀₀ —— DOLLARS

UB Union Bank

MEMO ———   *Tina Cordova*

⑆0720 ⑈01033⑆ 6171 5222⑈ 0896

In-Touch Electronics
612 Kent Avenue
Brooklyn, NY 11205

DATE: *May 23, 20--*   NO. *609*

SOLD TO   *Sue Ellen Scott*

CLERK	CASH	CHARGE	TERMS 2/10, n/30

QTY.	DESCRIPTION	UNIT PRICE	AMOUNT
1	Sony DVD player	$500.00	$500 00
		SUBTOTAL	$500 00
		SALES TAX	25 00
		TOTAL	$525 00

Thank You!

In-Touch Electronics
612 Kent Avenue
Brooklyn, NY 11205

DATE: May 24, 20--

NAME: Mark McCormick
ADDRESS: 2724 S. 1st Street
Brooklyn, NY 11205

Mark McCormick
CUSTOMER SIGNATURE

CREDIT MEMORANDUM   NO. 55

ORIGINAL SALES DATE	ORIGINAL SALES SLIP	APPROVAL	X MDSE RET
May 12, 20--	605	TC	

QTY.	DESCRIPTION	AMOUNT
1	Clarion accessory	$100.00

REASON FOR RETURN: *wrong model*

	SUB TOTAL	$100.00
THE TOTAL SHOWN AT THE RIGHT WILL BE CREDITED TO YOUR ACCOUNT.	SALES TAX	5.00
	TOTAL	$105.00

In-Touch Electronics   **897**
612 Kent Avenue
Brooklyn, NY 11205   74-103 / 720

DATE May 25   20--

PAY TO THE ORDER OF Surfside Insurance Co. $ 1,600.00

One thousand six hundred and 00/100 ———— DOLLARS

UB Union Bank

MEMO annual insurance   Tina Cordova

⑈0720 01033⑈ 6171 5222⑈ 0897

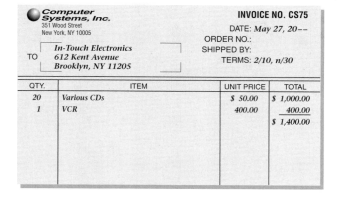

In-Touch Electronics   **MEMORANDUM 26**
612 Kent Avenue
Brooklyn, NY 11205

TO: Accounting Clerk
FROM: Senior Accountant
DATE: May 26, 20--
SUBJECT: Correcting entry

Please make the entry to correct the error in debiting Purchases rather than Transportation In last month for $50.00.

Computer Systems, Inc.   **INVOICE NO. CS75**
351 Wood Street
New York, NY 10005

DATE: May 27, 20--
ORDER NO.:
SHIPPED BY:
TERMS: 2/10, n/30

TO In-Touch Electronics
612 Kent Avenue
Brooklyn, NY 11205

QTY.	ITEM	UNIT PRICE	TOTAL
20	Various CDs	$ 50.00	$ 1,000.00
1	VCR	400.00	400.00
			$ 1,400.00

In-Touch Electronics
612 Kent Avenue
Brooklyn, NY 11205

DATE: May 28, 20--   NO. 610

SOLD TO: Marianne Martino

CLERK	CASH	CHARGE	TERMS 2/10, n/30

QTY.	DESCRIPTION	UNIT PRICE	AMOUNT
1	car stereo	$200.00	$200.00

	SUBTOTAL	$200.00
	SALES TAX	10.00
	TOTAL	$210.00

Thank You!

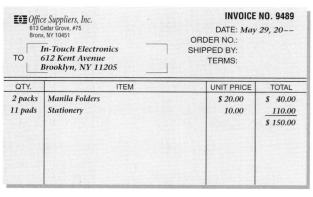

Office Suppliers, Inc.   **INVOICE NO. 9489**
613 Cedar Grove, #75
Bronx, NY 10451

DATE: May 29, 20--
ORDER NO.:
SHIPPED BY:
TERMS:

TO In-Touch Electronics
612 Kent Avenue
Brooklyn, NY 11205

QTY.	ITEM	UNIT PRICE	TOTAL
2 packs	Manila Folders	$ 20.00	$ 40.00
11 pads	Stationery	10.00	110.00
			$ 150.00

In-Touch Electronics   **898**
612 Kent Avenue
Brooklyn, NY 11205   74-103 / 720

DATE May 30   20--

PAY TO THE ORDER OF Green Realty $ 1,500.00

One thousand five hundred and 00/100 ———— DOLLARS

UB Union Bank

MEMO rent   Tina Cordova

⑈0720 01033⑈ 6171 5222⑈ 0898

Dec. 31
Tape 22

1,200.00	CA
60.00	ST
900.00	BCS
45.00	ST

---

552 ■ **Mini Practice Set 4**

$ __1,858.75__		No. 899
Date _5/31_		20 – –
To _Payroll Account_		
For _Payroll—May 31_		
	Dollars	Cents
Balance brought forward	8,093	20
less 12/31 bank svc. fee	25	00
12/31 bankcard fee	100	00
Total		
Less this check	1,858	75
Balance carried forward	6,109	45

PAYROLL REGISTER

PAY PERIOD ENDING _May 31_   20 – –   DATE OF PAYMENT _May 31, 20 – –_

EMPLOYEE NUMBER	NAME	MAR STATUS	ALLOW	TOTAL HOURS	RATE	EARNINGS			DEDUCTIONS							NET PAY	CK. NO.
						REGULAR	OVERTIME	TOTAL	SOC. SEC. TAX	MED. TAX	FED. INC. TAX	STATE INC. TAX	HOSP. INS.	OTHER	TOTAL		
24																	
25																	
	TOTALS					2500 00			155 00	36 25	400 00	50 00			641 25	1858 75	

Other Deductions: Write the appropriate code letter to the left of the amount: B—U.S. Savings Bonds; C—Credit Union; UD—Union Dues; UW—United Way.

In-Touch Electronics   **MEMORANDUM 27**
612 Kent Avenue
Brooklyn, NY 11205

TO: _Accounting Clerk_
FROM: _Payroll Dept._
DATE: _May 31, 20 – –_
SUBJECT: _Payroll Tax_

Please record employer payroll taxes for May 31 payroll.
FICA rate = 6.2%
Medicare rate = 1.45%
Fed. unemployment tax rate = 0.8%
State unemployment tax rate = 5.4%

Computerized Accounting Using Peachtree

Mini-Practice Set 4

INSTRUCTIONS

Beginning a Session

Step 1   Select the problem set: In-Touch Electronics (MP-4).

Step 2   Rename the company and set the system date to May 31, 2008.

Completing the Accounting Problem

Step 3   Review the transactions provided in your textbook (May 16–May 31).

 TIP: To save time entering transactions, group them by type and then enter the transactions in batches.

Sales and Cash Receipts

Step 4   Record the sales on account using the **Sales/Invoicing** option.

 TIP: Remember that you need to change the *GL Account* in the Receipts window to enter a cash receipt for the sale of an asset (e.g., supplies, office equipment).

Step 5   Enter and apply any sales returns.

Step 6   Record all of the cash receipts using the **Receipts** option.

Purchases and Cash Payments

Step 7   Enter the purchases on account using the **Purchases/Receive Inventory** option.

 TIP: Always verify the *GL Account* field when you enter a purchases on account transaction.

Step 8   Record and apply any purchases returns.

Step 9   Process all of the cash payments with the **Payments** option.

 TIP: Remember to use the **Payments** option to record any bank fees and charges.

Peachtree Guide

General Journal

Step 10 Use the **General Journal Entry** option to record the error discovered on May 26.

TIP: You can use the , and buttons to edit a multi-line general journal entry.

Step 11 Record the employer's payroll taxes using the **General Journal Entry** option.

Printing Reports and Proofing Your Work

Step 12 Print the following reports: General Journal, Purchases Journal, Cash Disbursements Journal, Sales Journal, and Cash Receipts Journal.

TIP: Double-click a report title in the Select a Report window to go directly to that report.

Step 13 Proof your work. Print updated reports, if necessary.
Step 14 Print the following reports: General Ledger, Vendor Ledgers, and Customer Ledgers.
Step 15 Print a General Ledger Trial Balance.

Ending the Session

Step 16 Click the **Close Problem** button and select the save option.
Step 17 If your teacher has asked you to check your solution, *select Check my answer to this problem*. Review, print, and close the report.

Analyzing Your Work

Step 18 Click the **Close Problem** button and select the save option.
Step 19 Answer the Analyze questions.

Mini Practice Set 4   (continued)

SALES JOURNAL

	DATE		SALES SLIP NO.	CUSTOMER'S ACCOUNT DEBITED	POST. REF.	SALES CREDIT	SALES TAX PAYABLE CREDIT	ACCOUNTS RECEIVABLE DEBIT	
1	20--								1
2	May	1	602	Sam Lorenzo	✓	1 8 0 00	9 00	1 8 9 00	2
3		3	603	Marianne Martino	✓	9 0 00	4 50	9 4 50	3
4		8	604	Tom Trout	✓	1 2 0 00	6 00	1 2 6 00	4
5		12	605	Mark McCormick	✓	2 0 0 00	1 0 00	2 1 0 00	5
6		13	606	Sue Ellen Scott	✓	3 0 0 00	1 5 00	3 1 5 00	6
7									7
8									8
9									9
10									10
11									11
12									12
13									13
14									14
15									15
16									16
17									17
18									18
19									19
20									20
21									21
22									22
23									23
24									24
25									25
26									26
27									27
28									28
29									29
30									30
31									31
32									32
33									33
34									34
35									35
36									36

Mini Practice Set 4 (continued)

CASH RECEIPTS JOURNAL

	DATE	DOC. NO.	ACCOUNT NAME	POST. REF.	GENERAL CREDIT	SALES CREDIT	SALES TAX PAYABLE CREDIT	ACCOUNTS RECEIVABLE CREDIT	SALES DISCOUNTS DEBIT	CASH IN BANK DEBIT
1	20--									
2	May 3	R350	Sue Ellen Scott	✓				35000	700	34300
3	5	R351	Sam Lorenzo	✓				30000	600	29400
4	7	T20	Cash Sales	—		210000	10500			220500
5	7	T20	Bankcard Sales	—		186000	9300			195300
6	8	R352	Marianne Martino	✓				37500		37500
7	12	R353	Tom Trout	✓				22500		22500
8	14	R354	Sam Lorenzo	✓				18900		18900
9	15	R355	Mark McCormick	✓				25000		25000
10	15	T21	Cash Sales	—		194000	9700			203700
11	15	T21	Bankcard Sales	—		166000	8300			174300
12										
13										
14										
15										
16										
17										
18										
19										
20										
21										
22										
23										
24										
25										
26										

Mini Practice Set 4   (continued)

PURCHASES JOURNAL

	DATE	INVOICE NO.	CREDITOR'S ACCOUNT CREDITED	POST. REF.	ACCOUNTS PAYABLE CREDIT	PURCHASES DEBIT	GENERAL ACCOUNT DEBITED	POST. REF.	GENERAL DEBIT	
	20--									
1	May 3	CS60	Computer Systems, Inc.	✓	60000	60000				1
2	4	HT88	Hi-Tech Elec. Outlet	✓	35000	35000				2
3	6	9451	Office Suppliers, Inc.	✓	77000		Office Equipment	155	77000	3
4	8	DW65	Desktop Wholesalers	✓	80000		Office Equipment	155	80000	4
5	10	601	Laser & Ink Jet Products	✓	100000	100000				5
6										6
7										7
8										8
9										9
10										10
11										11
12										12
13										13
14										14
15										15
16										16
17										17
18										18
19										19
20										20
21										21
22										22
23										23
24										24
25										25
26										26

Mini Practice Set 4   (continued)

CASH PAYMENTS JOURNAL

	DATE	DOC. NO.	ACCOUNT NAME	POST. REF.	GENERAL DEBIT	GENERAL CREDIT	ACCOUNTS PAYABLE DEBIT	PURCHASES DISCOUNTS CREDIT	CASH IN BANK CREDIT	
1	20--									1
2	May 1	887	Utilities Expense	635	12500				12500	2
3	2	888	Desktop Wholesalers	✓			90000		90000	3
4	4	889	Laser & Ink Jet Products	✓			75000	1500	73500	4
5	7	890	Office Suppliers, Inc.	✓			130000		130000	5
6	10	891	Transportation In	505	17500				17500	6
7										7
8										8
9										9
10										10
11										11
12										12
13										13
14										14
15										15
16										16
17										17
18										18
19										19
20										20
21										21
22										22
23										23
24										24
25										25
26										26
27										27
28										28

Mini Practice Set 4 (continued)

GENERAL JOURNAL

	DATE		DESCRIPTION	POST. REF.	DEBIT	CREDIT	
1	20--						1
2	May	5	Purchases	501	150000		2
3			Merchandise Inventory	110		150000	3
4			Memo 25				4
5							5
6							6
7							7
8							8
9							9
10							10
11							11
12							12
13							13
14							14
15							15
16							16
17							17
18							18
19							19
20							20
21							21
22							22
23							23
24							24
25							25
26							26
27							27
28							28
29							29
30							30
31							31
32							32
33							33
34							34
35							35
36							36
37							37
38							38

Mini Practice Set 4   (continued)

ACCOUNTS RECEIVABLE SUBSIDIARY LEDGER

Name *Sam Lorenzo*

Address *362 Oceanview, Miami, FL 33101*

DATE		DESCRIPTION	POST. REF.	DEBIT	CREDIT	BALANCE
20--						
May	1	Balance	✓			300 00
	1		S18	189 00		489 00
	5		CR15		300 00	189 00
	14		CR15		189 00	———

Name *Mark McCormick*

Address *14 Garden Place, Clearwater, FL 34618*

DATE		DESCRIPTION	POST. REF.	DEBIT	CREDIT	BALANCE
20--						
May	1	Balance	✓			250 00
	12		S18	210 00		460 00
	15		CR15		250 00	210 00

Name *Marianne Martino*

Address *92 Stafford Court, Fort Lauderdale, FL 33310*

DATE		DESCRIPTION	POST. REF.	DEBIT	CREDIT	BALANCE
20--						
May	1	Balance	✓			375 00
	3		S18	94 50		469 50
	8		CR15		375 00	94 50

Mini Practice Set 4   (continued)

Name *Sue Ellen Scott*

Address *302 Palm Drive, Jacksonville, FL 32203*

DATE		DESCRIPTION	POST. REF.	DEBIT	CREDIT	BALANCE
20--						
May	1	Balance	✓			3 5 0 00
	3		CR15		3 5 0 00	—
	13		S18	3 1 5 00		3 1 5 00

Name *Tom Trout*

Address *16 Del Mar, Boca Raton, FL 33431*

DATE		DESCRIPTION	POST. REF.	DEBIT	CREDIT	BALANCE
20--						
May	1	Balance	✓			2 2 5 00
	8		S18	1 2 6 00		3 5 1 00
	12		CR15		2 2 5 00	1 2 6 00

Mini Practice Set 4 (continued)

ACCOUNTS PAYABLE SUBSIDIARY LEDGER

Name _Computer Systems, Inc._

Address _Six Gulf Place, Hialeah, FL 33010_

DATE		DESCRIPTION	POST. REF.	DEBIT	CREDIT	BALANCE
20--						
May	1	Balance	✓			1 2000 00
	3		P12		6000 00	1 8000 00

Name _Desktop Wholesalers_

Address _Three Surfside, Palm Springs, FL 33460_

DATE		DESCRIPTION	POST. REF.	DEBIT	CREDIT	BALANCE
20--						
May	1	Balance	✓			9000 00
	2		CP14	9000 00		—
	8		P12		8000 00	8000 00

Name _Hi-Tech Electronics Outlet_

Address _Quadrangle Complex, Orlando, FL 32802_

DATE		DESCRIPTION	POST. REF.	DEBIT	CREDIT	BALANCE
20--						
May	1	Balance	✓			1 4000 00
	4		P12		3500 00	1 7500 00

Mini Practice Set 4   (continued)

Name *Laser & Ink Jet Products*

Address *32 Cypress Blvd., Tampa, FL 33602*

DATE		DESCRIPTION	POST. REF.	DEBIT	CREDIT	BALANCE
20--						
May	1	Balance	✓			75000
	4		CP14	75000		———
	10		P12		100000	100000

Name *Office Suppliers, Inc.*

Address *56 Sunset Blvd., Panama City, FL 32401*

DATE		DESCRIPTION	POST. REF.	DEBIT	CREDIT	BALANCE
20--						
May	1	Balance	✓			130000
	6		P12		77000	207000
	7		CP14	130000		77000

Mini Practice Set 4   (continued)

GENERAL LEDGER

ACCOUNT __Cash in Bank__   ACCOUNT NO. __101__

DATE		DESCRIPTION	POST. REF.	DEBIT	CREDIT	BALANCE DEBIT	BALANCE CREDIT
20--							
May	1	Balance	✓			7 5 0 0 00	

ACCOUNT __Accounts Receivable__   ACCOUNT NO. __105__

DATE		DESCRIPTION	POST. REF.	DEBIT	CREDIT	BALANCE DEBIT	BALANCE CREDIT
20--							
May	1	Balance	✓			1 5 0 0 00	

ACCOUNT __Merchandise Inventory__   ACCOUNT NO. __110__

DATE		DESCRIPTION	POST. REF.	DEBIT	CREDIT	BALANCE DEBIT	BALANCE CREDIT
20--							
May	1	Balance	✓			5 7 9 4 9 00	
	5		G7		1 5 0 0 00	5 6 4 4 9 00	

ACCOUNT __Supplies__   ACCOUNT NO. __115__

DATE		DESCRIPTION	POST. REF.	DEBIT	CREDIT	BALANCE DEBIT	BALANCE CREDIT
20--							
May	1	Balance	✓			5 0 0 00	

ACCOUNT __Prepaid Insurance__   ACCOUNT NO. __120__

DATE		DESCRIPTION	POST. REF.	DEBIT	CREDIT	BALANCE DEBIT	BALANCE CREDIT
20--							
May	1	Balance	✓			3 0 0 00	

Mini Practice Set 4 (continued)

ACCOUNT __Store Equipment__ ACCOUNT NO. __150__

DATE		DESCRIPTION	POST. REF.	DEBIT	CREDIT	BALANCE DEBIT	BALANCE CREDIT
20--							
May	1	Balance	✓			1300000	

ACCOUNT __Office Equipment__ ACCOUNT NO. __155__

DATE		DESCRIPTION	POST. REF.	DEBIT	CREDIT	BALANCE DEBIT	BALANCE CREDIT
20--							
May	1	Balance	✓			320000	
	6		P12	77000		397000	
	8		P12	80000		477000	

ACCOUNT __Accounts Payable__ ACCOUNT NO. __201__

DATE		DESCRIPTION	POST. REF.	DEBIT	CREDIT	BALANCE DEBIT	BALANCE CREDIT
20--							
May	1	Balance	✓				555000

ACCOUNT __Sales Tax Payable__ ACCOUNT NO. __205__

DATE		DESCRIPTION	POST. REF.	DEBIT	CREDIT	BALANCE DEBIT	BALANCE CREDIT
20--							
May	1	Balance	✓				41200

ACCOUNT __Employees' Federal Income Tax Payable__ ACCOUNT NO. __210__

DATE		DESCRIPTION	POST. REF.	DEBIT	CREDIT	BALANCE DEBIT	BALANCE CREDIT
20--							
May	1	Balance	✓				150000

Mini Practice Set 4   (continued)

ACCOUNT   *Employees' State Income Tax Payable*   ACCOUNT NO.   *211*

DATE		DESCRIPTION	POST. REF.	DEBIT	CREDIT	BALANCE DEBIT	BALANCE CREDIT
20--							
May	1	Balance	✓				2 1 5 00

ACCOUNT   *Social Security Tax Payable*   ACCOUNT NO.   *212*

DATE		DESCRIPTION	POST. REF.	DEBIT	CREDIT	BALANCE DEBIT	BALANCE CREDIT
20--							
May	1	Balance	✓				9 7 5 00

ACCOUNT   *Medicare Tax Payable*   ACCOUNT NO.   *213*

DATE		DESCRIPTION	POST. REF.	DEBIT	CREDIT	BALANCE DEBIT	BALANCE CREDIT
20--							
May	1	Balance	✓				1 8 0 00

ACCOUNT   *Federal Unemployment Tax Payable*   ACCOUNT NO.   *214*

DATE		DESCRIPTION	POST. REF.	DEBIT	CREDIT	BALANCE DEBIT	BALANCE CREDIT
20--							
May	1	Balance	✓				9 5 00

ACCOUNT   *State Unemployment Tax Payable*   ACCOUNT NO.   *215*

DATE		DESCRIPTION	POST. REF.	DEBIT	CREDIT	BALANCE DEBIT	BALANCE CREDIT
20--							
May	1	Balance	✓				1 3 0 00

Mini Practice Set 4   (continued)

ACCOUNT __Capital Stock__   ACCOUNT NO. ___301___

DATE	DESCRIPTION	POST. REF.	DEBIT	CREDIT	BALANCE DEBIT	BALANCE CREDIT
20--						
May 1	Balance	✓				3 500 000

ACCOUNT __Retained Earnings__   ACCOUNT NO. ___302___

DATE	DESCRIPTION	POST. REF.	DEBIT	CREDIT	BALANCE DEBIT	BALANCE CREDIT
20--						
May 1	Balance	✓				926 000

ACCOUNT __Income Summary__   ACCOUNT NO. ___303___

DATE	DESCRIPTION	POST. REF.	DEBIT	CREDIT	BALANCE DEBIT	BALANCE CREDIT

ACCOUNT __Sales__   ACCOUNT NO. ___401___

DATE	DESCRIPTION	POST. REF.	DEBIT	CREDIT	BALANCE DEBIT	BALANCE CREDIT
20--						
May 1	Balance	✓				6 000 000

ACCOUNT __Sales Discounts__   ACCOUNT NO. ___405___

DATE	DESCRIPTION	POST. REF.	DEBIT	CREDIT	BALANCE DEBIT	BALANCE CREDIT
20--						
May 1	Balance	✓			1 10 00	

Mini Practice Set 4   (continued)

ACCOUNT ___Sales Returns and Allowances___   ACCOUNT NO. ___410___

DATE		DESCRIPTION	POST. REF.	DEBIT	CREDIT	BALANCE DEBIT	BALANCE CREDIT
20--							
May	1	Balance	✓			37500	

ACCOUNT ___Purchases___   ACCOUNT NO. ___501___

DATE		DESCRIPTION	POST. REF.	DEBIT	CREDIT	BALANCE DEBIT	BALANCE CREDIT
20--							
May	1	Balance	✓			1200000	
	5		G7	150000		1350000	

ACCOUNT ___Transportation In___   ACCOUNT NO. ___505___

DATE		DESCRIPTION	POST. REF.	DEBIT	CREDIT	BALANCE DEBIT	BALANCE CREDIT
20--							
May	1	Balance	✓			60000	
	10		CP14	17500		77500	

ACCOUNT ___Purchases Discounts___   ACCOUNT NO. ___510___

DATE		DESCRIPTION	POST. REF.	DEBIT	CREDIT	BALANCE DEBIT	BALANCE CREDIT
20--							
May	1	Balance	✓				35000

ACCOUNT ___Purchases Returns and Allowances___   ACCOUNT NO. ___515___

DATE		DESCRIPTION	POST. REF.	DEBIT	CREDIT	BALANCE DEBIT	BALANCE CREDIT
20--							
May	1	Balance	✓				41200

ACCOUNT ___Advertising Expense___   ACCOUNT NO. ___605___

DATE		DESCRIPTION	POST. REF.	DEBIT	CREDIT	BALANCE DEBIT	BALANCE CREDIT
20--							
May	1	Balance	✓			51000	

Mini Practice Set 4 (continued)

ACCOUNT _Bankcard Fees Expense_   ACCOUNT NO. _610_

DATE		DESCRIPTION	POST. REF.	DEBIT	CREDIT	BALANCE DEBIT	BALANCE CREDIT
20--							
May	1	Balance	✓			600 00	

ACCOUNT _Miscellaneous Expense_   ACCOUNT NO. _615_

DATE		DESCRIPTION	POST. REF.	DEBIT	CREDIT	BALANCE DEBIT	BALANCE CREDIT
20--							
May	1	Balance	✓			75 00	

ACCOUNT _Payroll Tax Expense_   ACCOUNT NO. _620_

DATE		DESCRIPTION	POST. REF.	DEBIT	CREDIT	BALANCE DEBIT	BALANCE CREDIT
20--							
May	1	Balance	✓			1380 00	

ACCOUNT _Rent Expense_   ACCOUNT NO. _625_

DATE		DESCRIPTION	POST. REF.	DEBIT	CREDIT	BALANCE DEBIT	BALANCE CREDIT
20--							
May	1	Balance	✓			6000 00	

ACCOUNT _Salaries Expense_   ACCOUNT NO. _630_

DATE		DESCRIPTION	POST. REF.	DEBIT	CREDIT	BALANCE DEBIT	BALANCE CREDIT
20--							
May	1	Balance	✓			8000 00	

ACCOUNT _Utilities Expense_   ACCOUNT NO. _635_

DATE		DESCRIPTION	POST. REF.	DEBIT	CREDIT	BALANCE DEBIT	BALANCE CREDIT
20--							
May	1	Balance	✓			480 00	
	1		CP14	125 00		605 00	

Mini Practice Set 4   (continued)

In-Touch Electronics

Cash Proof

May 31, 20– –

Mini Practice Set 4   (continued)

In-Touch Electronics
Schedule of Accounts Receivable
May 31, 20--

In-Touch Electronics
Schedule of Accounts Payable
May 31, 20--

Mini Practice Set 4 (concluded)

Analyze: 1. _____

2. _____

3. _____

MINI PRACTICE SET 4

In-Touch Electronics

Audit Test

Directions: *Use your completed solutions to answer the following questions. Write the answer in the space to the left of each question.*

_____ **1.** How many accounts receivable customers does the business have?

_____ **2.** How many transactions were recorded in the sales journal for this period?

_____ **3.** What total amount was posted from the Sales Tax Payable credit column of the sales journal to the Sales Tax Payable account?

_____ **4.** What were the totals for debits and credits in the sales journal?

_____ **5.** What account was credited for May 18 transaction?

_____ **6.** What was the total of the Cash in Bank debit column for the cash receipts journal?

_____ **7.** Which account was debited for the second May 21 transaction?

_____ **8.** What was the total of the Purchases Debit column in the purchases journal?

_____ **9.** How many transactions were recorded in the purchases journal?

_____ **10.** How many transactions in the cash payments journal affected the Purchases account?

_____ **11.** How many transactions were recorded in the general journal in the month of May?

_____ **12.** For the payroll entry recorded on May 31, what amount was debited to Payroll Tax Expense?

13. What is the total of all accounts receivable subsidiary ledger accounts at month end?

14. What is the total of all accounts payable subsidiary ledger accounts at month end?

15. What is the balance of Cash in Bank at the end of the month?

16. What is the total of any payroll tax liabilities at May 31?

17. How many accounts are listed on the trial balance?

18. Which account has the largest balance on the trial balance?

19. Which customer owes In-Touch Electronics the most at the end of the month?

20. What is the balance of the Sales account at month end?

CHAPTER 20 Completing the Accounting Cycle for a Merchandising Corporation

Study Plan

Check Your Understanding

Section 1
Read Section 1 on pages 550–554 and complete the following exercises on page 555.
- ❏ Thinking Critically
- ❏ Communicating Accounting
- ❏ Problem 20-1 *Identifying Accounts Affected by Closing Entries*

Section 2
Read Section 2 on pages 556–561 and complete the following exercises on page 562.
- ❏ Thinking Critically
- ❏ Analyzing Accounting
- ❏ Problem 20-2 *Analyzing a Source Document*
- ❏ Problem 20-3 *Organizing the Steps in the Accounting Cycle*

Summary
Review the Chapter 20 Summary on page 563 in your textbook.
- ❏ Key Concepts

Review and Activities
Complete the following questions and exercises on pages 564–565 in your textbook.
- ❏ Using Key Terms
- ❏ Understanding Accounting Concepts and Procedures
- ❏ Case Study
- ❏ Conducting an Audit with Alex
- ❏ Internet Connection
- ❏ Workplace Skills

Computerized Accounting
Read the Computerized Accounting information on page 566 in your textbook.
- ❏ Making the Transition from a Manual to a Computerized System
- ❏ Closing the Accounting Period in Peachtree

Problems
Complete the following end-of-chapter problems for Chapter 20 in your textbook.
- ❏ Problem 20-4 *Journalizing Closing Entries*
- ❏ Problem 20-5 *Journalizing and Posting Closing Entries*
- ❏ Problem 20-6 *Identifying Accounts for Closing Entries*
- ❏ Problem 20-7 *Completing End-of-Period Activities*

Challenge Problem
- ❏ Problem 20-8 *Preparing Adjusting and Closing Entries*

Chapter Reviews and Working Papers
Complete the following exercises for Chapter 20 in your Chapter Reviews and Working Papers.
- ❏ Chapter Review
- ❏ Self-Test

CHAPTER 20 REVIEW
Completing the Accounting Cycle for a Merchandising Corporation

Part 1 Accounting Vocabulary (6 points)

Directions: *Using the terms from the following list, complete the sentences below. Write the letter of the term you have chosen in the space provided.*

Total Points **26**
Student's Score

A. adjusting entries	**D.** post-closing trial balance	**F.** "Closing Entries"
B. closing entries	**E.** temporary accounts	**G.** "Adjusting Entries"
C. permanent accounts		

___G___ **0.** _____ is written in the Description column of the general journal before recording the entries that update the general ledger accounts.

_____ **1.** After the closing entries have been posted, there will be zero balances in the _____.

_____ **2.** To transfer the balances of temporary accounts to a permanent account, _____ are prepared.

_____ **3.** _____ update the general ledger accounts at the end of a period.

_____ **4.** The report prepared at the end of the period to test the equality of the general ledger after all adjusting and closing entries have been posted is the _____.

_____ **5.** Once the post-closing trial balance is prepared, the _____ should be in balance.

_____ **6.** _____ is written in the Description column of the general journal before recording the entries that bring all temporary accounts to zero balances.

Part 2 Rules for Adjusting and Closing Entries (7 points)

Directions: *Read each of the following statements to determine whether the statement is true or false. Write your answer in the space provided.*

___False___ **0.** The work sheet adjustments update the general ledger accounts.

_____ **1.** The sum of all the debit balances of the contra revenue, cost of merchandise, and expense accounts is the amount credited to Income Summary when recording the closing entry.

_____ **2.** The basic steps in the accounting cycle are not the same for all businesses.

_____ **3.** The account balances that appear on the post-closing trial balance are the same as those on the balance sheet.

_____ **4.** Four closing entries are required to close the temporary accounts for a merchandising business organized as a corporation.

_____ **5.** The source of information for the closing entries is the Balance Sheet section of the work sheet.

_____ **6.** After the closing entries have been posted, all of the permanent accounts will have zero balances.

_____ **7.** The temporary revenue account is closed by crediting it for its balance.

Part 3   The End-of-Period Steps in the Accounting Cycle (4 points)

Directions: *For each of the following statements, select the answer that best completes the sentence. Write your answer in the space provided.*

    __C__   **0.** The source of information for the adjusting entries is the
         (A) Adjusted Trial Balance section of the work sheet.
         (B) Income Statement section of the work sheet.
         (C) Adjustments section of the work sheet.
         (D) Trial Balance section of the work sheet.

    _____   **1.** Which of the following accounts is not a temporary account?
         (A) Rent Expense   (C) Sales Discounts
         (B) Cash in Bank   (D) Purchases Returns and Allowances

    _____   **2.** Which of the following accounts would not be affected by a closing entry?
         (A) Capital Stock   (C) Retained Earnings
         (B) Sales   (D) Income Summary

    _____   **3.** For a corporation, the balance of the Income Summary account is closed into
         (A) Capital Stock.   (C) Sales.
         (B) Cash in Bank.   (D) Retained Earnings.

    _____   **4.** The first closing entry is made to close the
         (A) contra revenue, cost of merchandise, and expense accounts with debit balances into Income Summary.
         (B) temporary revenue and contra cost of merchandise accounts with credit balances into Income Summary.
         (C) balance of the Withdrawals account into the Retained Earnings account.
         (D) balance of the Income Summary account into the Retained Earnings account.

Part 4   Accounting Cycle for a Merchandising Business (9 points)

Directions: *In the space provided before the explanation of each step of the accounting cycle, write the letter of the step that matches the explanation.*

A. Prepare trial balance	**F.** Prepare financial statements
B. Analyze transactions	**G.** Prepare post-closing trial balance
C. Complete the work sheet	**H.** Journalize and post adjusting entries
D. Journalize business transactions	**I.** Journalize and post closing entries
E. Collect and verify data from business	**J.** Post to general and subsidiary ledgers

    __E__   **0.** Receive and check the source documents for verification and accuracy.
    _____   **1.** Examine the source documents to determine the accounts to be debited and credited.
    _____   **2.** Record the information from source documents in the appropriate journals.
    _____   **3.** Transfer the information in a journal entry to an individual account.
    _____   **4.** Prove the equality of total debits and credits in the general ledger before making adjustments.
    _____   **5.** Gather information for use in preparing the end-of-period financial statements and journal entries.
    _____   **6.** Report the changes that have taken place during the period and the financial condition of the business at the end of the period.
    _____   **7.** Update the general ledger accounts at the end of a period.
    _____   **8.** Transfer the temporary account balances to a permanent account.
    _____   **9.** Prove that the permanent general ledger accounts are in balance at the close of the accounting period.

Working Papers *for Section Problems*

Problem 20-1 Identifying Accounts Affected by Closing Entries

Account	Is the account affected by a closing entry?	During closing, is the account debited or credited?	During closing, is Income Summary debited or credited?
Accounts Receivable			
Bankcard Fees Expense			
Capital Stock			
Cash in Bank			
Equipment			
Federal Corporate Income Tax Expense			
Federal Corporate Income Tax Payable			
Income Summary			
Insurance Expense			
Merchandise Inventory			
Miscellaneous Expense			
Prepaid Insurance			
Purchases			
Purchases Discounts			
Purchases Returns and Allowances			
Retained Earnings			
Sales			
Sales Discounts			
Sales Returns and Allowances			
Sales Tax Payable			
Supplies			
Supplies Expense			
Transportation In			
Utilities Expense			

Problem 20-2   Analyzing a Source Document

CASH PAYMENTS JOURNAL

PAGE _____

DATE	DOC. NO.	ACCOUNT NAME	POST. REF.	GENERAL DEBIT	GENERAL CREDIT	ACCOUNTS PAYABLE DEBIT	PURCHASES DISCOUNTS CREDIT	CASH IN BANK CREDIT	
									1
									2
									3
									4

Your Backpack Inc.
29000 White Road
Cold Springs, TX 77282-4513

MEMORANDUM 42

TO: Robert Chan, Chief Accountant
FROM: James Perkins, President
DATE: July 12, 20––
SUBJECT: New Storage Facility Rent

Would you please make a check out to Warehouse Inc. for $750. The check is for the new storage facility we are renting. Please mail the check to:

   Mr. James Skiller, Controller
   Warehouse Inc.
   7576 County Line Highway
   Crossplains, TX 77361-8411

Problem 20-3 Organizing the Steps in the
Accounting Cycle

1. _____

2. _____

3. _____

4. _____

5. _____

6. _____

7. _____

8. _____

9. _____

10. _____

Computerized Accounting Using Peachtree

Software Objectives

When you have completed this chapter, you will be able to use Peachtree to:

1. Perform the year-end closing for a corporation.
2. Print a post-closing trial balance.

Problem 20-4   Journalizing Closing Entries

INSTRUCTIONS

Beginning a Session

- **Step 1**   Select the problem set: Sunset Surfwear (Prob. 20-4).
- **Step 2**   Rename the company and set the system date to December 31, 2008.

Completing the Accounting Problem

- **Step 3**   Review the information in your textbook.
- **Step 4**   Since Peachtree records closing entries automatically, choose **System** from the *Tasks* menu and then select **Year-End Wizard** to close the fiscal year for Sunset Surfwear.

 TIP: If you need help performing the closing process, refer to the instructions in Problem 10-4 (page 197). The closing process for a merchandising corporation is the same as the procedure to close a sole proprietorship.

- **Step 5**   Print a Post-Closing Trial Balance.
- **Step 6**   Answer the Analyze question.

Ending the Session

- **Step 7**   Click the **Close Problem** button in the Glencoe Smart Guide window.

Problem 20-5   Journalizing and Posting Closing Entries

INSTRUCTIONS

Beginning a Session

- **Step 1**   Select the problem set: Shutterbug Cameras (Prob. 20-5).
- **Step 2**   Rename the company and set the system date to December 31, 2008.

DO YOU HAVE A QUESTION ?

Q. *Are the steps to close a sole proprietorship any different than those to close a merchandising corporation?*

A. The steps to close a merchandising corporation are the same as those to close a sole proprietorship. Peachtree automatically closes all of the temporary accounts to a permanent account. In the case of a corporation, Peachtree closes all of the temporary accounts to the **Retained Earnings** account.

Notes

Peachtree requires you to make a backup before performing the year-end closing.

Notes

Print the General Ledger Trial Balance whenever you are instructed to print a post-closing trial balance. You can change the name of the report if you want to rename it.

Completing the Accounting Problem

Step 3   Review the information in your textbook.

Step 4   Choose **System** from the *Tasks* menu and then select **Year-End Wizard** to close the fiscal year.

Step 5   Print a Post-Closing Trial Balance.

Step 6   Answer the Analyze question.

Ending the Session

Step 7   Click the **Close Problem** button in the Glencoe Smart Guide window.

Notes

Peachtree automatically creates the closing entries and posts them to the general ledger accounts when you close the fiscal year.

Problem 20-6   Identifying Accounts for Closing Entries

INSTRUCTIONS

Beginning a Session

Step 1   Select the problem set: Shutterbug Cameras (Prob. 20-6).

Step 2   Rename the company and set the system date to December 31, 2008.

Completing the Accounting Problem

Step 3   Review the information in your textbook.

Step 4   Print a Chart of Accounts report. (Do **not** use the General Ledger accounts in your text.) List or highlight all the accounts that will be debited when closed. Next, list all the accounts that will be credited when closed.

Step 5   Answer the Analyze question.

Ending the Session

Step 6   Click the **Close Problem** button in the Glencoe Smart Guide window.

Problem 20-8   Preparing Adjusting and Closing Entries

INSTRUCTIONS

Beginning a Session

Step 1   Select the problem set: Buzz Newsstand (Prob. 20-8).

Step 2   Rename the company and set the system date to December 31, 2008.

Completing the Accounting Problem

Step 3   Review the information in your textbook.

Step 4   Print a Working Trial Balance report and use it to prepare the adjustments.

Step 5   Record the adjusting entries.

Step 6   Print a General Journal report and proof your work.

Step 7   Answer the Analyze question.

Step 8   Click the **Save Pre-closing Balances** button in the Glencoe Smart Guide window.

Step 9   Close the fiscal year.

Step 10   Print a Post-Closing Trial Balance.

Ending the Session

Step 11   Click the **Close Problem** button in the Glencoe Smart Guide window.

Computerized Accounting Using Spreadsheets

Problem 20-7   Completing End-of-Period Activities

Completing the Spreadsheet

Step 1 Read the instructions for Problem 20-7 in your textbook. This problem involves preparing a ten-column work sheet and the end-of-period financial statements for River's Edge Canoe & Kayak.

Step 2 Open the Glencoe Accounting: Electronic Learning Center software.

Step 3 From the Program Menu, click on the **Peachtree Complete® Accounting Software and Spreadsheet Applications** icon.

Step 4 Log onto the Management System by typing your user name and password.

Step 5 Under the **Problems & Tutorials** tab, select template 20-7 from the Chapter 20 drop-down menu. The template should look like the one shown below.

```
PROBLEM 20-7
COMPLETING END-OF-PERIOD ACTIVITIES

(name)
(date)


RIVER'S EDGE CANOE & KAYAK
WORK SHEET
FOR THE YEAR ENDED DECEMBER 31, 20--

ACCOUNT NUMBER       ACCOUNT NAME              TRIAL BALANCE              ADJUSTMENTS        > <        BALANCE SHEET
                                           DEBIT        CREDIT      DEBIT        CREDIT  > <   DEBIT          CREDIT
     101        Cash in Bank               22,236.57                                     > <  22,236.57
     115        Accounts Receivable         7,400.00                                     > <   7,400.00
     130        Merchandise Inventory      25,000.00                            AMOUNT   > <       0.00
     135        Supplies                    4,100.00                            AMOUNT   > <       0.00
     140        Prepaid Insurance           3,000.00                            AMOUNT   > <       0.00
     145        Delivery Truck             67,900.00                                     > <  67,900.00
     201        Accounts Payable                        13,000.00                        > <                 13,000.00
     210        Federal Corp. Income Tax Payable          ------            AMOUNT        > <                      0.00
     215        Sales Tax Payable                          526.57                         > <                    526.57
     301        Capital Stock                           40,000.00                        > <                 40,000.00
     305        Retained Earnings                       25,400.00                        > <                 25,400.00
     310        Income Summary               ------      ------   AMOUNT                  > <
     401        Sales                                  175,000.00                        > <
     405        Sales Discounts             3,775.00                                     > <
     410        Sales Returns & Allowances  2,500.00                                     > <
     501        Purchases                  75,300.00                                     > <
     505        Transportation In           5,000.00                                     > <
     510        Purchases Discounts                      2,300.00                        > <
     515        Purchases Returns & Allowances           5,600.00                        > <
     605        Bankcard Fees Expense       3,515.00                                     > <
     625        Federal Corporate Income Tax Expense  2,940.00             AMOUNT         > <
     635        Insurance Expense            ------                        AMOUNT         > <
     655        Miscellaneous Expense       5,960.00                                     > <
     660        Rent Expense                9,000.00                                     > <
     665        Salaries Expense           22,200.00                                     > <
     670        Supplies Expense             ------                        AMOUNT         > <
     680        Utilities Expense           2,000.00                                     > <
                                          261,826.57  261,826.57     0.00        0.00    > <  97,536.57      78,926.57
                Net Income                                                                > <                 53,650.00
                                                                                         > <  97,536.57     132,576.57
```

Step 6 Key your name and today's date in the cells containing the *(name)* and *(date)* placeholders.

Step 7 The trial balance amounts in the work sheet are given for you. The first adjustment that must be made is to adjust beginning merchandise inventory of $25,000 to an ending balance of $20,000. To make this adjustment, you must debit Income Summary and credit Merchandise Inventory for the difference between the beginning and ending merchandise inventory amounts. Enter the Income Summary adjustment in cell E25 and the Merchandise Inventory adjustment in cell F16. Remember, it is not necessary to include a comma or the decimal point and two zeroes as part of the amount. Notice that, as you enter the adjustments, the balances for the affected accounts in the adjusted trial balance change accordingly.

Step 8 Enter the remaining adjustments into the Adjustments section of the work sheet. When you have entered all of the adjustments, move the cell pointer into the Adjusted Trial Balance, Income Statement, and Balance Sheet sections of the work sheet. Notice that the amounts for the Adjusted Trial Balance, Income Statement, and Balance Sheet are automatically entered. The program also calculates the column totals and the net income for River's Edge Canoe & Kayak.

Step 9 Now scroll down below the work sheet and look at the income statement, statement of retained earnings, and balance sheet for River's Edge Canoe & Kayak. Notice the financial statements are already completed. This is because the spreadsheet template includes formulas that automatically pull information from the filled-in work sheet to complete the financial statements.

Step 10 Now scroll down below the balance sheet and complete the closing entries in the general journal. The account names and posting references are given for you.

Step 11 Scroll down below the closing entries and look at the post-closing trial balance. The amounts are automatically calculated using formulas.

Step 12 Save the spreadsheet using the **Save** option from the *File* menu. You should accept the default location for the save as this is handled by the management system.

Step 13 Print the completed spreadsheet.

TIP: If your spreadsheet is too wide to fit on an 8.5-inch wide piece of paper, you can change your print settings to print the worksheet *landscape*. Landscape means that the worksheet will be printed broadside on the page. Some spreadsheet applications also allow you to choose a "print to page" option. This function will reduce the width and/or depth of the worksheet to fit on one page.

TIP: When printing a long spreadsheet with multiple parts, you may want to insert page breaks between the sections so that each one begins printing at the top of a new page. Page breaks have already been entered into this spreadsheet template. Check your program's Help file for instructions on how to enter page breaks.

Step 14 Exit the spreadsheet program.

Step 15 In the Close Options box, select the location where you would like to save your work.

Step 16 Answer the Analyze question from your textbook for this problem.

What-If Analysis

If unexpired insurance on December 31 were $2,500, what adjustments would be made? What would net income be? How would this affect ending Retained Earnings?

Working Papers *for End-of-Chapter Problems*

Problem 20-4   Journalizing Closing Entries

GENERAL JOURNAL   PAGE _____

	DATE	DESCRIPTION	POST. REF.	DEBIT	CREDIT	
1						1
2						2
3						3
4						4
5						5
6						6
7						7
8						8
9						9
10						10
11						11
12						12
13						13
14						14
15						15
16						16
17						17
18						18
19						19
20						20
21						21
22						22
23						23
24						24
25						25
26						26
27						27
28						28
29						29
30						30

Analyze: _____

_____

Problem 20-5   Journalizing and Posting Closing Entries

(1)

GENERAL JOURNAL   PAGE _____

	DATE	DESCRIPTION	POST. REF.	DEBIT	CREDIT	
1						1
2						2
3						3
4						4
5						5
6						6
7						7
8						8
9						9
10						10
11						11
12						12
13						13
14						14
15						15
16						16
17						17

(2)

GENERAL LEDGER (PARTIAL)

ACCOUNT _Retained Earnings_   ACCOUNT NO. ___305___

DATE		DESCRIPTION	POST. REF.	DEBIT	CREDIT	BALANCE DEBIT	BALANCE CREDIT
20--							
Dec.	1	Balance	✓				20 41 00 00

ACCOUNT _Income Summary_   ACCOUNT NO. ___310___

DATE		DESCRIPTION	POST. REF.	DEBIT	CREDIT	BALANCE DEBIT	BALANCE CREDIT
20--							
Dec.	31	Adjusting Entry	G13	4 00 00 00		4 00 00 00	

Problem 20-5 (continued)

ACCOUNT **Sales**   ACCOUNT NO. **401**

DATE		DESCRIPTION	POST. REF.	DEBIT	CREDIT	BALANCE DEBIT	BALANCE CREDIT
20--							
Dec.	1	Balance	✓				15 000 00

ACCOUNT **Sales Returns and Allowances**   ACCOUNT NO. **410**

DATE		DESCRIPTION	POST. REF.	DEBIT	CREDIT	BALANCE DEBIT	BALANCE CREDIT
20--							
Dec.	1	Balance	✓			5 000 00	

ACCOUNT **Purchases**   ACCOUNT NO. **501**

DATE		DESCRIPTION	POST. REF.	DEBIT	CREDIT	BALANCE DEBIT	BALANCE CREDIT
20--							
Dec.	1	Balance	✓			9 000 00	

ACCOUNT **Transportation In**   ACCOUNT NO. **505**

DATE		DESCRIPTION	POST. REF.	DEBIT	CREDIT	BALANCE DEBIT	BALANCE CREDIT
20--							
Dec.	1	Balance	✓			5 000 00	

ACCOUNT **Purchases Discounts**   ACCOUNT NO. **510**

DATE		DESCRIPTION	POST. REF.	DEBIT	CREDIT	BALANCE DEBIT	BALANCE CREDIT
20--							
Dec.	1	Balance	✓				1 000 00

Problem 20-5   (concluded)

ACCOUNT   *Purchases Returns and Allowances*   ACCOUNT NO.   515

DATE		DESCRIPTION	POST. REF.	DEBIT	CREDIT	BALANCE	
						DEBIT	CREDIT
20--							
Dec.	1	Balance	✓				1 50000

ACCOUNT   *Federal Corporate Income Tax Expense*   ACCOUNT NO.   620

DATE		DESCRIPTION	POST. REF.	DEBIT	CREDIT	BALANCE	
						DEBIT	CREDIT
20--							
Dec.	1	Balance	✓			3 90000	
	31	Adjusting Entry	G13	80000		4 70000	

ACCOUNT   *Miscellaneous Expense*   ACCOUNT NO.   645

DATE		DESCRIPTION	POST. REF.	DEBIT	CREDIT	BALANCE	
						DEBIT	CREDIT
20--							
Dec.	1	Balance	✓			30000	

ACCOUNT   *Rent Expense*   ACCOUNT NO.   650

DATE		DESCRIPTION	POST. REF.	DEBIT	CREDIT	BALANCE	
						DEBIT	CREDIT
20--							
Dec.	1	Balance	✓			6 00000	

ACCOUNT   *Supplies Expense*   ACCOUNT NO.   660

DATE		DESCRIPTION	POST. REF.	DEBIT	CREDIT	BALANCE	
						DEBIT	CREDIT
20--							
Dec.	31	Adjusting Entry	G13	1 63000		1 63000	

ACCOUNT   *Utilities Expense*   ACCOUNT NO.   670

DATE		DESCRIPTION	POST. REF.	DEBIT	CREDIT	BALANCE	
						DEBIT	CREDIT
20--							
Dec.	1	Balance	✓			3 00000	

Analyze: _____

_____

Problem 20-6   Identifying Accounts for Closing Entries

Accounts Debited	**Accounts Credited**
_____	_____
_____	_____
_____	_____
_____	_____
_____	_____
_____	_____
_____	_____
_____	_____
_____	_____
_____	_____

Analyze: _____

_____

Problem 20-7   Completing End-of-Period Activities
(1), (2)

River's Edge

Work

For the Year Ended

	ACCT. NO.	ACCOUNT NAME	TRIAL BALANCE		ADJUSTMENTS	
			DEBIT	CREDIT	DEBIT	CREDIT
1	101	Cash in Bank	2223657			
2	115	Accounts Receivable	740000			
3	130	Merchandise Inventory	2500000			
4	135	Supplies	410000			
5	140	Prepaid Insurance	300000			
6	145	Delivery Truck	6790000			
7	201	Accounts Payable		1300000		
8	210	Fed. Corporate Income Tax Pay.		—		
9	215	Sales Tax Payable		52657		
10	301	Capital Stock		4000000		
11	305	Retained Earnings		2540000		
12	310	Income Summary	—	—		
13	401	Sales		17500000		
14	405	Sales Discounts	377500			
15	410	Sales Returns and Allowances	250000			
16	501	Purchases	7530000			
17	505	Transportation In	500000			
18	510	Purchases Discounts		230000		
19	515	Purchases Returns and Allow.		560000		
20	605	Bankcard Fees Expense	351500			
21	625	Fed. Corporate Income Tax Exp.	294000			
22	635	Insurance Expense	—			
23	655	Miscellaneous Expense	596000			
24	660	Rent Expense	900000			
25	665	Salaries Expense	2220000			
26	670	Supplies Expense	—			
27	680	Utilities Expense	200000			
28						
29						
30						
31						
32						

Canoe & Kayak _____

Sheet _____

December 31, 20– –

ADJUSTED TRIAL BALANCE		INCOME STATEMENT		BALANCE SHEET		
DEBIT	CREDIT	DEBIT	CREDIT	DEBIT	CREDIT	
						1
						2
						3
						4
						5
						6
						7
						8
						9
						10
						11
						12
						13
						14
						15
						16
						17
						18
						19
						20
						21
						22
						23
						24
						25
						26
						27
						28
						29
						30
						31
						32

Problem 20-7   (continued)
(3)

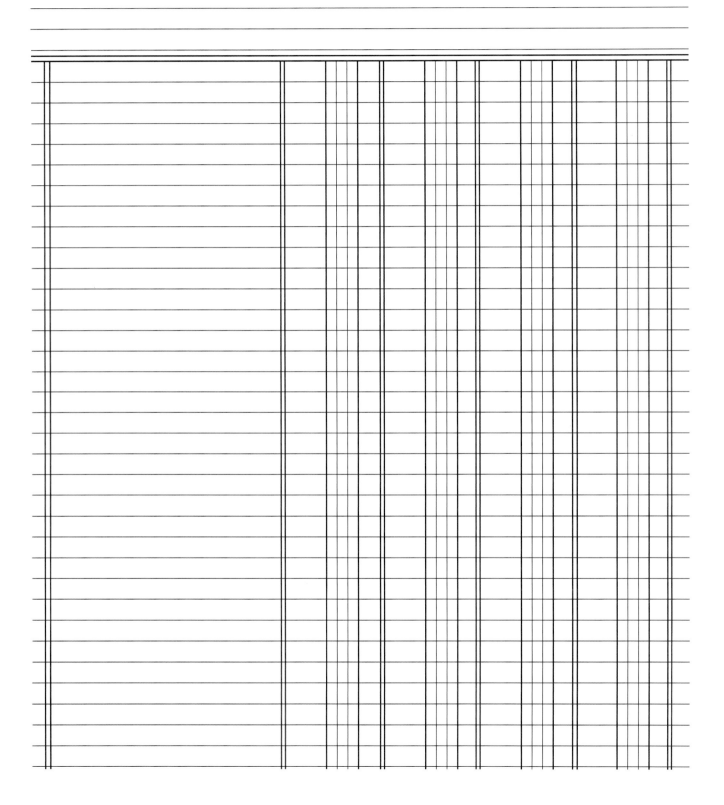

Problem 20-7   (continued)

(4)

(5)

Problem 20-7 (continued)

(6), (7)

GENERAL JOURNAL   PAGE _____

	DATE	DESCRIPTION	POST. REF.	DEBIT	CREDIT	
1						1
2						2
3						3
4						4
5						5
6						6
7						7
8						8
9						9
10						10
11						11
12						12
13						13
14						14
15						15
16						16
17						17
18						18
19						19
20						20
21						21
22						22
23						23
24						24
25						25
26						26
27						27
28						28
29						29
30						30
31						31
32						32
33						33
34						34
35						35
36						36

Problem 20-7 (continued)

(7)

GENERAL LEDGER

ACCOUNT __Cash in Bank__   ACCOUNT NO. __101__

DATE		DESCRIPTION	POST. REF.	DEBIT	CREDIT	BALANCE DEBIT	BALANCE CREDIT
20--							
Dec.	1	Balance	✓			2048297	
	31		CR19	1715170		3763467	
	31		CP22		1539810	2223657	

ACCOUNT __Accounts Receivable__   ACCOUNT NO. __115__

DATE		DESCRIPTION	POST. REF.	DEBIT	CREDIT	BALANCE DEBIT	BALANCE CREDIT
20--							
Dec.	1	Balance	✓			1119438	
	7		G12		17500	1101938	
	31		S17	695162		1797100	
	31		CR19		1057100	740000	

ACCOUNT __Merchandise Inventory__   ACCOUNT NO. __130__

DATE		DESCRIPTION	POST. REF.	DEBIT	CREDIT	BALANCE DEBIT	BALANCE CREDIT
20--							
Dec.	1	Balance	✓			2500000	

ACCOUNT __Supplies__   ACCOUNT NO. __135__

DATE		DESCRIPTION	POST. REF.	DEBIT	CREDIT	BALANCE DEBIT	BALANCE CREDIT
20--							
Dec.	1	Balance	✓			396650	
	10		P16	10600		407250	
	21		P16	2750		410000	

ACCOUNT __Prepaid Insurance__   ACCOUNT NO. __140__

DATE		DESCRIPTION	POST. REF.	DEBIT	CREDIT	BALANCE DEBIT	BALANCE CREDIT
20--							
Dec.	1	Balance	✓			300000	

Problem 20-7 (continued)

ACCOUNT __Delivery Truck__   ACCOUNT NO. __145__

DATE		DESCRIPTION	POST. REF.	DEBIT	CREDIT	BALANCE DEBIT	BALANCE CREDIT
20--							
Dec.	1	Balance	✓			6790000	

ACCOUNT __Accounts Payable__   ACCOUNT NO. __201__

DATE		DESCRIPTION	POST. REF.	DEBIT	CREDIT	BALANCE DEBIT	BALANCE CREDIT
20--							
Dec.	1	Balance	✓				1303050
	19		G12	25600			1277450
	31		P16		943350		2220800
	31		CP22	920800			1300000

ACCOUNT __Federal Corporate Income Tax Payable__   ACCOUNT NO. __210__

DATE		DESCRIPTION	POST. REF.	DEBIT	CREDIT	BALANCE DEBIT	BALANCE CREDIT
20--							

ACCOUNT __Sales Tax Payable__   ACCOUNT NO. __215__

DATE		DESCRIPTION	POST. REF.	DEBIT	CREDIT	BALANCE DEBIT	BALANCE CREDIT
20--							
Dec.	1	Balance	✓				46300
	14		CP22	46300			—
	31		S17		26737		26737
	31		CR19		25920		52657

ACCOUNT __Capital Stock__   ACCOUNT NO. __301__

DATE		DESCRIPTION	POST. REF.	DEBIT	CREDIT	BALANCE DEBIT	BALANCE CREDIT
20--							
Dec.	1	Balance	✓				4000000

Problem 20-7   (continued)

ACCOUNT **Retained Earnings**   ACCOUNT NO. **305**

DATE		DESCRIPTION	POST. REF.	DEBIT	CREDIT	BALANCE DEBIT	BALANCE CREDIT
20--							
Dec.	1	Balance	✓				25400 00

ACCOUNT **Income Summary**   ACCOUNT NO. **310**

DATE		DESCRIPTION	POST. REF.	DEBIT	CREDIT	BALANCE DEBIT	BALANCE CREDIT

ACCOUNT **Sales**   ACCOUNT NO. **401**

DATE		DESCRIPTION	POST. REF.	DEBIT	CREDIT	BALANCE DEBIT	BALANCE CREDIT
20--							
Dec.	1	Balance	✓				161835 75
	31		S17		6684 25		168520 00
	31		CR19		6480 00		175000 00

ACCOUNT **Sales Discounts**   ACCOUNT NO. **405**

DATE		DESCRIPTION	POST. REF.	DEBIT	CREDIT	BALANCE DEBIT	BALANCE CREDIT
20--							
Dec.	1	Balance	✓			3616 50	
	31		CR19	158 50		3775 00	

ACCOUNT **Sales Returns and Allowances**   ACCOUNT NO. **410**

DATE		DESCRIPTION	POST. REF.	DEBIT	CREDIT	BALANCE DEBIT	BALANCE CREDIT
20--							
Dec.	1	Balance	✓			2325 00	
	7		G12	175 00		2500 00	

Problem 20-7 (continued)

ACCOUNT __Purchases__   ACCOUNT NO. __501__

DATE		DESCRIPTION	POST. REF.	DEBIT	CREDIT	BALANCE DEBIT	BALANCE CREDIT
20--							
Dec.	1	Balance	✓			64385 00	
	11		CP22	1615 00		66000 00	
	31		P16	9300 00		75300 00	

ACCOUNT __Transportation In__   ACCOUNT NO. __505__

DATE		DESCRIPTION	POST. REF.	DEBIT	CREDIT	BALANCE DEBIT	BALANCE CREDIT
20--							
Dec.	1	Balance	✓			4949 00	
	21		CP22	51 00		5000 00	

ACCOUNT __Purchases Discounts__   ACCOUNT NO. __510__

DATE		DESCRIPTION	POST. REF.	DEBIT	CREDIT	BALANCE DEBIT	BALANCE CREDIT
20--							
Dec.	1	Balance	✓				2162 00
	31		CP22		138 00		2300 00

ACCOUNT __Purchases Returns and Allowances__   ACCOUNT NO. __515__

DATE		DESCRIPTION	POST. REF.	DEBIT	CREDIT	BALANCE DEBIT	BALANCE CREDIT
20--							
Dec.	1	Balance	✓				5344 00
	19		G12		256 00		5600 00

ACCOUNT __Bankcard Fees Expense__   ACCOUNT NO. __605__

DATE		DESCRIPTION	POST. REF.	DEBIT	CREDIT	BALANCE DEBIT	BALANCE CREDIT
20--							
Dec.	1	Balance	✓			3253 00	
	27		CP22	262 00		3515 00	

Problem 20-7   (continued)

ACCOUNT   *Federal Corporate Income Tax Expense*   ACCOUNT NO.   **625**

DATE		DESCRIPTION	POST. REF.	DEBIT	CREDIT	BALANCE DEBIT	BALANCE CREDIT
20--							
Dec.	1	Balance	✓			2 2 0 5 00	
	15		CP22	7 3 5 00		2 9 4 0 00	

ACCOUNT   *Insurance Expense*   ACCOUNT NO.   **635**

DATE	DESCRIPTION	POST. REF.	DEBIT	CREDIT	BALANCE DEBIT	BALANCE CREDIT

ACCOUNT   *Miscellaneous Expense*   ACCOUNT NO.   **655**

DATE		DESCRIPTION	POST. REF.	DEBIT	CREDIT	BALANCE DEBIT	BALANCE CREDIT
20--							
Dec.	1	Balance	✓			5 5 2 0 00	
	6		CP22	4 2 5 00		5 9 4 5 00	
	27		CP22	1 5 00		5 9 6 0 00	

ACCOUNT   *Rent Expense*   ACCOUNT NO.   **660**

DATE		DESCRIPTION	POST. REF.	DEBIT	CREDIT	BALANCE DEBIT	BALANCE CREDIT
20--							
Dec.	1	Balance	✓			8 2 5 0 00	
	1		CP22	7 5 0 00		9 0 0 0 00	

ACCOUNT   *Salaries Expense*   ACCOUNT NO.   **665**

DATE		DESCRIPTION	POST. REF.	DEBIT	CREDIT	BALANCE DEBIT	BALANCE CREDIT
20--							
Dec.	1	Balance	✓			2 0 3 5 0 00	
	12		CP22	1 8 5 0 00		2 2 2 0 0 00	

Problem 20-7 (concluded)

ACCOUNT **Supplies Expense**   ACCOUNT NO. **670**

DATE	DESCRIPTION	POST. REF.	DEBIT	CREDIT	BALANCE DEBIT	BALANCE CREDIT

ACCOUNT **Utilities Expense**   ACCOUNT NO. **680**

DATE	DESCRIPTION	POST. REF.	DEBIT	CREDIT	BALANCE DEBIT	BALANCE CREDIT
20--						
Dec. 1	Balance	✓			1 8 3 7 90	
9		CP22	5 7 50		1 8 9 5 40	
23		CP22	1 0 4 60		2 0 0 0 00	

(8)

Analyze: _____

Problem 20-8   Preparing Adjusting and Closing Entries
(1)

GENERAL JOURNAL   PAGE _____

	DATE	DESCRIPTION	POST. REF.	DEBIT	CREDIT	
1						1
2						2
3						3
4						4
5						5
6						6
7						7
8						8
9						9
10						10
11						11
12						12
13						13
14						14
15						15
16						16
17						17
18						18
19						19
20						20
21						21
22						22
23						23
24						24
25						25
26						26
27						27
28						28
29						29
30						30
31						31
32						32
33						33
34						34
35						35
36						36

Problem 20-8   (concluded)

(2)

Analyze:

CHAPTER 20   Completing the Accounting Cycle for a Merchandising Corporation

Self-Test

Part A   True or False

Directions: *Circle the letter* T *in the Answer column if the statement is true; circle the letter* F *if the statement is false.*

Answer

T	F	**1.** The information for closing entries can be found on the Income Statement columns of the work sheet.
T	F	**2.** Revenue accounts are closed when they are credited.
T	F	**3.** All temporary accounts are closed into Income Summary.
T	F	**4.** The Retained Earnings account is closed into the Capital account.
T	F	**5.** Adjustments are prepared and posted before closing the books.
T	F	**6.** After closing the books, the Income Summary account has a credit balance.
T	F	**7.** Expense accounts are closed by crediting them.
T	F	**8.** To close Purchases Returns and Allowances and Sales Discounts will require a debit.
T	F	**9.** After closing the ledger, only the permanent accounts have balances.
T	F	**10.** The post-closing trial balance is a list of temporary accounts.

Part B   Matching

Directions: *The steps in the accounting cycle are listed below. Put the list in proper order by placing the correct letter next to the numbers 1–10.*

Answer

1. _____

2. _____

3. _____

4. _____

5. _____

6. _____

7. _____

8. _____

9. _____

10. _____

A. Prepare the trial balance

B. Analyze source documents

C. Complete the work sheet

D. Journalize transactions from source documents

E. Collect and verify source documents

F. Prepare financial statements

G. Prepare post-closing trial balance

H. Journalize and post closing entries

I. Journalize and post adjusting entries

J. Post to the general and subsidiary ledgers

CHAPTER 21   Accounting for Publicly Held Corporations

Study Plan

Check Your Understanding

Section 1	*Read Section 1 on pages 574–577 and complete the following exercises on page 578.* ❑ Thinking Critically ❑ Communicating Accounting ❑ Problem 21-1 *Examining Capital Stock Transactions*
Section 2	*Read Section 2 on pages 579–582 and complete the following exercises on page 583.* ❑ Thinking Critically ❑ Computing in the Business World ❑ Problem 21-2 *Distributing Corporate Earnings* ❑ Problem 21-3 *Analyzing a Source Document*
Section 3	*Read Section 3 on pages 585–587 and complete the following exercises on page 588.* ❑ Thinking Critically ❑ Analyzing Accounting ❑ Problem 21-4 *Examining the Statement of Stockholders' Equity*
Summary	*Review the Chapter 21 Summary on page 589 in your textbook.* ❑ Key Concepts
Review and Activities	*Complete the following questions and exercises on pages 590–591 in your textbook.* ❑ Using Key Terms ❑ Understanding Accounting Concepts and Procedures ❑ Case Study ❑ Conducting an Audit with Alex ❑ Internet Connection ❑ Workplace Skills
Computerized Accounting	*Read the Computerized Accounting information on page 592 in your textbook.* ❑ *Making the Transition from a Manual to a Computerized System* ❑ *Customizing Financial Reports in Peachtree*
Problems	*Complete the following end-of-chapter problems for Chapter 21 in your textbook.* ❑ Problem 21-5 *Distributing Corporate Earnings* ❑ Problem 21-6 *Journalizing the Issue of Stock* ❑ Problem 21-7 *Journalizing Common and Preferred Stock Dividend Transactions* ❑ Problem 21-8 *Preparing Corporate Financial Statements*
Challenge Problem	❑ Problem 21-9 *Recording Stockholders' Equity Transactions*
Chapter Reviews and Working Papers	*Complete the following exercises for Chapter 21 in your Chapter Reviews and Working Papers.* ❑ Chapter Review ❑ Self-Test

CHAPTER 21 REVIEW — Accounting for Publicly Held Corporations

Part 1   Accounting Vocabulary (10 points)

Total Points	43
Student's Score	

Directions: *Using terms from the following list, complete the sentences below. Write the letter of the term you have chosen in the space provided.*

A. authorized capital stock	**E.** dividend	**I.** proxy
B. board of directors	**F.** paid-in capital in excess of par	**J.** publicly held corporation
C. closely held corporation	**G.** par value	**K.** statement of stockholders'
D. common stock	**H.** preferred stock	equity

_____ **K** _____   **0.** The financial statement that reports the changes that have taken place in all of the stockholders' equity accounts during the period is the _____.

_____   **1.** The maximum number of shares of stock that a corporation may issue is called its _____.

_____   **2.** The stock that has certain privileges over common stock is called _____.

_____   **3.** When a corporation issues only one type of stock, the stock is called _____.

_____   **4.** The stockholders' equity account that is used to record the sale of stock at higher than par value is the _____ account.

_____   **5.** A corporation owned by a few persons or by a family and whose stock is not sold to the general public is called a(n) _____.

_____   **6.** A group who governs and is responsible for the affairs of the corporation is called a(n) _____.

_____   **7.** A(n) _____ is one whose stock is widely held, has a large market, and is traded on a stock exchange.

_____   **8.** _____ is the amount assigned to each share of stock and printed as a dollar amount on the stock certificates.

_____   **9.** A(n) _____ is a return on the money invested by the stockholder.

_____   **10.** A document that gives the stockholder's voting rights to someone else when the stockholder cannot attend a stockholders' meeting is called a(n) _____.

Part 2   Analyzing Transactions for a Corporation (10 points)

Directions: *Using the following account titles, analyze the transactions below. Determine the account(s) to be debited and credited. Write your answers in the space provided.*

A. Cash in Bank	**D.** Dividends—Preferred	**G.** $5 Preferred Stock
B. Common Stock	**E.** Dividends Payable—Common	**H.** Paid-in Capital in
C. Dividends—Common	**F.** Dividends Payable—Preferred	Excess of Par

Debit	Credit	
A	*B*	**0.** Issued 10,000 shares of common stock at $6 per share.
_____	_____	**1.** Issued 2,000 shares of $6 par common stock at $7.50 per share.
_____	_____	**2.** Declared a cash dividend of 85¢ per share on the 12,000 shares.
_____	_____	**3.** Issued a check for the payment of the dividend declared.
_____	_____	**4.** Issued 500 shares of $5 preferred stock, $100 par, at $100 per share.
_____	_____	**5.** Declared a cash dividend on the 500 shares of $5 preferred stock issued.

Part 3   Corporations (15 points)

Directions: *Read each of the following statements to determine whether the statement is true or false. Write your answer in the space provided.*

    True   **0.** The Dividends account is increased on the debit side.

    _____   **1.** An ownership certificate is proof of ownership in a corporation and lists the name of the stockholder, the number of shares issued, and the date they were issued.

    _____   **2.** When a dividend is declared, the method used by all corporations is to debit the amount of the dividend directly into the Retained Earnings account.

    _____   **3.** The stated dividend rate for preferred stock is an annual rate.

    _____   **4.** Publicly held corporations cannot prepare a statement of retained earnings and must therefore prepare a statement of stockholders' equity.

    _____   **5.** The usual preference that preferred stockholders have is the right to receive dividends before they are paid to common stockholders.

    _____   **6.** Even though a corporation is authorized to issue two types of stock, only one account is set up to record the two types of stock.

    _____   **7.** Forming a corporation is less costly than forming a sole proprietorship or a partnership.

    _____   **8.** Amounts that decrease account balances are enclosed in parentheses on the statement of stockholders' equity.

    _____   **9.** The board of directors has the duty to hire professional managers to operate the corporation.

    _____   **10.** Both the net income earned and the dividends declared by a corporation increase the Retained Earnings account.

    _____   **11.** A privately held corporation is usually owned by a small family, and a publicly held corporation is usually owned by a very large family.

    _____   **12.** There are more corporations than there are sole proprietorships and partnerships in the United States.

    _____   **13.** A corporation may enter into contracts, borrow money, acquire property, and sue in the courts in the same manner as a person.

    _____   **14.** The only types of information reported on a statement of stockholders' equity are the number of any shares of stock issued and the total amount received for those shares.

    _____   **15.** Organization costs incurred by a corporation when getting started may include attorneys' fees for legal services and payments to promoters to sell stock.

Part 4   Advantages and Disadvantages of Corporations (8 points)

Directions: *In the space provided below, indicate whether the statement expresses an advantage or a disadvantage of a corporation. Place an "A" in the space for an advantage or a "D" for a disadvantage.*

    D   **0.** Filing of numerous reports

    _____   **1.** Limited liability of the owners

    _____   **2.** Close regulation by state and federal governments

    _____   **3.** Risk limited to individual investment

    _____   **4.** Sale of stock between stockholders without approval of other stockholders

    _____   **5.** Continuous existence of the corporation

    _____   **6.** Separate legal entity

Working Papers *for Section Problems*

Problem 21-1   Examining Capital Stock Transactions

1. _____

_____

_____

_____

_____

2. _____

_____

_____

_____

_____

3. _____

_____

_____

_____

_____

_____

Problem 21-2   Distributing Corporate Earnings

1. _____

_____

2. _____

_____

Problem 21-3   Analyzing a Source Document

GENERAL JOURNAL   PAGE _____

	DATE		DESCRIPTION	POST. REF.	DEBIT	CREDIT	
1							1
2							2
3							3
4							4
5							5
6							6
7							7
8							8
9							9
10							10
11							11
12							12
13							13
14							14
15							15
16							16
17							17
18							18
19							19
20							20
21							21
22							22

Problem 21-4   **Examining the Statement of Stockholders' Equity**

Transaction	Reported on Statement of Stockholders' Equity? (Yes/No)
1	
2	
3	
4	
5	
6	
7	
8	

Computerized Accounting Using Peachtree

Software Objectives

When you have completed this chapter, you will be able to use Peachtree to:

1. Record journal entries to record the issue of stock.
2. Record journal entries to record the distribution of earnings to owners.
3. Print a balance sheet for a publicly held corporation.

Problem 21-6   Journalizing the Issue of Stock

INSTRUCTIONS

Beginning a Session

Step 1   Select the problem set: InBeat CD Shop (Prob. 21-6).
Step 2   Rename the company and set the system date to June 30, 2008.

Completing the Accounting Problem

Step 3   Review the information in your textbook.
Step 4   Record all of the transactions using the **General Journal Entry** option.

TIP: To save time, just enter the day of the month when you record the date for a general journal entry.

Step 5   Print a General Journal report and proof your work.

TIP: You can use the report design features to widen the *Transaction Description* column on a General Journal report.

Step 6   Answer the Analyze question.

Ending the Session

Step 7   Click the **Close Problem** button in the Glencoe Smart Guide window.

Mastering Peachtree

Print a Chart of Accounts report. What account types are assigned to the capital accounts? On a separate sheet of paper, discuss the significance of each account type.

DO YOU HAVE A QUESTION

Q. *Do you have to use the General Journal to record the issue of stock?*

A. Although your textbook demonstrates how to use the General Journal to record the issue of stock, you could use the **Receipts** option instead of the **General Journal Entry** option. You can record any transaction involving a cash receipt using the **Receipts** option. For consistency, however, the instructions provided below explain how to use the **General Journal Entry** option.

Peachtree Guide

Problem 21-7   Journalizing Common and Preferred Stock Dividend Transactions

INSTRUCTIONS

Beginning a Session

Step 1 Select the problem set: Shutterbug Cameras (Prob. 21-7).
Step 2 Rename the company and set the system date to December 31, 2008.

Completing the Accounting Problem

Step 3 Review the information in your textbook.
Step 4 Record all of the transactions using the **General Journal Entry** option. Make sure each transaction is entered in the correct accounting period. When entering the October transaction, verify the accounting period in the lower right hand corner is Period 10—10/1/08 to 10/31/08. When entering the November transaction, change the accounting period to Nov. 1, 2008 to Nov. 30, 2008.

TIP: To change the accounting period, select **System** from the *Tasks* menu, and then choose **Change Accounting Period**.

Step 5 Print a General Journal report and proof your work.

TIP: If you need to edit a general journal entry, click the button. If you don't see the entry, click the **Show** drop-down to select the correct period.

NOTE: You must set the date filter option to **This Quarter** to print a General Journal report that includes the entries for the entire quarter.

Step 6 Print a General Ledger report.

NOTE: Change the General Ledger report format to *Summary by Transaction* from the Options Filter tab, and choose *Period 10* through *Period 12* for the time frame. You must set the time frame because you entered transactions across multiple accounting periods.

Step 7 Answer the Analyze question.

Checking Your Work and Ending the Session

Step 8 Verify that the accounting period displayed in the lower right hand corner is Period 12—12/1/08 to 12/31/08. Click the **Close Problem** button in the Glencoe Smart Guide window.
Step 9 If your teacher has asked you to check your solution, select *Check my answer to this problem*. Review, print, and close the report.
Step 10 Click the **Close Problem** button and select a save option.

Problem 21-9   Recording Stockholders' Equity Transactions

INSTRUCTIONS

Beginning a Session

Step 1   Select the problem set: Buzz Newsstand (Prob. 21-9).

Step 2   Rename the company and set the system date to December 31, 2008.

Completing the Accounting Problem

Step 3   Review the information in your textbook.

Step 4   Record all of the transactions using the **General Journal Entry** option. When entering the March transaction, set the accounting period to Mar. 1, 2008 to Mar. 31, 2008. When entering the April transaction, set the accounting period to Apr. 1, 2008 to Apr. 30, 2008, and so forth.

Step 5   Print a General Journal report and proof your work.

TIP: Set the date filter option to print a General Journal report for the period March 1, 2008 to December 31, 2008.

Step 6   Print a Balance Sheet report.

TIP: Choose *Range* for the time frame and select *Period 1* through *Period 12* to print a Balance Sheet for the entire year.

Step 7   Answer the Analyze question.

TIP: Add net income to **Retained Earnings** and subtract the dividends to determine the ending **Retained Earnings** account balance.

Checking Your Work and Ending the Session

Step 8   Verify that the accounting period displayed in the lower right hand corner is Period 11—11/1/08 to 11/30/08. Click the **Close Problem** button in the Glencoe Smart Guide window.

Step 9   If your teacher has asked you to check your solution, select *Check my answer to this problem*. Review, print, and close the report.

Step 10   Click the **Close Problem** button and select a save option.

Mastering Peachtree

Change capital section title on the Balance Sheet to *Stockholders' Equity*. Save the customized report layout and print the revised Balance Sheet.

Peachtree Guide

FAQs

Why doesn't Peachtree show all of the transactions when you click the Open button?

Peachtree shows only the transactions entered in the current period. Suppose you are currently in Period 10—10/1/08 to 10/31/08 and you enter a transaction dated November 15. If you chose to edit a record, this transaction would not appear in the list. Click the **Show** drop-down menu in the Select General Journal Entry window and choose *All Transactions* or select the desired period to display the transactions.

The reports (e.g., General Journal, General Ledger, Balance Sheet, etc.) do not reflect all of the transactions recorded.

Peachtree allows you to enter transactions that occur over multiple accounting periods. Although this does not cause any difficulties when you enter transactions, you could encounter some anomalies when printing reports.

By default, Peachtree uses the default accounting period for most reports. Suppose you are currently in Period 10—10/1/08 to 10/31/08 and you enter transactions on Oct. 15, Nov. 15, and Dec. 15. When you print the General Journal report using the default options, only the October 15 transaction will appear because it is the only transaction for period 10. You must set the date range to include all of the periods in which you entered transactions.

Peachtree displays the error, "00/00/00 is not a valid date in this period."

If the current period is set to Period 10—10/1/08 to 10/31/08, Peachtree will allow you to enter transactions for October, November, and December. However, Peachtree will not permit you to enter a transaction with a date earlier than the current period (e.g., August or September).

If you receive this (or a similar) error message, choose **System** from the *Tasks* menu and then select **Change Accounting Period**. Select the accounting period in which you want to enter a transaction.

Computerized Accounting Using Spreadsheets

Problem 21-8   Preparing Corporate Financial Statements

Completing the Spreadsheet

Step 1   Read the instructions for Problem 21-8 in your textbook. This problem involves preparing a statement of stockholders' equity and a balance sheet.

Step 2   Open the Glencoe Accounting: Electronic Learning Center software.

Step 3   From the Program Menu, click on the **Peachtree Complete®
Accounting Software and Spreadsheet Applications** icon.

Step 4   Log onto the Management System by typing your user name and password.

Step 5   Under the **Problems & Tutorials** tab, select template 21-8 from the Chapter 21 drop-down menu. The template should look like the one shown below.

```
PROBLEM 21-8
PREPARING CORPORATE FINANCIAL STATEMENTS

(name)
(date)

RIVER'S EDGE CANOE & KAYAK
STATEMENT OF STOCKHOLDERS' EQUITY
FOR THE YEAR ENDED DECEMBER 31, 20--

                                    $100 PAR      $5 PAR       PAID-IN
                                    9% PREFERRED  COMMON       CAPITAL IN    RETAINED
                                    STOCK         STOCK        EXCESS OF PAR EARNINGS    TOTAL
Balance, January 1, 20--            AMOUNT        AMOUNT       AMOUNT        AMOUNT       0.00
Stock Issued:
  500 Shares of 9% Preferred at Par AMOUNT                                               0.00
  25,000 Shares of Common at $9                   AMOUNT       AMOUNT                     0.00
Net Income                                                                   AMOUNT       0.00
Cash Dividends:
  Preferred Stock                                                           AMOUNT       0.00
  Common Stock                                                              AMOUNT       0.00
Balance, December 31, 20--          AMOUNT        AMOUNT       AMOUNT        0.00         0.00
```

Step 6   Key your name and today's date in the cells containing the *(name)* and *(date)* placeholders.

Step 7   Enter the missing amounts in the statement of stockholders' equity in the cells containing the AMOUNT placeholders. Remember, it is not necessary to add a comma or the decimal point and ending zeroes. The total for each line will be automatically computed.

TIP: To enter a negative number, precede the amount entered by a minus sign. The spreadsheet will automatically format the number as negative using parentheses.

Step 8   Now scroll down below the statement of stockholders' equity and look at the balance sheet. Note the amounts have already been computed. The spreadsheet automatically pulls information from the statement of stockholders' equity to complete the balance sheet.

Step 9 Save the spreadsheet using the **Save** option from the *File* menu. You should accept the default location for the save as this is handled by the management system.

Step 10 Print the completed spreadsheet.

TIP: If your spreadsheet is too wide to fit on an 8.5-inch wide piece of paper, you can change your print settings to print the worksheet *landscape*. Landscape means that the worksheet will be printed broadside on the page. Some spreadsheet applications also allow you to choose a "print to page" option. This function will reduce the width and/or depth of the worksheet to fit on one page.

Step 11 Exit the spreadsheet program.

Step 12 In the Close Options box, select the location where you would like to save your work.

Step 13 Answer the Analyze question from your textbook for this problem.

What-If Analysis

If net income were $511,000, how would this affect retained earnings?

_____

_____

Working Papers *for End-of-Chapter Problems*

Problem 21-5 Distributing Corporate Earnings

1. _____

2. _____

Analyze: _____

_____

_____

Problem 21-6 Journalizing the Issue of Stock

GENERAL JOURNAL   PAGE _____

	DATE	DESCRIPTION	POST. REF.	DEBIT	CREDIT	
1						1
2						2
3						3
4						4
5						5
6						6
7						7
8						8
9						9
10						10
11						11
12						12
13						13
14						14
15						15
16						16
17						17
18						18
19						19
20						20
21						21
22						22

Analyze: _____

_____

_____

Problem 21-7   **Journalizing Common and Preferred Stock Dividend Transactions**

(1)

GENERAL JOURNAL   PAGE _____

	DATE	DESCRIPTION	POST. REF.	DEBIT	CREDIT	
1						1
2						2
3						3
4						4
5						5
6						6
7						7
8						8
9						9
10						10
11						11
12						12
13						13
14						14
15						15
16						16
17						17
18						18
19						19
20						20
21						21

Problem 21-7   (concluded)

(2)

GENERAL LEDGER (PARTIAL)

ACCOUNT   *Cash in Bank*   ACCOUNT NO.   *101*

DATE	DESCRIPTION	POST. REF.	DEBIT	CREDIT	BALANCE DEBIT	BALANCE CREDIT
20--						
Dec. 1	*Balance*	✓			9 8 6 5 0 00	

ACCOUNT   *Dividends Payable—Preferred*   ACCOUNT NO.   *203*

DATE	DESCRIPTION	POST. REF.	DEBIT	CREDIT	BALANCE DEBIT	BALANCE CREDIT

ACCOUNT   *Dividends Payable—Common*   ACCOUNT NO.   *204*

DATE	DESCRIPTION	POST. REF.	DEBIT	CREDIT	BALANCE DEBIT	BALANCE CREDIT

Analyze: _____

_____

_____

Problem 21-8   Preparing Corporate Financial Statements

(1)

Problem 21-8   (concluded)

(2)

Analyze: _____

_____

Problem 21-9   Source Documents

Instructions: *Use the following source documents to record the transactions for this problem.*

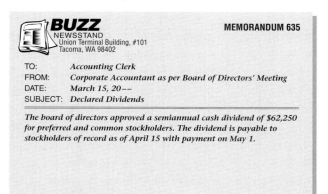

BUZZ NEWSSTAND
Union Terminal Building, #101
Tacoma, WA 98402

MEMORANDUM 635

TO: *Accounting Clerk*
FROM: *Corporate Accountant as per Board of Directors' Meeting*
DATE: *March 15, 20––*
SUBJECT: *Declared Dividends*

The board of directors approved a semiannual cash dividend of $62,250 for preferred and common stockholders. The dividend is payable to stockholders of record as of April 15 with payment on May 1.

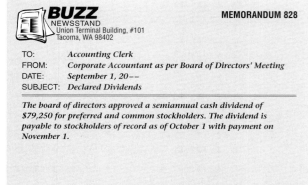

BUZZ NEWSSTAND
Union Terminal Building, #101
Tacoma, WA 98402

MEMORANDUM 828

TO: *Accounting Clerk*
FROM: *Corporate Accountant as per Board of Directors' Meeting*
DATE: *September 1, 20––*
SUBJECT: *Declared Dividends*

The board of directors approved a semiannual cash dividend of $79,250 for preferred and common stockholders. The dividend is payable to stockholders of record as of October 1 with payment on November 1.

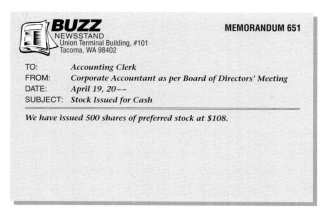

BUZZ NEWSSTAND
Union Terminal Building, #101
Tacoma, WA 98402

MEMORANDUM 651

TO: *Accounting Clerk*
FROM: *Corporate Accountant as per Board of Directors' Meeting*
DATE: *April 19, 20––*
SUBJECT: *Stock Issued for Cash*

We have issued 500 shares of preferred stock at $108.

$ 79,250.00   No. 2451
Date *November 1*   20––
To *Dividends Checking Account*
For *Dividends Payable*

	Dollars	Cents
Balance brought forward	323,908	00
Add deposits		
Total	323,908	00
Less this check	79,250	00
Balance carried forward	244,658	00

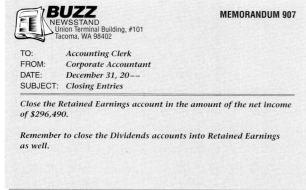

BUZZ NEWSSTAND
Union Terminal Building, #101
Tacoma, WA 98402

MEMORANDUM 907

TO: *Accounting Clerk*
FROM: *Corporate Accountant*
DATE: *December 31, 20––*
SUBJECT: *Closing Entries*

Close the Retained Earnings account in the amount of the net income of $296,490.

Remember to close the Dividends accounts into Retained Earnings as well.

$ 62,250.00   No. 1256
Date *May 1*   20––
To *Dividends Checking Account*
For *Dividends Payable*

	Dollars	Cents
Balance brought forward	357,225	00
Add deposits		
Total	357,225	00
Less this check	62,250	00
Balance carried forward	294,975	00

Problem 21-9   Recording Stockholders' Equity Transactions
(1)

GENERAL JOURNAL   PAGE _____

	DATE	DESCRIPTION	POST. REF.	DEBIT	CREDIT	
1						1
2						2
3						3
4						4
5						5
6						6
7						7
8						8
9						9
10						10
11						11
12						12
13						13
14						14
15						15
16						16
17						17
18						18
19						19
20						20
21						21
22						22
23						23
24						24
25						25
26						26
27						27
28						28
29						29
30						30
31						31
32						32
33						33
34						34
35						35
36						36

Problem 21-9   (concluded)

(2)

Analyze: _____

CHAPTER 21   Accounting for Publicly Held Corporations

Self-Test

Part A   Fill in the Missing Term

Directions: *Using terms from the following list, complete the sentences below. Write the letter of the term you have selected in the space provided.*

A. authorized capital stock	**F.** paid-in capital in excess of par	**I.** proxy
B. board of directors		**J.** publicly held corporation
C. closely held corporation	**G.** par value	**K.** statement of stockholders' equity
D. common stock	**H.** preferred stock	
E. dividend		

Answer

_____ 1. A document that gives the stockholder's voting rights to someone else when the stockholder does not want to attend the stockholders' meeting is called a(n) _____.

_____ 2. The stock that has certain privileges over common stock is called _____.

_____ 3. The financial report that shows changes that have taken place in all of the stockholders' equity accounts is called the _____.

_____ 4. The maximum number of shares of stock that a corporation may issue is called its _____.

_____ 5. A(n) _____ is one whose stock is widely held, has a large market, and is traded on a stock exchange.

_____ 6. A corporation owned by a few persons or by a family and whose stock is not sold to the general public is called a(n) _____.

_____ 7. _____ is the amount assigned to each share of stock and printed as a dollar amount on the stock certificates.

_____ 8. A group who governs and is responsible for the affairs of the corporation is called a(n) _____.

_____ 9. The stockholders' equity account that is used to record the sale of stock at higher than par value is the _____ account.

_____ 10. When a corporation issues only one type of stock, the stock is called _____.

_____ 11. A(n) _____ is a cash return on the money invested by the stockholder.

Part B   True or False

Directions: *Read each of the following statements to determine whether the statement is true or false. Write your answer in the space provided.*

Answer

_____   **1.** The stockholders of a corporation have unlimited liability.

_____   **2.** Changes in stockholders' ownership end the life of a corporation.

_____   **3.** A corporation may own property in the name of the corporation.

_____   **4.** The Dividend account in a corporation is similar to the Withdrawals account in a sole proprietorship.

_____   **5.** Declaring a cash dividend requires formal action by the board of directors.

_____   **6.** The Paid-in Capital in Excess of Par account has a debit balance.

_____   **7.** Preferred stockholders have the right to vote for the board of directors.

_____   **8.** Stockholders who own stock at the date of record will receive a dividend if declared.

_____   **9.** The earnings of a corporation are taxed.

_____   **10.** Paid-in Capital in Excess of Par is a current asset.

Part C   Analyzing Transactions

Directions: *Record each of the following unrelated transactions in the general journal forms provided.*

1. May 1. Issued 500 shares of $5 par value common stock for $5.00 per share.

	DATE		DESCRIPTION	POST. REF.	DEBIT	CREDIT	
1							1
2							2
3							3
4							4

2. May 15. Declared a cash dividend of $3.15 per share on 2,000 shares of common stock.

	DATE		DESCRIPTION	POST. REF.	DEBIT	CREDIT	
1							1
2							2
3							3
4							4

CHAPTER 22 Cash Funds

Study Plan

Check Your Understanding

Section 1	Read Section 1 on pages 600–603 and complete the following exercises on page 604.
	❑ Thinking Critically
	❑ Communicating Accounting
	❑ Problem 22-1 *Preparing a Cash Proof*
	❑ Problem 22-2 *Recording a Cash Overage*
Section 2	Read Section 2 on pages 605–613 and complete the following exercises on page 614.
	❑ Thinking Critically
	❑ Computing in the Business World
	❑ Problem 22-3 *Analyzing a Source Document*

Summary

Review the Chapter 22 Summary on page 615 in your textbook.
❑ Key Concepts

Review and Activities

Complete the following questions and exercises on pages 616–617 in your textbook.
❑ Using Key Terms
❑ Understanding Accounting Concepts and Procedures
❑ Case Study
❑ Conducting an Audit with Alex
❑ Internet Connection
❑ Workplace Skills

Computerized Accounting

Read the Computerized Accounting information on page 618 in your textbook.
❑ *Making the Transition from a Manual to a Computerized System*
❑ *Maintaining Cash Funds in Peachtree*

Problems

Complete the following end-of-chapter problems for Chapter 22 in your textbook.
❑ Problem 22-4   *Establishing a Change Fund*
❑ Problem 22-5   *Establishing and Replenishing a Petty Cash Fund*
❑ Problem 22-6   *Establishing and Replenishing a Petty Cash Fund*
❑ Problem 22-7   *Using a Petty Cash Register*
❑ Problem 22-8   *Handling a Petty Cash Fund*

Challenge Problem

❑ Problem 22-9   *Locating Errors in a Petty Cash Register*

Chapter Reviews and Working Papers

Complete the following exercises for Chapter 22 in your Chapter Reviews and Working Papers.
❑ Chapter Review
❑ Self-Test

CHAPTER 22 REVIEW   Cash Funds

Part 1   Accounting Vocabulary (6 points)

Total Points	41
Student's Score	

Directions: *Using terms from the following list, complete the sentences below. Write the letter of the term you have chosen in the space provided.*

A. change fund	**D.** petty cashier	**F.** petty cash requisition
B. petty cash disbursement	**E.** petty cash register	**G.** petty cash voucher
C. petty cash fund		

_____F_____   **0.** The form used for requesting money to replenish the petty cash fund is a _____.

_____   **1.** A _____ consists of varying denominations of bills and coins and is used to make change in cash transactions.

_____   **2.** Cash that is kept on hand by a business for making small, incidental cash payments is called a _____.

_____   **3.** A _____ is a proof of payment from the petty cash fund.

_____   **4.** The person responsible for handling the petty cash fund is the _____.

_____   **5.** Any payment from the petty cash fund is called a _____.

_____   **6.** A _____ is a record of all disbursements made from the petty cash fund.

Part 2   Examining Cash Funds (8 points)

Directions: *For each of the following, select the choice that is the most suitable. Write your answer in the space provided.*

_____A_____   **0.** The Change Fund account is listed on the chart of accounts as a(n)
    (A) asset.   (C) revenue account.
    (B) liability.   (D) expense.

_____   **1.** Which of the following businesses would probably **not** have a change fund?
    (A) drugstore.   (C) lawyer's office.
    (B) supermarket.   (D) newsstand.

_____   **2.** A cash proof for the change fund should be prepared at the end of the
    (A) accounting period.   (C) day.
    (B) month.   (D) week.

_____   **3.** Cash Short & Over is classified as a(n)
    (A) liability.   (C) expense.
    (B) temporary owner's equity account.   (D) revenue account.

_____   **4.** The Petty Cash Fund account is debited
    (A) every time the fund is replenished.
    (B) when the fund is established.
    (C) when the amount of money in the fund is increased.
    (D) B and C.

_____   **5.** The petty cash fund is replenished
    (A) when its balance reaches the minimum amount.
    (B) at the end of each day.
    (C) at the end of the fiscal period.
    (D) A and C.

_____   **6.** The Petty Cash Fund account is a(n)
    (A) expense account.   (C) asset account.
    (B) revenue account.   (D) owner's equity account.

_____   **7.** In business, cash overages are
    (A) revenue.   (C) assets.
    (B) expenses.   (D) liabilities.

8. Cash shortages or overages are recorded in the
  (A) cash payments journal.
  (B) cash receipts journal.
  (C) general journal.
  (D) A and B.

Part 3   Accounting for Cash Funds (15 points)

Directions: *Read each of the following statements to determine whether the statement is true or false. Write your answer in the space provided.*

True   **0.** The size of the petty cash fund is determined by the needs of the business.

_____ **1.** The amount of cash sales for the day is taken from the cash register tape and entered on the cash proof form.

_____ **2.** At the end of the fiscal period, the balance of the Cash Short & Over account is closed directly into Retained Earnings.

_____ **3.** Replenishing the petty cash fund increases the fund's original balance.

_____ **4.** The salesclerk usually counts the cash in the drawer, verifies its accuracy, and signs the cash proof form.

_____ **5.** Cash shortages are liabilities and are debited to the Cash Short & Over account.

_____ **6.** The size of the petty cash fund will not change unless the business finds that it needs more or less than its original estimate.

_____ **7.** The Change Fund account is debited each time the fund is replenished.

_____ **8.** Some businesses that have a petty cash fund do not use a petty cash register.

_____ **9.** A cash proof is prepared to verify that the amount of cash in the cash register drawer is equal to the total cash sales for the day.

_____ **10.** Cash overages are revenue and are recorded as credits to the Sales account.

_____ **11.** Businesses that use a petty cash envelope for recording petty cash disbursements use a new petty cash envelope for each period's disbursements.

_____ **12.** The amounts paid out of the petty cash fund must be journalized and recorded in the appropriate general ledger accounts when the petty cash fund is replenished.

_____ **13.** The petty cash register is considered an accounting journal because all amounts are posted from this register to general ledger accounts.

_____ **14.** The balance of the cash from the cash register drawer is deposited in the business's checking account after the cash in the cash register drawer is counted and the change fund is set aside.

_____ **15.** The entry to establish the petty cash fund is recorded in the cash payments journal.

Part 4   Analyzing Cash Funds Transactions (12 points)

Directions: *Using the following list of account titles, determine the account titles to be debited and credited for the transactions below.*

A. Cash in Bank	**D.** Supplies	**F.** Delivery Expense
B. Change Fund	**E.** Cash Short & Over	**G.** Miscellaneous Expense
C. Petty Cash Fund		

Debit	Credit	
B	A	**0.** Established a change fund.
___	___	**1.** Recorded a cash overage in the change fund.
___	___	**2.** Established the petty cash fund.
___	___	**3.** Replenished the petty cash fund; petty cash vouchers were for supplies, delivery expenses, and miscellaneous expenses.
___	___	**4.** Increased the petty cash fund.
___	___	**5.** Recorded a cash shortage in the petty cash fund.
___	___	**6.** Decreased the change fund.

Working Papers *for Section Problems*

Problem 22-1   Preparing a Cash Proof

(1)

CASH PROOF

Date _____

Cash Register No. _____

Total cash sales (from cash register tape)		$ _____
Cash in drawer	$ _____	
Less change fund	_____	
Net cash received		$ _____
Cash short		_____
Cash over		_____

Salesclerk _____

Supervisor _____

(2)

GENERAL JOURNAL   PAGE _____

	DATE	DESCRIPTION	POST. REF.	DEBIT	CREDIT	
1						1
2						2
3						3
4						4
5						5
6						6
7						7
8						8
9						9
10						10
11						11
12						12

Problem 22-2   Recording a Cash Overage
(1)

CASH PROOF

Date _____

Cash Register No. _____

Total cash sales (from cash register tape)		$ _____
Cash in drawer	$ _____	
Less change fund	_____	
Net cash received		$ _____
Cash short		_____
Cash over		_____

Salesclerk _____

Supervisor _____

(2)

GENERAL JOURNAL   PAGE _____

	DATE	DESCRIPTION	POST. REF.	DEBIT	CREDIT	
1						1
2						2
3						3
4						4
5						5
6						6
7						7
8						8
9						9
10						10
11						11
12						12

Problem 22-3   Analyzing a Source Document

	Dollars	Cents
$ _____ No. 973		
Date _____ 20 ___		
To _____		
For _____		
Balance brought forward	77,432	86
Add deposits		
Total		
Less this check		
Balance carried forward		

Riddle's Card Shop
1500 Main Street
Concord, MA 01742

973

53-215
———
113

DATE _____ 20 ___

PAY TO THE
ORDER OF _____ $ _____

_____ DOLLARS

Patriot Bank
CONCORD, MASSACHUSETTS

MEMO _____

⑆011302153⑆ 331 234 9⑈ 0973

Computerized Accounting Using Peachtree

Software Objectives

When you have completed this chapter, you will be able to use Peachtree to:
1. Record the entry to establish a change fund.
2. Record the entries to establish and replenish a petty cash fund.

Problem 22-4   Establishing a Change Fund

INSTRUCTIONS

Beginning a Session

Step 1   Select the problem set: Sunset Surfwear (Prob. 22-4).
Step 2   Rename the company by adding your initials, e.g., Sunset (Prob. 22-4: XXX).
Step 3   Set the system date to February 29, 2008.

TIP: The year 2008 is a leap year. Therefore, the end of the month is February 29.

Completing the Accounting Problem

Step 4   Review the information provided in your textbook.
Step 5   Record the entry to establish the change fund using the **General Journal Entry** option.

TIP: As a shortcut, you can type only the day of the month when you enter a transaction date.

Step 6   Manually prepare a cash proof.
Step 7   Record the cash sales using the **General Journal Entry** option.
Step 8   Print a General Journal report and proof your work.
Step 9   Answer the Analyze question.

Ending the Session

Step 10 Click the **Close Problem** button in the Glencoe Smart Guide window to save your work.

Mastering Peachtree

Explore how to export data from Peachtree into another application such as a spreadsheet program. Demonstrate your understanding of this feature by exporting the chart of accounts. Include only the general ledger ID and account description. Import the file into a spreadsheet or word processor, and then print the data.

DO YOU HAVE A QUESTION

Q. *Can you use the General Journal Entry option for transactions involving cash?*

A. You have learned how to use the **Payments** and **Receipts** options to record cash payments and cash receipts, respectively. However, you can always use the **General Journal Entry** option to record transactions involving cash. The only disadvantage to using the **General Journal Entry** option is that Peachtree will not print checks for any cash payments transactions you record. However, this is not an issue if you manually write checks and use Peachtree only to record the transactions.

The Peachtree instructions for this chapter suggest that you record the various cash funds transactions using the **General Journal Entry** option since your textbook also uses general journal transactions. You can, however, use the **Payments** and **Receipts** options if you are more comfortable using these options.

Problem 22-5 Establishing and Replenishing a Petty Cash Fund

INSTRUCTIONS

Beginning a Session

Step 1 Select the problem set: InBeat CD Shop (Prob. 22-5).
Step 2 Rename the company, and set the system date to February 29, 2008.

Completing the Accounting Problem

Step 3 Review the information provided in your textbook.
Step 4 Record the entry to establish the petty cash fund using the **General Journal Entry** option.

TIP: Click the [Open] button if you need to edit a general journal entry you already posted.

DO YOU HAVE A QUESTION

Q. *Does Peachtree include any features to prepare a petty cash register?*

A. Peachtree does not provide any features to prepare a petty cash register. Although you must manually prepare this document, you can use the **General Journal Entry** option to record the transactions to establish and replenish a petty cash fund.

Step 5 Record the entry to replenish the petty cash fund using the **General Journal Entry** option.
Step 6 Print a General Journal report and proof your work.
Step 7 Answer the Analyze question.

Checking Your Work and Ending the Session

Step 8 Click the **Close Problem** button in the Glencoe Smart Guide window.
Step 9 If your teacher has asked you to check your solution, select *Check my answer to this problem.* Review, print, and close the report.
Step 10 Click the **Close Problem** button and select a save option.

Problem 22-6 Establishing and Replenishing a Petty Cash Fund

INSTRUCTIONS

Beginning a Session

Step 1 Select the problem set: Shutterbug Cameras (Prob. 22-6).
Step 2 Rename the company, and set the system date to February 29, 2008.

Completing the Accounting Problem

Step 3 Review the information provided in your textbook.
Step 4 Record the entry to establish the petty cash fund.

TIP: You can use the [Row], [Add] and [Row], [Remove] buttons to edit a multi-line general journal entry.

Step 5 Make a list of the petty cash vouchers and manually prepare a petty cash requisition.
Step 6 Record the entry to replenish the petty cash fund.

Step 7   Print a General Journal report and proof your work.

Step 8   Answer the Analyze question.

Ending the Session

Step 9   Click the **Close Problem** button in the Glencoe Smart Guide window to save your work.

Problem 22-7   Using a Petty Cash Register

INSTRUCTIONS

Beginning a Session

Step 1   Select the problem set: Cycle Tech Bicycles (Prob. 22-7).

Step 2   Rename the company, and set the system date to February 29, 2008.

Completing the Accounting Problem

Step 3   Review the information provided in your textbook.

Step 4   Record the entry to establish the petty cash fund.

TIP: Use the **General Journal Entry** option to record the cash fund transactions.

Step 5   Manually record the petty cash disbursements.

Step 6   Prepare a petty cash requisition.

Step 7   Record the entry to replenish the petty cash fund.

Step 8   Print a General Journal report and proof your work.

TIP: Double-click a report title to go directly to the report, skipping the report options.

Step 9   Answer the Analyze question.

Checking Your Work and Ending the Session

Step 10   Click the **Close Problem** button in the Glencoe Smart Guide window.

Step 11   If your teacher has asked you to check your solution, select *Check my answer to this problem*. Review, print, and close the report.

Step 12   Click the **Close Problem** button and select a save option.

Problem 22-8   Handling a Petty Cash Fund

INSTRUCTIONS

Beginning a Session

Step 1   Select the problem set: River's Edge Canoe & Kayak (Prob. 22-8).

Step 2   Rename the company, and set the system date to February 29, 2008.

Completing the Accounting Problem

Step 3   Review the information provided in your textbook.

Step 4   Record the entry to establish the petty cash fund.

> **TIP:** Use the **General Journal Entry** option to record the cash fund transactions.

Step 5   Manually record the petty cash disbursements.

Step 6   Reconcile the petty cash register and prepare a petty cash requisition. Update the petty cash register to record the replenishment information.

Step 7   Record the entry to replenish the petty cash fund.

Step 8   Record the entry to increase the petty cash fund.

Step 9   Print a General Journal report and proof your work.

Step 10 Answer the Analyze question.

Ending the Session

Step 11 Click the **Close Problem** button in the Glencoe Smart Guide window to save your work.

FAQs

Can you use the General Journal Entry option for cash fund transactions?

You have learned how to use the **Payments** and **Receipts** options to record cash payments and cash receipts, respectively. However, you can always use the **General Journal Entry** option to record transactions involving cash. The only disadvantage to using the **General Journal Entry** option is that Peachtree will not print checks for any cash payments transactions you record. However, this is not an issue if you manually write checks and use Peachtree only to record the transactions.

You can record various cash funds transactions using the **General Journal Entry** option, but you can always work with the **Payments** and **Receipts** options if you are more comfortable using these options.

Does Peachtree include any features to prepare a petty cash register?

Peachtree does not provide any features to prepare a petty cash register. Although you must manually prepare this document, you can use the **General Journal Entry** option to record the transactions to establish and replenish a petty cash fund.

Working Papers *for End-of-Chapter Problems*

Problem 22-4   Establishing a Change Fund
(1), (3)

GENERAL JOURNAL   PAGE _____

	DATE	DESCRIPTION	POST. REF.	DEBIT	CREDIT	
1						1
2						2
3						3
4						4
5						5
6						6
7						7
8						8
9						9
10						10

Analyze: _____

_____

(2)

CASH PROOF

Date _____

Cash Register No. ____*1*____

Total cash sales
(from cash register tape) $ _____

Cash in drawer $ _____

Less change fund   _____

Net cash received $ _____

Cash short   _____

Cash over   _____

Salesclerk _____

Supervisor _____

Problem 22-5 Establishing and Replenishing a Petty Cash Fund

(1)

GENERAL JOURNAL   PAGE _____

	DATE	DESCRIPTION	POST. REF.	DEBIT	CREDIT	
1						1
2						2
3						3
4						4
5						5
6						6
7						7
8						8
9						9
10						10

(2)

GENERAL JOURNAL   PAGE _____

	DATE	DESCRIPTION	POST. REF.	DEBIT	CREDIT	
1						1
2						2
3						3
4						4
5						5
6						6
7						7
8						8
9						9
10						10

Analyze: _____

_____

Problem 22-6 Establishing and Replenishing a Petty Cash Fund
(1)

GENERAL JOURNAL   PAGE _____

	DATE	DESCRIPTION	POST. REF.	DEBIT	CREDIT	
1						1
2						2
3						3
4						4
5						5
6						6
7						7
8						8
9						9
10						10

(5)

GENERAL JOURNAL   PAGE _____

	DATE	DESCRIPTION	POST. REF.	DEBIT	CREDIT	
1						1
2						2
3						3
4						4
5						5
6						6
7						7
8						8
9						9
10						10

Analyze: _____

_____

Problem 22-6   (concluded)

(2)

Voucher No.	Account Name	Amount
101	Supplies	$ _____
102	Advertising Expense	_____
103	Miscellaneous Expense	_____
104	Delivery Expense	_____
105	Supplies	_____
106	Miscellaneous Expense	_____
107	Delivery Expense	_____
108	Miscellaneous Expense	_____
109	Supplies	_____
110	Delivery Expense	_____
111	Advertising Expense	_____
112	Miscellaneous Expense	_____
113	Advertising Expense	_____
	TOTAL	$ _____

(3)

Supplies	$ _____
Advertising Expense	_____
Delivery Expense	_____
Miscellaneous Expense	_____

(4)

PETTY CASH REQUISITION

Accounts for which payments were made:

	Amount
_____	$ _____
_____	_____
_____	_____
_____	_____
_____	_____

TOTAL CASH NEEDED TO REPLENISH FUND: $ _____

Requested by: _____ (PETTY CASHIER)   Date _____

Approved by: _____ (ACCOUNTANT)   Date _____

Check No. _____

Analyze: _____
_____
_____
_____
_____

Problem 22-7   Source Documents

Instructions: *Use the following source documents to record the transactions for this problem.*

PETTY CASH VOUCHER	No. 0001
DATE *February 2* 20 --	
PAID TO *Silver City Star*	$ *9.25*
FOR *Newspaper ad*	
ACCOUNT *Advertising Expense*	
APPROVED BY	PAYMENT RECEIVED BY
	Dudley Hartel

PETTY CASH VOUCHER	No. 0005
DATE *February 19* 20 --	
PAID TO *National Express*	$ *15.00*
FOR *Parts delivered*	
ACCOUNT *Delivery Expense*	
APPROVED BY	PAYMENT RECEIVED BY
	Jessica Kirby

PETTY CASH VOUCHER	No. 0002
DATE *February 5* 20 --	
PAID TO *Maxwell Office Supplies*	$ *5.00*
FOR *Pens and pencils*	
ACCOUNT *Supplies*	
APPROVED BY	PAYMENT RECEIVED BY
	Joshua Maxwell

PETTY CASH VOUCHER	No. 0006
DATE *February 20* 20 --	
PAID TO *Postmaster*	$ *3.90*
FOR *Postage stamps*	
ACCOUNT *Miscellaneous Expense*	
APPROVED BY	PAYMENT RECEIVED BY
	Sam Haygood

PETTY CASH VOUCHER	No. 0003
DATE *February 9* 20 --	
PAID TO *Tumbleweed Florist*	$ *12.50*
FOR *Flowers for employee's birthday*	
ACCOUNT *Miscellaneous Expense*	
APPROVED BY	PAYMENT RECEIVED BY
	Louise Wicker

PETTY CASH VOUCHER	No. 0007
DATE *February 22* 20 --	
PAID TO *Krystal Clean*	$ *16.00*
FOR *Show window cleaned*	
ACCOUNT *Miscellaneous Expense*	
APPROVED BY	PAYMENT RECEIVED BY
	Krystal Adams

PETTY CASH VOUCHER	No. 0004
DATE *February 12* 20 --	
PAID TO *Maxwell Office Supplies*	$ *3.95*
FOR *Cash register tape*	
ACCOUNT *Supplies*	
APPROVED BY	PAYMENT RECEIVED BY
	Joshua Maxwell

PETTY CASH VOUCHER	No. 0008
DATE *February 24* 20 --	
PAID TO *Silver City Star*	$ *11.00*
FOR *Newspaper ad*	
ACCOUNT *Advertising Expense*	
APPROVED BY	PAYMENT RECEIVED BY
	Dudley Hartel

Problem 22-7 (continued)

PETTY CASH VOUCHER	No. 0009
DATE *February 25* 20 --	
PAID TO *Maxwell Office Supplies* $ *10.00*	
FOR *Stationery*	
ACCOUNT *Supplies*	
APPROVED BY _____ PAYMENT RECEIVED BY *Joshua Maxwell*	

PETTY CASH VOUCHER	No. 0013
DATE *February 28* 20 --	
PAID TO *Lara Allen* $ *4.00*	
FOR *Tip for daily newspaper delivery*	
ACCOUNT *Miscellaneous Expense*	
APPROVED BY _____ PAYMENT RECEIVED BY *Lara Allen*	

PETTY CASH VOUCHER	No. 0010
DATE *February 26* 20 --	
PAID TO *National Express* $ *8.25*	
FOR *Packages delivered*	
ACCOUNT *Delivery Expense*	
APPROVED BY _____ PAYMENT RECEIVED BY *Jessica Kirby*	

PETTY CASH VOUCHER	No. 0014
DATE *February 28* 20 --	
PAID TO *Postmaster* $ *6.80*	
FOR *Postage stamps*	
ACCOUNT *Miscellaneous Expense*	
APPROVED BY _____ PAYMENT RECEIVED BY *Sam Haygood*	

PETTY CASH VOUCHER	No. 0011
DATE *February 27* 20 --	
PAID TO *Maxwell Office Supplies* $ *8.00*	
FOR *Supplies*	
ACCOUNT	
APPROVED BY _____ PAYMENT RECEIVED BY	

VOID

PETTY CASH VOUCHER	No. 0015
DATE *February 28* 20 --	
PAID TO *Silver City Star* $ *10.00*	
FOR *Newspaper ad*	
ACCOUNT *Advertising Expense*	
APPROVED BY _____ PAYMENT RECEIVED BY *Dudley Hartel*	

PETTY CASH VOUCHER	No. 0012
DATE *February 27* 20 --	
PAID TO *Maxwell Office Supplies* $ *3.00*	
FOR *Memo pads*	
ACCOUNT *Supplies*	
APPROVED BY _____ PAYMENT RECEIVED BY *Joshua Maxwell*	

Problem 22-7   Using a Petty Cash Register
(1)

GENERAL JOURNAL

PAGE _____

	DATE	DESCRIPTION	POST. REF.	DEBIT	CREDIT	
1						1
2						2
3						3
4						4
5						5
6						6
7						7
8						8
9						9
10						10

(7)

GENERAL JOURNAL

PAGE _____

	DATE	DESCRIPTION	POST. REF.	DEBIT	CREDIT	
1						1
2						2
3						3
4						4
5						5
6						6
7						7
8						8
9						9
10						10

Problem 22-7   (continued)   **(2), (3), (4), (5)**

PAGE _____

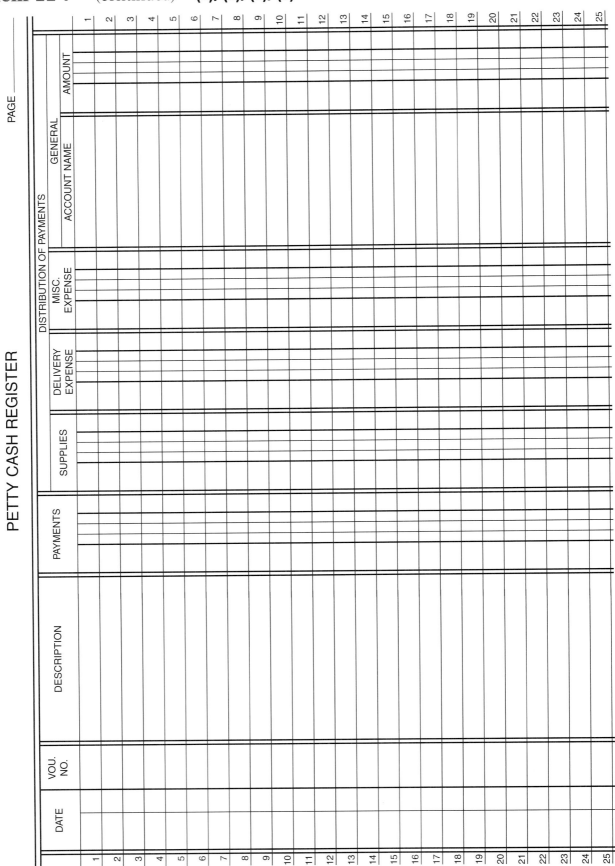

PETTY CASH REGISTER

Problem 22-7   (concluded)
(6)

PETTY CASH REQUISITION

Accounts for which
payments were made: Amount

_____ $ _____

_____   _____

_____   _____

_____   _____

_____   _____

_____   _____

TOTAL CASH NEEDED TO REPLENISH FUND: $ _____

Requested by: _____ Date _____
            PETTY CASHIER

Approved by: _____ Date _____
            ACCOUNTANT

                       Check No. _____

Analyze: _____

_____

Problem 22-8   Handling a Petty Cash Fund
(1)

GENERAL JOURNAL PAGE _____

	DATE	DESCRIPTION	POST. REF.	DEBIT	CREDIT	
1						1
2						2
3						3
4						4
5						5
6						6
7						7

(7)

GENERAL JOURNAL PAGE _____

	DATE	DESCRIPTION	POST. REF.	DEBIT	CREDIT	
1						1
2						2
3						3
4						4
5						5
6						6
7						7
8						8
9						9
10						10

(9)

GENERAL JOURNAL PAGE _____

	DATE	DESCRIPTION	POST. REF.	DEBIT	CREDIT	
1						1
2						2
3						3
4						4
5						5
6						6
7						7

Problem 22-8 (continued) (2), (3), (4), (5), (8)

PAGE _____

PETTY CASH REGISTER

DATE	VOU. NO.	DESCRIPTION	PAYMENTS	DISTRIBUTION OF PAYMENTS			GENERAL	
				SUPPLIES	DELIVERY EXPENSE	MISC. EXPENSE	ACCOUNT NAME	AMOUNT
								1
								2
								3
								4
								5
								6
								7
								8
								9
								10
								11
								12
								13
								14
								15
								16
								17
								18
								19
								20
								21
								22
								23
								24

Problem 22-8   (concluded)

(6)

PETTY CASH REQUISITION

Accounts for which
payments were made: Amount

_____ $_____

_____   _____

_____   _____

_____   _____

_____   _____

_____   _____

TOTAL CASH NEEDED TO REPLENISH FUND: $_____

Requested by:_____ Date _____
                PETTY CASHIER

Approved by: _____ Date _____
                ACCOUNTANT

                   Check No. _____

Analyze: _____

_____

Problem 22-9   Locating Errors in a Petty Cash Register

Petty Cash Disbursements

Item	Date	Amt.	Item	Date	Amt.
Delivery charge	Feb. 4	$ 7.60	Order forms	Feb. 5	$ 9.45
Delivery charge	14	9.65	Memo pads	6	12.14
Delivery charge	27	8.75	Writing tablets	9	9.43
			Wrapping paper	9	8.49
Newspaper ad	Feb. 7	$10.00	Pens/pencils	20	7.24
Newspaper ad	10	8.50	Coffee filters	25	3.14
			Fax paper	28	9.30
Stamps	Feb. 2	$ 8.25	Gasoline	28	11.42
Stamps	11	9.00	Gasket	28	6.28
			Telephone directory	28	4.30

Analyze: _____

_____

_____

Problem 22-9 (concluded)   (1), (2), (3), (4)

PETTY CASH REGISTER

PAGE _____

DATE		VOU. NO.	DESCRIPTION	PAYMENTS	SUPPLIES	DELIVERY EXPENSE	MISC. EXPENSE	GENERAL ACCOUNT NAME	AMOUNT
20--									
Feb.	1	—	Est. fund $150 Ck. 948						
	2	101	Postage stamps	825			825		
	4	102	Delivery charge	670			670		
	5	103	Order forms	945			945		
	6	104	Memo pads	1214	1214				
	7	105	Newspaper ad	1000				Advertising Expense	1000
	9	106	Writing tablets	943	934				
	9	107	Wrapping paper	849	849				
	10	108	Newspaper ad	852			850		
	11	109	Stamps	900				Postage	900
	14	110	VOID						
	14	111	Delivery charge	965		965			
	20	112	Pens/pencils	724	724				
	25	113	Coffee filters	314			314		
	27	114	Delivery charge	875		675			
	28	115	Fax paper	930	930				
	28	116	Gasoline	1142				Gasoline Expense	1142
	28	117	Gasket	628				Maintenance Expense	628
	28	118	Telephone directory	430			430		
	28		Totals	13890	4651	1646	3834		3670
			Reconciled Balance	706					
			Cash Over	104					
			Replenishment Check	14190					
			Total	15000					

CHAPTER 22 Cash Funds

Self-Test

Part A Fill in the Missing Term

Directions: *In the Answer column, write the letter of the word or phrase that best completes the sentence.*

A. cash proof	**E.** petty cash fund	**H.** petty cash voucher
B. change fund	**F** petty cash register	**I.** petty cashier
C. over	**G.** petty cash requisition	**J.** short
D. petty cash disbursement		

Answer

_____ **1.** A(n) _____ is any payment made from the petty cash fund.

_____ **2.** A(n) _____ is a form prepared by the petty cashier to request cash for replenishing the petty cash fund.

_____ **3.** If the change fund was established for $100.00 and there are $296.00 in cash receipts and the cash balance is $400.00, cash for the day is said to be _____.

_____ **4.** Every payment made from the petty cash fund must be supported by a(n) _____.

_____ **5.** A(n) _____ is usually used in retail stores such as supermarkets for making change for cash customers.

_____ **6.** The _____ is responsible for making payments from the petty cash fund.

_____ **7.** Many businesses use a(n) _____ for paying small cash payments.

_____ **8.** If the petty cash fund was established for $50.00 and there is $4.00 in cash with $45.00 in vouchers, cash is said to be _____.

_____ **9.** All payments made from the petty cash fund are recorded in the _____.

_____ **10.** A(n) _____ is prepared to verify that the amount of cash in the drawer is equal to the total cash sales for the day plus the change fund cash.

Part B   True or False

Directions: *Circle the letter* T *in the Answer column if the statement is true;*
circle the letter F *if the statement is false.*

Answer

T   F   **1.** The size of the change fund does not change unless the business finds it needs more or less than its original estimate.

T   F   **2.** Change funds are classified as assets and are listed directly below Cash in Bank in the chart of accounts.

T   F   **3.** Once a business establishes a petty cash fund for a set amount, the Internal Revenue Service does not permit the size of the fund to be changed.

T   F   **4.** The purpose of a change fund is to enable the cashier using the cash register to have enough coins and currency to conduct business for the day.

T   F   **5.** Salesclerks are usually required to sign the cash proof to indicate they have counted the cash in the drawer and verified its accuracy.

T   F   **6.** If there is a cash shortage at the end of the day, the amount of cash recorded on the cash register tape for the day's cash sales will be less than the amount of cash in the drawer.

T   F   **7.** The account Cash Short & Over is a temporary equity account.

T   F   **8.** The normal balance of the Cash Short & Over account is a credit balance.

T   F   **9.** Cash shortages are losses to the business; cash overages are gains or revenues.

T   F   **10.** If cash shortages occur more frequently than cash overages, the account Cash Short & Over will have a credit balance.

T   F   **11.** Businesses use a petty cash fund because writing checks for small amounts is costly, time consuming, and impractical.

T   F   **12.** For internal control reasons, at least two people should be responsible for the petty cash fund, and the petty cash box should be kept in a locked desk drawer or safe.

T   F   **13.** The petty cash fund is debited each time the fund is replenished.

T   F   **14.** Each time a payment is made from the petty cash fund, a prenumbered voucher must be prepared to verify proof of payment.

T   F   **15.** Replenishing the petty cash fund restores the fund to its original balance.

CHAPTER 23   Plant Assets and Depreciation

Study Plan

Check Your Understanding

Section 1   *Read Section 1 on pages 626–628 and complete the exercises on page 629.*
- ❏ Thinking Critically
- ❏ Communicating Accounting
- ❏ Problem 23-1   *Classifying Asset Accounts*

Section 2   *Read Section 2 on pages 630–631 and complete the exercises on page 632.*
- ❏ Thinking Critically
- ❏ Computing in the Business World
- ❏ Problem 23-2   *Calculating Depreciation Expense*
- ❏ Problem 23-3   *Completing a Plant Asset Record*

Section 3   *Read Section 3 on pages 633–640 and complete the exercises on page 641.*
- ❏ Thinking Critically
- ❏ Communicating Accounting
- ❏ Problem 23-4   *Analyzing a Source Document*
- ❏ Problem 23-5   *Preparing a Depreciation Schedule and Journalizing the Depreciation Adjusting Entry*

Summary   *Review the Chapter 23 Summary on page 643 in your textbook.*
- ❏ Key Concepts

Review and Activities   *Complete the following questions and exercises on pages 644–645.*
- ❏ Using Key Terms
- ❏ Understanding Accounting Concepts and Procedures
- ❏ Case Study
- ❏ Conducting an Audit with Alex
- ❏ Internet Connection
- ❏ Workplace Skills

Computerized Accounting   *Read the Computerized Accounting information on page 646.*
- ❏ *Making the Transition from a Manual to a Computerized System*
- ❏ *Recording Depreciation in Peachtree*

Problems   *Complete the following end-of-chapter problems for Chapter 23.*
- ❏ Problem 23-6   *Opening a Plant Asset Record*
- ❏ Problem 23-7   *Recording Adjusting Entries for Depreciation*
- ❏ Problem 23-8   *Reporting Depreciation Expense on the Work Sheet and Financial Statements*
- ❏ Problem 23-9   *Calculating and Recording Depreciation Expense*
- ❏ Problem 23-10   *Calculating and Recording Adjustments*

Challenge Problem
- ❏ Problem 23-11   *Examining Depreciation Adjustments*

Chapter Reviews and Working Papers   *Complete the following exercises for Chapter 23 in your Chapter Reviews and Working Papers.*
- ❏ Chapter Review
- ❏ Self-Test

CHAPTER 23 REVIEW   Plant Assets and Depreciation

Part 1   Accounting Vocabulary (6 points)

Total Points	26
Student's Score	

Directions: *Using terms from the following list, complete the sentences below. Write the letter of the term you have chosen in the space provided.*

A. accumulated depreciation	**D.** depreciation	**F.** plant assets
B. book value	**E.** disposal value	**G.** straight-line depreciation
C. current assets		

_____E_____ **0.** The estimated value of a plant asset at its replacement time is called _____.

_____ **1.** _____ are long-lived assets that are used in the production or sale of other assets or services over several accounting periods.

_____ **2.** _____ is a method of equally distributing the depreciation expense on a plant asset over its estimated useful life.

_____ **3.** Allocating the cost of a plant asset over the asset's useful life is called _____.

_____ **4.** The original cost of a plant asset less its accumulated depreciation is the _____.

_____ **5.** _____ is the total amount of depreciation for a plant asset that has been recorded up to a specific point in time.

_____ **6.** _____ are either used up or converted to cash during a one-year accounting period.

Part 2   Accounting for Depreciation (11 points)

Directions: *Read each of the following statements to determine whether the statement is true or false. Write your answer in the space provided.*

___False___ **0.** Depreciation Expense is reported on the income statement in the cost of goods sold section.

_____ **1.** The amount of depreciation taken for a plant asset is usually recorded in the accounting records at the beginning of the fiscal period.

_____ **2.** All plant assets, including land, depreciate in value.

_____ **3.** The cost of a plant asset is the price paid for the asset plus taxes, installation charges, and delivery charges.

_____ **4.** Depreciation amounts are estimates of the decrease in value or usefulness of a plant asset over a period of time.

_____ **5.** Accumulated Depreciation is reported on the balance sheet as a liability.

_____ **6.** Depreciation Expense is recorded by an adjusting entry made in the general journal.

_____ **7.** The adjusting entry for depreciation affects two accounts for each type of plant asset: Depreciation Expense and Delivery Equipment.

_____ **8.** Delivery equipment, office equipment, buildings, and land are long-lived assets because they are expected to produce benefits for the business for more than one year.

_____ **9.** The plant asset record does not list the accumulated depreciation of an asset or its book value at the end of each year.

_____ **10.** Accumulated Depreciation is classified as a contra plant asset account.

_____ **11.** Current assets include cash, merchandise, equipment, and accounts receivable.

Part 3   Analyzing the Depreciation of Equipment (9 points)

Directions: *On January 3, Washington Delivery Service purchased a new delivery truck for $20,000. The delivery truck has an estimated disposal value of $3,200 and an estimated useful life of seven years. Using this information, select the answer that best completes each of the following statements. Write your answer in the space provided.*

B   **0.** The amount that will be debited to Delivery Equipment is
- (A) $16,800.
- (B) $20,000.
- (C) $3,200.
- (D) $23,200.

1. The estimated annual depreciation amount using the straight-line method is
- (A) $3,200.
- (B) $16,800.
- (C) $2,857.
- (D) $2,400.

2. The estimated depreciation amount will be recorded as a debit to
- (A) Accumulated Depreciation—Delivery Equipment.
- (B) Delivery Expense.
- (C) Depreciation Expense—Delivery Equipment.
- (D) Delivery Equipment.

3. At the end of the fiscal period, the adjusting entry for the depreciation is a
- (A) debit to Depreciation Expense—Delivery Equipment and a credit to Delivery Equipment.
- (B) debit to Depreciation Expense—Delivery Equipment and a credit to Accumulated Depreciation—Delivery Equipment.
- (C) debit to Accumulated Depreciation—Delivery Equipment and a credit to Depreciation Expense—Delivery Equipment.
- (D) debit to Delivery Equipment and a credit to Accumulated Depreciation—Delivery Equipment.

4. After the adjusting entries are posted to the general ledger, Depreciation Expense—Delivery Equipment will have
- (A) a zero balance.
- (B) either a debit or credit balance.
- (C) a debit balance.
- (D) a credit balance.

5. After the adjusting entries are posted, the Delivery Equipment account will have a
- (A) debit balance equal to the original purchase price less the amount of the accumulated depreciation.
- (B) debit balance for the amount of the depreciation.
- (C) debit balance equal to the original purchase price.
- (D) credit balance for the amount of the accumulated depreciation.

6. After the closing entry is posted, the Depreciation Expense—Delivery Equipment account will have
- (A) a zero balance.
- (B) a debit balance.
- (C) a credit balance.
- (D) either a debit or credit balance.

7. At the beginning of the next fiscal period, Accumulated Depreciation—Delivery Equipment will have
- (A) a debit balance.
- (B) a credit balance.
- (C) either a debit or credit balance.
- (D) a zero balance.

8. The book value of the delivery truck at the end of four years is
- (A) $9,600.
- (B) $12,800.
- (C) $10,400.
- (D) $7,200.

9. If the Delivery Equipment had been purchased on October 1 instead of January 3, the estimated depreciation for the first year would be
- (A) $400.
- (B) $1,200.
- (C) $200.
- (D) $600.

Working Papers *for Section Problems*

Problem 23-1   Classifying Asset Accounts

Asset	Current Asset	Plant Asset
Accounts Receivable	✓	
Building		
Cash in Bank		
Change Fund		
Delivery Equipment		
Land		
Merchandise Inventory		
Office Equipment		
Office Furniture		
Petty Cash Fund		
Prepaid Insurance		
Store Equipment		
Supplies		

Problem 23-2   Calculating Depreciation Expense

	Asset	(1) Amount to be Depreciated	Depreciation Rate	(2) Annual Depreciation	Months Owned	(3) First Year Depreciation
1	Cash register	$		$		$
2	Computer	$		$		$
3	Conference table	$		$		$
4	Delivery truck	$		$		$
5	Desk	$		$		$

Problem 23-3 Completing a Plant Asset Record

PLANT ASSET RECORD

ITEM _____ GENERAL LEDGER ACCOUNT _____

SERIAL NUMBER _____ MANUFACTURER _____

PURCHASED FROM _____ EST. DISPOSAL VALUE _____

ESTIMATED LIFE _____ LOCATION _____

DEPRECIATION
METHOD _____ DEPRECIATION
PER YEAR _____

DATE	EXPLANATION	ASSET			ACCUMULATED DEPRECIATION			BOOK VALUE
		DEBIT	CREDIT	BALANCE	DEBIT	CREDIT	BALANCE	

Problem 23-4 Analyzing a Source Document

(1)

GENERAL JOURNAL   PAGE _____

	DATE	DESCRIPTION	POST. REF.	DEBIT	CREDIT	
1						1
2						2
3						3

(2)

GENERAL JOURNAL   PAGE _____

	DATE	DESCRIPTION	POST. REF.	DEBIT	CREDIT	
1						1
2						2
3						3

(3)

GENERAL JOURNAL   PAGE _____

	DATE	DESCRIPTION	POST. REF.	DEBIT	CREDIT	
1						1
2						2
3						3

Problem 23-5 **Preparing a Depreciation Schedule and Journalizing the Depreciation Adjusting Entry**

(1)

Date	Cost	Annual Depreciation	Accumulated Depreciation	Book Value
January 7	$2,360	———	———	$2,360
First year				
Second year				
Third year				
Fourth year				
Fifth year				

(2), (3)

GENERAL JOURNAL   PAGE ___40___

	DATE		DESCRIPTION	POST. REF.	DEBIT	CREDIT	
1							1
2							2
3							3
4							4
5							5
6							6
7							7
8							8
9							9
10							10
11							11
12							12

(4)

_____

_____

Computerized Accounting Using Peachtree

Software Objectives

When you have completed this chapter, you will be able to use Peachtree to:

1. Record adjusting entries for depreciation.
2. Prepare financial statements that include accumulated depreciation and depreciation expenses.

Problem 23-7   Recording Adjusting Entries for Depreciation

INSTRUCTIONS

Beginning a Session

Step 1   Select the problem set: InBeat CD Shop (Prob. 23-7).
Step 2   Rename the company by adding your initials, e.g., InBeat (Prob. 23-7: XXX).
Step 3   Set the system date to December 31, 2008.

Completing the Accounting Problem

Step 4   Review the information provided in your textbook.
Step 5   Record the adjusting entries for depreciation using the **General Journal Entry** option.

 TIP: As a shortcut, you can use the General Ledger navigation aid shown at the bottom of the Peachtree window to access the General Journal Entry window.

Step 6   Print a General Journal report and proof your work.
Step 7   Answer the Analyze question.

Checking Your Work and Ending the Session

Step 8   Click the **Close Problem** button in the Glencoe Smart Guide window.
Step 9   If your teacher has asked you to check your solution, select *Check my answer to this problem*. Review, print, and close the report.
Step 10 Click the **Close Problem** button and select a save option.

Mastering Peachtree

On a separate sheet of paper, list the account types assigned to each of the following general ledger accounts: **Store Equipment, Accum. Depr.—Store Equipment,** and **Depr. Exp.—Store Equipment**.

DO YOU HAVE A QUESTION

Q. *Does Peachtree include any options to track plant assets?*

A. No, Peachtree does not include any options to track plant assets. However, there are specialized programs designed to help a company keep track of its assets and prepare depreciation schedules. As another alternative, you could use a spreadsheet program for this task.

Notes

Peachtree automatically posts general journal entries to the corresponding general ledger accounts.

Problem 23-8   Reporting Depreciation Expense on the Work Sheet and Financial Statements

INSTRUCTIONS

Beginning a Session

Step 1   Select the problem set: Shutterbug Cameras (Prob. 23-8).

Step 2   Rename the company by adding your initials.

Step 3   Set the system date to December 31, 2008.

Completing the Accounting Problem

Step 4   Review the information provided in your textbook.

Step 5   Record the adjusting entries for depreciation using the **General Journal Entry** option.

Notes

Peachtree does not include an option to complete a work sheet.

TIP: Set the date range from 12/31/08 to 12/31/08 to print only the adjusting entries.

Step 6   Print a General Journal report and proof your work.

Step 7   Print an Income Statement, a Statement of Retained Earnings, and a Balance Sheet.

TIP: Use the standard financial statements when instructed to print an Income Statement, a Statement of Retained Earnings, or Balance Sheet.

Step 8   Answer the Analyze question.

Ending the Session

Step 9   Click the **Close Problem** button in the Glencoe Smart Guide window to save your work.

Problem 23-9   Calculating and Recording Depreciation Expense

INSTRUCTIONS

Beginning a Session

Step 1   Select the problem set: Cycle Tech Bicycles (Prob. 23-9).
Step 2   Rename the company by adding your initials.
Step 3   Set the system date to December 31, 2008.

Completing the Accounting Problem

Step 4   Review the information provided in your textbook.
Step 5   Calculate the depreciation for the first two years.
Step 6   Record the depreciation adjustment for the first year using the **General Journal Entry** option.
Step 7   Print a General Journal report and proof your work.
Step 8   Print a General Ledger report for the manufacturing equipment (accumulated depreciation and depreciation expense) general ledger accounts.

TIP: Set the report filter options to control which accounts appear on the General Ledger report.

Step 9   Answer the Analyze question.

Checking Your Work and Ending the Session

Step 10   Click the **Close Problem** button in the Glencoe Smart Guide window.
Step 11   If your teacher has asked you to check your solution, select *Check my answer to this problem*. Review, print, and close the report.
Step 12   Click the **Close Problem** button and select a save option.

Peachtree Guide

Problem 23-10   Calculating and Recording Adjustments

INSTRUCTIONS

Beginning a Session

Step 1   Select the problem set: River's Edge Canoe & Kayak (Prob. 23-10).
Step 2   Rename the company by adding your initials.
Step 3   Set the system date to December 31, 2008.

Completing the Accounting Problem

Step 4   Review the information provided in your textbook.
Step 5   Record all of the adjusting entries.

TIP: Remember to use the **Inventory Adjustment** account to record the merchandise inventory adjustment.

Step 6   Print a General Journal report and proof your work.
Step 7   Print an Income Statement and a Balance Sheet.
Step 8   Click the **Save Pre-closing Balances** button in the Glencoe Smart Guide window.
Step 9   Close the fiscal year.

TIP: To close the fiscal year, choose **System** from the *Tasks* menu and then select **Year-End Wizard.**

Step 10   Print a Post-Closing Trial Balance.
Step 11   Answer the Analyze question.

Ending the Session

Step 12   Click the **Close Problem** button in the Glencoe Smart Guide window.

FAQs

Can you edit a general journal entry from an earlier period after you close the fiscal year?

No, you cannot edit a general journal entry after you close the fiscal year. You must make a correcting entry, or restore a backup and then make the correction.

Working Papers *for End-of-Chapter Problems*

Problem 23-6   Opening a Plant Asset Record

PLANT ASSET RECORD

ITEM _____   GENERAL LEDGER ACCOUNT _____

SERIAL NUMBER _____   MANUFACTURER _____

PURCHASED FROM _____   EST. DISPOSAL VALUE _____

ESTIMATED LIFE _____   LOCATION _____

DEPRECIATION
METHOD _____   DEPRECIATION
                                 PER YEAR _____

DATE	EXPLANATION	ASSET			ACCUMULATED DEPRECIATION			BOOK VALUE
		DEBIT	CREDIT	BALANCE	DEBIT	CREDIT	BALANCE	

Analyze: _____

_____

Problem 23-7   Recording Adjusting Entries for Depreciation

GENERAL JOURNAL   PAGE _____

	DATE	DESCRIPTION	POST. REF.	DEBIT	CREDIT	
1						1
2						2
3						3
4						4
5						5
6						6
7						7

Analyze: _____

_____

Problem 23-8   Reporting Depreciation Expense on the Work Sheet and Financial Statements

(1), (2)

Shutterbug

Work

For the Year Ended

	ACCT. NO.	ACCOUNT NAME	TRIAL BALANCE DEBIT	TRIAL BALANCE CREDIT	ADJUSTMENTS DEBIT	ADJUSTMENTS CREDIT
1	101	Cash in Bank	9 30 00 00			
2	105	Change Fund	5 00 00			
3	110	Petty Cash Fund	2 00 00			
4	115	Accounts Receivable	1 20 00 0			
5	125	Merchandise Inventory	50 00 00 0			(a) 3 69 00 00
6	130	Supplies	4 00 00 0			(c) 2 96 00 00
7	135	Prepaid Insurance	2 40 00 0			(b) 1 20 00 00
8	140	Office Equipment	26 00 00 0			
9	142	Accum. Depr.—Office Equip.		7 50 00 0		
10	145	Store Equipment	19 20 00 0			
11	147	Accum. Depr.—Store Equip.		3 60 00 0		
12	201	Accounts Payable		1 51 00 0		
13	207	Fed. Corporate Income Tax Pay.	—	—		(d) 60 00 0
14	215	Sales Tax Payable		1 32 00 0		
15	301	Capital Stock		51 36 50 0		
16	305	Retained Earnings		24 00 00 0		
17	310	Income Summary	—	—	(a) 3 69 00 00	
18	401	Sales		66 94 00 0		
19	410	Sales Returns and Allowances	8 75 00			
20	501	Purchases	21 00 00 0			
21	505	Transportation In	3 10 00			
22	510	Purchases Discounts		2 15 00		
23	515	Purchases Returns and Allow.		1 60 00 0		
24	601	Advertising Expense	3 00 00 0			
25	610	Cash Short & Over	5 00 0			
26	615	Depr. Exp.—Office Equip.	—			
27	617	Depr. Exp.—Store Equip.	—			
28	620	Fed. Corporate Income Tax Exp.	5 00 00 0		(d) 60 00 0	
29		Carry Forward	143 03 50 0	158 05 00 0		
30						
31						
32						
33						

Cameras _____

Sheet _____

December 31, 20-- _____

ADJUSTED TRIAL BALANCE		INCOME STATEMENT		BALANCE SHEET		
DEBIT	CREDIT	DEBIT	CREDIT	DEBIT	CREDIT	
						1
						2
						3
						4
						5
						6
						7
						8
						9
						10
						11
						12
						13
						14
						15
						16
						17
						18
						19
						20
						21
						22
						23
						24
						25
						26
						27
						28
						29
						30
						31
						32
						33

Problem 23-8 (continued)

	ACCT. NO.	ACCOUNT NAME	TRIAL BALANCE		ADJUSTMENTS	
			DEBIT	CREDIT	DEBIT	CREDIT
1		**Brought Forward**	143 035 00	158 050 00		
2						
3	630	**Insurance Expense**	—		(b) 1 200 00	
4	640	**Maintenance Expense**	1 400 00			
5	645	**Miscellaneous Expense**	425 00			
6	650	**Rent Expense**	6 000 00			
7	655	**Salaries Expense**	6 040 00			
8	660	**Supplies Expense**	—		(c) 2 960 00	
9	670	**Utilities Expense**	1 150 00			
10			158 050 00	158 050 00		
11		**Net Income**				
12						
13						
14						
15						
16						
17						
18						
19						
20						
21						
22						
23						
24						
25						
26						
27						
28						
29						
30						
31						
32						

Cameras

(continued)

December 31, 20– –

ADJUSTED TRIAL BALANCE		INCOME STATEMENT		BALANCE SHEET		
DEBIT	CREDIT	DEBIT	CREDIT	DEBIT	CREDIT	
						1
						2
						3
						4
						5
						6
						7
						8
						9
						10
						11
						12
						13
						14
						15
						16
						17
						18
						19
						20
						21
						22
						23
						24
						25
						26
						27
						28
						29
						30
						31
						32

Problem 23-8   (continued)
(3)

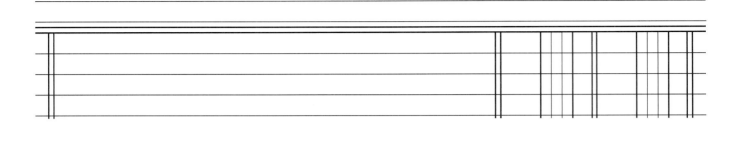

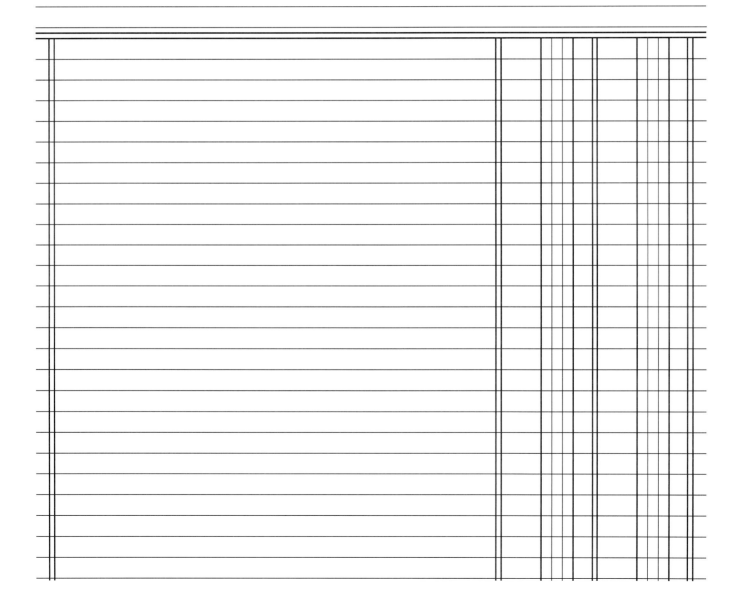

Problem 23-8 (continued)

Analyze: _____

_____ _____

_____ _____

_____ _____

Problem 23-9 Calculating and Recording Depreciation Expense

(1)

Date	Cost	Annual Depreciation	Accumulated Depreciation	Book Value
Purchased Aug. 1	$410,000	——	——	$410,000
First year				
Second year				

(2)

Depreciation Expense—
Manufacturing Equipment
Account 623

Accumulated Depreciation—
Manufacturing Equipment
Account 152

(3)

GENERAL JOURNAL   PAGE _____

	DATE	DESCRIPTION	POST. REF.	DEBIT	CREDIT	
1						1
2						2
3						3
4						4
5						5
6						6
7						7

Problem 23-9 (concluded)

(4)

GENERAL LEDGER

ACCOUNT _____ ACCOUNT NO. _____

DATE	DESCRIPTION	POST. REF.	DEBIT	CREDIT	BALANCE	
					DEBIT	CREDIT

ACCOUNT _____ ACCOUNT NO. _____

DATE	DESCRIPTION	POST. REF.	DEBIT	CREDIT	BALANCE	
					DEBIT	CREDIT

Analyze: _____

_____

Problem 23-10   Calculating and Recording Adjustments
(1), (2)

River's Edge

Work

For the Year Ended

	ACCT. NO.	ACCOUNT NAME	TRIAL BALANCE		ADJUSTMENTS	
			DEBIT	CREDIT	DEBIT	CREDIT
1	101	Cash in Bank	6 69 0 00			
2	105	Change Fund	20 0 00			
3	115	Accounts Receivable	1 2 40 0 00			
4	130	Merchandise Inventory	1 6 30 0 00			
5	135	Supplies	4 85 0 00			
6	140	Prepaid Insurance	6 00 0 00			
7	145	Delivery Truck	3 2 00 0 00			
8	147	Accum. Depr.—Delivery Truck		1 6 00 0 00		
9	150	Store Equipment	1 3 00 0 00			
10	152	Accum. Depr.—Store Equip.		6 00 0 00		
11	155	Building	1 6 0 00 0 00			
12	157	Accum. Depr.—Building		6 0 00 0 00		
13	160	Land	5 5 00 0 00			
14	201	Accounts Payable		2 30 0 00		
15	204	Fed. Corporate Income Tax Pay.	—	—		
16	215	Sales Tax Payable		1 62 0 00		
17	301	Capital Stock		1 4 0 00 0 00		
18	305	Retained Earnings		2 0 52 0 00		
19	310	Income Summary	—			
20	401	Sales		1 8 0 00 0 00		
21	501	Purchases	9 6 00 0 00			
22	505	Transportation In	5 00 0 00			
23	601	Advertising Expense	3 50 0 00			
24	615	Depr. Exp.—Store Equip.	—			
25	620	Depr. Exp.—Delivery Truck	—			
26	622	Depr. Exp.—Building	—			
27	625	Fed. Corporate Income Tax Exp.	2 40 0 00			
28		Carry Forward	4 1 3 34 0 00	4 2 6 44 0 00		
29						
30						
31						
32						
33						

Canoe & Kayak

Sheet

December 31, 20– –

	ADJUSTED TRIAL BALANCE		INCOME STATEMENT		BALANCE SHEET		
	DEBIT	CREDIT	DEBIT	CREDIT	DEBIT	CREDIT	
							1
							2
							3
							4
							5
							6
							7
							8
							9
							10
							11
							12
							13
							14
							15
							16
							17
							18
							19
							20
							21
							22
							23
							24
							25
							26
							27
							28
							29
							30
							31
							32
							33

Problem 23-10 (continued)

	ACCT. NO.	ACCOUNT NAME	TRIAL BALANCE DEBIT	TRIAL BALANCE CREDIT	ADJUSTMENTS DEBIT	ADJUSTMENTS CREDIT
1		**Brought Forward**	4133400 0	4264400 0		
2						
3	630	**Gas Expense**	14000 0			
4	635	**Insurance Expense**	————			
5	655	**Miscellaneous Expense**	54000 0			
6	670	**Supplies Expense**	————			
7	680	**Utilities Expense**	63000 0			
8			4264400 0	4264400 0		
9		**Net Income**				
10						
11						
12						
13						
14						
15						
16						
17						
18						
19						
20						
21						
22						
23						
24						
25						
26						
27						
28						
29						
30						
31						
32						

Canoe & Kayak

(continued)

December 31, 20––

ADJUSTED TRIAL BALANCE		INCOME STATEMENT		BALANCE SHEET		
DEBIT	CREDIT	DEBIT	CREDIT	DEBIT	CREDIT	
						1
						2
						3
						4
						5
						6
						7
						8
						9
						10
						11
						12
						13
						14
						15
						16
						17
						18
						19
						20
						21
						22
						23
						24
						25
						26
						27
						28
						29
						30
						31
						32

Problem 23-10   (continued)
(3), (4)

GENERAL JOURNAL   PAGE _____

	DATE	DESCRIPTION	POST. REF.	DEBIT	CREDIT	
1						1
2						2
3						3
4						4
5						5
6						6
7						7
8						8
9						9
10						10
11						11
12						12
13						13
14						14
15						15
16						16
17						17
18						18
19						19
20						20
21						21
22						22
23						23
24						24
25						25
26						26
27						27
28						28
29						29
30						30
31						31
32						32
33						33
34						34
35						35
36						36

Problem 23-10   (continued)

GENERAL LEDGER (PARTIAL)

ACCOUNT ___Merchandise Inventory___ ACCOUNT NO. ___130___

DATE		DESCRIPTION	POST. REF.	DEBIT	CREDIT	BALANCE	
						DEBIT	CREDIT
20--							
Dec.	31	Balance	✓			16 300 00	

ACCOUNT ___Supplies___ ACCOUNT NO. ___135___

DATE		DESCRIPTION	POST. REF.	DEBIT	CREDIT	BALANCE	
						DEBIT	CREDIT
20--							
Dec.	31	Balance	✓			4 850 00	

ACCOUNT ___Prepaid Insurance___ ACCOUNT NO. ___140___

DATE		DESCRIPTION	POST. REF.	DEBIT	CREDIT	BALANCE	
						DEBIT	CREDIT
20--							
Dec.	31	Balance	✓			6 000 00	

ACCOUNT ___Accumulated Depreciation—Delivery Truck___ ACCOUNT NO. ___147___

DATE		DESCRIPTION	POST. REF.	DEBIT	CREDIT	BALANCE	
						DEBIT	CREDIT
20--							
Dec.	31	Balance	✓				16 000 00

Problem 23-10   (continued)

ACCOUNT __Accumulated Depreciation—Store Equipment__   ACCOUNT NO. __152__

DATE	DESCRIPTION	POST. REF.	DEBIT	CREDIT	BALANCE DEBIT	BALANCE CREDIT
20--						
Dec. 31	Balance	✓				6 000 00

ACCOUNT __Accumulated Depreciation—Building__   ACCOUNT NO. __157__

DATE	DESCRIPTION	POST. REF.	DEBIT	CREDIT	BALANCE DEBIT	BALANCE CREDIT
20--						
Dec. 31	Balance	✓				60 000 00

ACCOUNT __Federal Corporate Income Tax Payable__   ACCOUNT NO. __204__

DATE	DESCRIPTION	POST. REF.	DEBIT	CREDIT	BALANCE DEBIT	BALANCE CREDIT

ACCOUNT __Retained Earnings__   ACCOUNT NO. __305__

DATE	DESCRIPTION	POST. REF.	DEBIT	CREDIT	BALANCE DEBIT	BALANCE CREDIT
20--						
Dec. 31	Balance	✓				20 520 00

Problem 23-10   (continued)

ACCOUNT __Income Summary__   ACCOUNT NO. ___310___

DATE		DESCRIPTION	POST. REF.	DEBIT	CREDIT	BALANCE	
						DEBIT	CREDIT

ACCOUNT __Sales__   ACCOUNT NO. ___401___

DATE		DESCRIPTION	POST. REF.	DEBIT	CREDIT	BALANCE	
						DEBIT	CREDIT
20--							
Dec.	31	Balance	✓				18000000

ACCOUNT __Purchases__   ACCOUNT NO. ___501___

DATE		DESCRIPTION	POST. REF.	DEBIT	CREDIT	BALANCE	
						DEBIT	CREDIT
20--							
Dec.	31	Balance	✓			9600000	

ACCOUNT __Transportation In__   ACCOUNT NO. ___505___

DATE		DESCRIPTION	POST. REF.	DEBIT	CREDIT	BALANCE	
						DEBIT	CREDIT
20--							
Dec.	31	Balance	✓			500000	

Problem 23-10 (continued)

ACCOUNT _Advertising Expense_____ ACCOUNT NO. ___601___

DATE		DESCRIPTION	POST. REF.	DEBIT	CREDIT	BALANCE	
						DEBIT	CREDIT
20--							
Dec.	31	Balance	✓			3 5 0 0 00	

ACCOUNT _Depreciation Expense—Store Equipment_____ ACCOUNT NO. ___615___

DATE	DESCRIPTION	POST. REF.	DEBIT	CREDIT	BALANCE	
					DEBIT	CREDIT

ACCOUNT _Depreciation Expense—Delivery Truck_____ ACCOUNT NO. ___620___

DATE	DESCRIPTION	POST. REF.	DEBIT	CREDIT	BALANCE	
					DEBIT	CREDIT

ACCOUNT _Depreciation Expense—Building_____ ACCOUNT NO. ___622___

DATE	DESCRIPTION	POST. REF.	DEBIT	CREDIT	BALANCE	
					DEBIT	CREDIT

Problem 23-10 (continued)

ACCOUNT __Federal Corporate Income Tax Expense__ ACCOUNT NO. __625__

DATE		DESCRIPTION	POST. REF.	DEBIT	CREDIT	BALANCE	
						DEBIT	CREDIT
20--							
Dec.	31	Balance	✓			2400 00	

ACCOUNT __Gas Expense__ ACCOUNT NO. __630__

DATE		DESCRIPTION	POST. REF.	DEBIT	CREDIT	BALANCE	
						DEBIT	CREDIT
20--							
Dec.	31	Balance	✓			1400 00	

ACCOUNT __Insurance Expense__ ACCOUNT NO. __635__

DATE		DESCRIPTION	POST. REF.	DEBIT	CREDIT	BALANCE	
						DEBIT	CREDIT

ACCOUNT __Miscellaneous Expense__ ACCOUNT NO. __655__

DATE		DESCRIPTION	POST. REF.	DEBIT	CREDIT	BALANCE	
						DEBIT	CREDIT
20--							
Dec.	31	Balance	✓			5400 00	

Problem 23-10   (concluded)

ACCOUNT   *Supplies Expense*   ACCOUNT NO.   *670*

DATE	DESCRIPTION	POST. REF.	DEBIT	CREDIT	BALANCE DEBIT	BALANCE CREDIT

ACCOUNT   *Utilities Expense*   ACCOUNT NO.   *680*

DATE	DESCRIPTION	POST. REF.	DEBIT	CREDIT	BALANCE DEBIT	BALANCE CREDIT
20--						
Dec. 31	*Balance*	✓			6 30 0 00	

Analyze: _____

_____

_____

_____

_____

Problem 23-11   Examining Depreciation Adjustments

1. _____

_____

_____

_____

2. _____

_____

_____

Analyze: _____

_____

_____

_____

CHAPTER 23   Plant Assets and Depreciation

Self-Test

Part A   True or False

Directions: *Circle the letter* T *in the Answer column if the statement is true; circle the letter* F *if the statement is false.*

Answer

T   F   **1.** If estimated annual depreciation is $3,600, a plant asset that was owned only four months during the year would have a depreciation expense of $1,200.

T   F   **2.** For accounting purposes, the cost of land is not depreciated.

T   F   **3.** The normal balance of Accumulated Depreciation—Computer Equipment is a debit.

T   F   **4.** Depreciation Expense is classified as an expense, has a normal credit balance, and is reported on the income statement.

T   F   **5.** The cost of a plant asset is the price paid for the asset when it is purchased, plus any installation and delivery charges.

T   F   **6.** The book value of a plant asset is determined by subtracting the accumulated depreciation to date from the initial cost of the asset.

T   F   **7.** Accumulated Depreciation—Store Equipment is a contra asset account and is reported on the income statement.

T   F   **8.** A company will often decide to replace, sell, or discard a plant asset while the asset still has some monetary value.

T   F   **9.** The straight-line method of depreciation allocates the same amount of depreciation expense each year over the estimated useful life of the asset.

T   F   **10.** The estimated disposal value of a plant asset is the part of the asset's cost that the business expects to get back at the end of the asset's useful life.

Part B   Multiple Choice

Directions: *Only one of the choices given with each of the following statements is correct. Write the letter of the correct answer in the Answer column.*

Answer

_____ **1.** Among the following accounts, the only non-plant asset is
(A) Supplies.
(B) Computer Equipment.
(C) Land.
(D) Accumulated Depreciation—Office Equipment.

_____ **2.** The cost of a plant asset is determined by the purchase price for the asset and all of the following *except*
(A) sales taxes.
(B) delivery charges.
(C) estimated disposal value of the asset.
(D) installation charges.

_____ **3.** Each of the following is one of the four factors that affect the depreciation estimate *except*
(A) the cost of the plant asset.
(B) the estimated useful life of the asset.
(C) the depreciation method used.
(D) the location of the asset.

Directions: *Using the information below, choose the correct answers for questions 4 though 6.*

A computer was purchased on January 4 at a cost of $3,800. The computer has an estimated disposal value of $200 and an estimated useful life of six years.

Answer

_____ **4.** The estimated amount to be depreciated on this computer is
(A) $3,800.   (C) $600.
(B) $3,600.   (D) $200.

_____ **5.** The estimated annual depreciation expense is
(A) $633.33.   (C) $600.
(B) $3,600.   (D) $200.

_____ **6.** If the computer had been purchased on August 4 instead of in January, the estimated depreciation expense for the partial year would be
(A) $200.   (C) $600.
(B) $3,600.   (D) $250.

CHAPTER 24   Uncollectible Accounts Receivable

Study Plan

Check Your Understanding

Section 1	*Read Section 1 on pages 652–656 and complete the following exercises on page 657.*
	❏ Thinking Critically
	❏ Communicating Accounting
	❏ Problem 24-1 *Using the Direct Write-Off Method*
Section 2	*Read Section 2 on pages 658–665 and complete the following exercises on page 666.*
	❏ Thinking Critically
	❏ Analyzing Accounting
	❏ Problem 24-2 *Writing Off Accounts Under the Allowance Method*
Section 3	*Read Section 3 on pages 667–669 and complete the following exercises on page 670.*
	❏ Thinking Critically
	❏ Computing in the Business World
	❏ Problem 24-3 *Estimating Uncollectible Accounts Expense Using the Percentage of Net Sales Method*

Summary	*Review the Chapter 24 Summary on page 671 in your textbook.*
	❏ Key Concepts
Review and Activities	*Complete the following questions and exercises on pages 672–673.*
	❏ Using Key Terms
	❏ Understanding Accounting Concepts and Procedures
	❏ Case Study
	❏ Conducting an Audit with Alex
	❏ Internet Connection
	❏ Workplace Skills
Computerized Accounting	*Read the Computerized Accounting information on page 674.*
	❏ *Making the Transition from a Manual to a Computerized System*
	❏ *Accounts Receivable in Peachtree*
Problems	*Complete the following end-of-chapter problems for Chapter 24.*
	❏ Problem 24-4 *Using the Direct Write-Off Method*
	❏ Problem 24-5 *Calculating and Recording Estimated Uncollectible Accounts Expense*
	❏ Problem 24-6 *Writing Off Accounts Under the Allowance Method*
	❏ Problem 24-7 *Estimating Uncollectible Accounts Expense*
	❏ Problem 24-8 *Reporting Uncollectible Amounts on the Financial Statements*
Challenge Problem	❏ Problem 24-9 *Using the Allowance Method for Write-Offs*
Chapter Reviews and Working Papers	*Complete the following exercises for Chapter 24 in your Chapter Reviews and Working Papers.*
	❏ Chapter Review
	❏ Self-Test

CHAPTER 24 REVIEW Uncollectible Accounts Receivable

Part 1 Accounting Vocabulary (5 points)

Total Points	39
Student's Score	

Directions: *Using terms from the following list, complete the sentences below. Write the letter of the term you have chosen in the space provided.*

A. aging of accounts receivable method	C. book value of accounts receivable	E. percentage of net sales method
B. allowance method	D. direct write-off method	F. uncollectible accounts

___F___ **0.** Accounts receivable accounts that cannot be collected are called _____.

_____ **1.** The amount a business can reasonably expect to receive from all its charge customers is called the _____.

_____ **2.** When a business determines that an actual amount is uncollectible, the _____ is used to remove the uncollectible amount from the accounts receivable subsidiary ledger and the controlling account in the general ledger.

_____ **3.** When the _____ is used to estimate the uncollectible amount, each customer's account is examined and classified according to its due date.

_____ **4.** When using the _____ of estimating uncollectible accounts expense, a business assumes that a certain percentage of each year's net sales will be uncollectible.

_____ **5.** The _____ of accounting for uncollectible accounts matches potential bad debts expenses with sales made during the same fiscal period.

Part 2 Accounting for Uncollectible Accounts Receivable (16 points)

Directions: *Using the following list, analyze the transactions below. Determine the account(s) to be debited and credited. Write your answers in the space provided.*

General Ledger Accounts		Subsidiary Ledger Accounts
A. Cash in Bank	D. Sales Tax Payable	G. Jim Wright
B. Accounts Receivable	E. Sales	H. Ti Yong
C. Allowance for Uncollectible Accounts	F. Uncollectible Accounts Expense	

Debit	Credit	
B,H	**D,E**	**0.** Sold merchandise on account, plus sales tax, to Ti Yong.
_____	_____	**1.** Wrote off the account of Ti Yong as uncollectible using the allowance method.
_____	_____	**2.** Sold merchandise on account, plus sales tax, to Jim Wright.
_____	_____	**3.** Reinstated Ti Yong's account.
_____	_____	**4.** Received a check for payment on account from Ti Yong.
_____	_____	**5.** Wrote off Jim Wright's account as uncollectible using the direct write-off method.
_____	_____	**6.** Reinstated the account of Jim Wright.
_____	_____	**7.** Received a check from Jim Wright for payment of his account.
_____	_____	**8.** Estimated that 3% of the net sales would be uncollectible.

Part 3   Extending Credit (10 points)

Directions: *Read each of the following statements to determine whether the statement is true or false. Write your answer in the space provided.*

__True__ **0.** Two common methods used to estimate bad debts expense are the percentage of net sales method and the aging of accounts receivable method.

_____ **1.** When the direct write-off method is used for writing off an uncollectible account, Uncollectible Accounts Expense is the account debited for the amount of the loss.

_____ **2.** The two accounts affected by the adjusting entry for the allowance method of accounting for uncollectible accounts are Uncollectible Accounts Expense and Accounts Receivable.

_____ **3.** Before reinstating a charge customer's account, the receipt of cash to pay off the amount owed must be journalized.

_____ **4.** The book value of accounts receivable is the difference between the balance of Accounts Receivable and the balance of Allowance for Uncollectible Accounts.

_____ **5.** Businesses that sell on credit usually expect to sell more than if they accepted only cash.

_____ **6.** Allowance for Uncollectible Accounts is classified as a contra asset account and appears on the balance sheet as a deduction from Cash in Bank.

_____ **7.** Charge customers' accounts that are declared uncollectible become a liability to the business.

_____ **8.** Allowance for Uncollectible Accounts usually has a zero balance at the end of a fiscal period.

_____ **9.** Uncollectible accounts are sometimes paid at a later date by the customer whose account was written off.

_____ **10.** The two general ledger accounts affected by the direct write-off method of accounting for uncollectible accounts are Uncollectible Accounts Expense and Allowance for Uncollectible Accounts.

Part 4   Estimating Uncollectible Accounts Expense (8 points)

Directions: *The Ramona Estevez Company uses the allowance method of handling uncollectible accounts. Before any adjustments, the ledger contains the following balances:*

Sales	$400,000
Accounts Receivable	120,000
Sales Discounts	20,000
Allowance for Uncollectible Accounts	500
Sales Returns and Allowances	10,000
Uncollectible Accounts Expense	0

The Ramona Estevez Company estimates that the uncollectible accounts for the year will be 2% of the net sales. Using this information, answer the following questions.

____$370,000____ **0.** The net sales for the fiscal year are _____.

_____ **1.** The estimated percentage of net sales that will be uncollectible is _____.

_____ **2.** The estimated uncollectible account expense for the fiscal year is _____.

_____ **3.** The account to be debited for the estimated uncollectible amount is _____.

_____ **4.** After the adjusting entry is posted, the balance of the Allowance for Uncollectible Accounts account is _____.

_____ **5.** The account to be credited for the estimated uncollectible amount is _____.

_____ **6.** After posting the adjusting entry, the balance of the Uncollectible Accounts Expense account is _____.

_____ **7.** The book value of accounts receivable after the adjusting entry is posted is _____.

_____ **8.** The financial statement on which Uncollectible Accounts Expense is reported is the _____.

Working Papers _for Section Problems_

Problem 24-1   Using the Direct Write-Off Method

(1)

GENERAL JOURNAL   PAGE _____

	DATE	DESCRIPTION	POST. REF.	DEBIT	CREDIT	
1						1
2						2
3						3
4						4
5						5
6						6
7						7
8						8
9						9
10						10
11						11
12						12
13						13
14						14
15						15
16						16
17						17
18						18
19						19
20						20
21						21
22						22
23						23
24						24
25						25
26						26
27						27
28						28
29						29
30						30

Problem 24-1   (continued)

(2)

GENERAL LEDGER

ACCOUNT __Cash in Bank__   ACCOUNT NO. __101__

DATE		DESCRIPTION	POST. REF.	DEBIT	CREDIT	BALANCE DEBIT	BALANCE CREDIT
20--							
Apr.	1	Balance	✓			9 4 2 8 00	

ACCOUNT __Accounts Receivable__   ACCOUNT NO. __115__

DATE		DESCRIPTION	POST. REF.	DEBIT	CREDIT	BALANCE DEBIT	BALANCE CREDIT
20--							
Apr.	1	Balance	✓			7 2 9 0 00	

ACCOUNT __Sales Tax Payable__   ACCOUNT NO. __215__

DATE		DESCRIPTION	POST. REF.	DEBIT	CREDIT	BALANCE DEBIT	BALANCE CREDIT
20--							
Apr.	1	Balance	✓				2 4 8 00

ACCOUNT __Sales__   ACCOUNT NO. __401__

DATE		DESCRIPTION	POST. REF.	DEBIT	CREDIT	BALANCE DEBIT	BALANCE CREDIT
20--							
Apr.	1	Balance	✓				2 4 1 6 0 00

ACCOUNT __Uncollectible Accounts Expense__   ACCOUNT NO. __680__

DATE		DESCRIPTION	POST. REF.	DEBIT	CREDIT	BALANCE DEBIT	BALANCE CREDIT
20--							
Apr.	1	Balance	✓			9 2 8 00	

Problem 24-1   (concluded)

(2)

ACCOUNTS RECEIVABLE SUBSIDIARY LEDGER

Name *Sonya Dickson* _____

Address _____

DATE		DESCRIPTION	POST. REF.	DEBIT	CREDIT	BALANCE

Problem 24-2   Writing Off Accounts Under the Allowance Method

(1)

GENERAL JOURNAL   PAGE _____

	DATE		DESCRIPTION	POST. REF.	DEBIT	CREDIT	
1							1
2							2
3							3
4							4
5							5
6							6
7							7
8							8
9							9
10							10
11							11
12							12
13							13
14							14
15							15
16							16
17							17
18							18
19							19
20							20
21							21
22							22
23							23
24							24
25							25
26							26
27							27
28							28
29							29
30							30

Problem 24-2 (continued)

(2)

ACCOUNTS RECEIVABLE SUBSIDIARY LEDGER

Name _Jack Bowers_ _____

Address _____

DATE		DESCRIPTION	POST. REF.	DEBIT	CREDIT	BALANCE
20--						
May	1	Balance	✓			1 05 0 00

GENERAL LEDGER (PARTIAL)

ACCOUNT _Cash in Bank_ _____ ACCOUNT NO. _101_

DATE		DESCRIPTION	POST. REF.	DEBIT	CREDIT	BALANCE DEBIT	BALANCE CREDIT
20--							
Nov.	1	Balance	✓			9 42 0 00	

ACCOUNT _Accounts Receivable_ _____ ACCOUNT NO. _115_

DATE		DESCRIPTION	POST. REF.	DEBIT	CREDIT	BALANCE DEBIT	BALANCE CREDIT
20--							
May	1	Balance	✓			20 40 0 00	

Problem 24-2 (concluded)

ACCOUNT ___Allowance for Uncollectible Accounts___ ACCOUNT NO. ___117___

DATE	DESCRIPTION	POST. REF.	DEBIT	CREDIT	BALANCE DEBIT	BALANCE CREDIT
20-- May 1	Balance	✓				1 200 00

ACCOUNT ___Uncollectible Accounts Expense___ ACCOUNT NO. ___675___

DATE	DESCRIPTION	POST. REF.	DEBIT	CREDIT	BALANCE DEBIT	BALANCE CREDIT

(3)

Problem 24-3　Estimating Uncollectible Accounts Expense Using the Percentage of Net Sales Method

(1)

Company	Net Sales	Percentage of Net Sales	Uncollectible Accounts Expense
Andrews Co.		2%	
The Book Nook		1%	
Cable, Inc.		$1\frac{1}{2}$%	
Davis, Inc.		2%	
Ever-Sharp Co.		$1\frac{1}{4}$%	

(2)

GENERAL JOURNAL　　　　　　　　　　　PAGE ___21___

	DATE	DESCRIPTION	POST. REF.	DEBIT	CREDIT	
1						1
2						2
3						3
4						4
5						5
6						6
7						7
8						8
9						9
10						10
11						11
12						12

Computerized Accounting Using Peachtree

Software Objectives

When you have completed this chapter, you will be able to use Peachtree to:

1. Record a transaction to write off an account using the direct write-off method.
2. Record estimated uncollectible accounts expense.
3. Record a transaction to write off an account using the allowance method.
4. Print financial statements that include uncollectible amounts.

Problem 24-4   Using the Direct Write-Off Method

INSTRUCTIONS

Beginning a Session

Step 1   Select the problem set: Sunset Surfwear (Prob. 24-4).
Step 2   Rename the company by adding your initials.
Step 3   Set the system date to June 30, 2008.

Completing the Accounting Problem

Step 4   Review the information provided in your textbook.
Step 5   Record the June 1 transaction using the **Receipts** option.

> ### *June 1 Wrote off the $288.75 account of Alex Hamilton as uncollectible, Memo 223.*

To record a transaction to write off an account:

- Choose **Receipts** from the *Tasks* menu.
- Type **6/1/08** in the *Deposit ticket ID* field.
- Enter **HAM** in the *Customer ID* field to select the record for Alex Hamilton.
- Type **Memo 223** in the *Reference* field.
- Enter **6/1/08** in the *Date* field.
- Move to the *Cash Account* field shown below the *Payment Method* field.

TIP: You must identify the **Uncollectible Accounts Expense** as the Cash Account to record a transaction to write off an invoice. If the account does not appear, choose **Global** from the *Options* menu. Change the setting to show the general ledger accounts for Accounts Receivable.

- Enter **670** (Uncollectible Accounts Expense) in the *Cash Account* field.

  Changing the Cash Account to **Uncollectible Accounts Expense** is required to write off an account. When you mark an invoice as "Paid," Peachtree will debit **Uncollectible Accounts Expense** instead of **Cash**. Peachtree will also credit **Accounts Receivable.**

- Click the **Pay** check box next to the outstanding invoice.
- Compare the information on your screen to the completed **Receipts** windows shown in Figure 24-4A. If necessary, make any changes before you continue.

DO YOU HAVE A QUESTION

Q. *Can you use the* **General Journal Entry** *task option to record a transaction to write off an uncollectible account?*

A. No, you cannot use the **General Journal Entry** option to record a transaction to write off an uncollectible account. As you may remember, Peachtree does not provide a way to identify an accounts receivable subsidiary ledger account. Therefore, you must use the **Receipts** task option.

Notes

You must use the **Receipts** *task option to record a transaction to write off an account as uncollectible.*

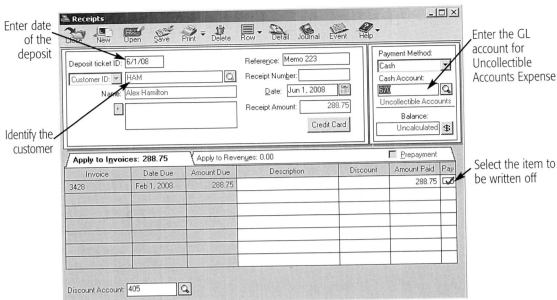

Enter date of the deposit

Identify the customer

Enter the GL account for Uncollectible Accounts Expense

Select the item to be written off

Figure 24-4A *Accounts Write-Off Transaction (June 1)*

- Click the [Journal] button to display the Accounting Behind the Screens window. (See Figure 24-4B.)

    As you can see, **Uncollectible Accounts Expense** is debited $288.75 and **Accounts Receivable** is credited $288.75.

- Close the Accounting Behind the Screens window.
- Click [Save] to post the transaction.

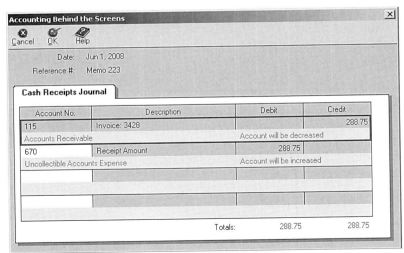

Figure 24-4B *Completed Accounts Write-Off Transaction (June 1)*

Step 6 Enter the remaining transactions to write off the other uncollectible accounts. (June 4, June 14, and June 29).

Step 7 Enter the June 22 transaction to record the cash receipt from a customer whose account was previously written off.

June 22, Received $288.75 from Alex Hamilton in full payment of his account, which was written off previously, Memorandum 298 and Receipt 944.

To reinstate an account and record the cash receipt:

- Use the **Sales/Invoicing** task option to enter a new sales invoice for $288.75.

  Enter **Memo 298** for the invoice number. In the Apply to Sales tab, make sure that you enter **670** (Uncollectible Accounts Expense) in the *GL Account* field. Also, remember to change the sales tax to 0.00. (See Figure 24-4C.)

- Choose the **Receipts** option to record the cash receipt.

  Record the receipt just as you would record any other receipt on account. **IMPORTANT:** Be sure to change the *Cash Account* field back to the **Cash in Bank** (101) account when you enter a cash receipt.

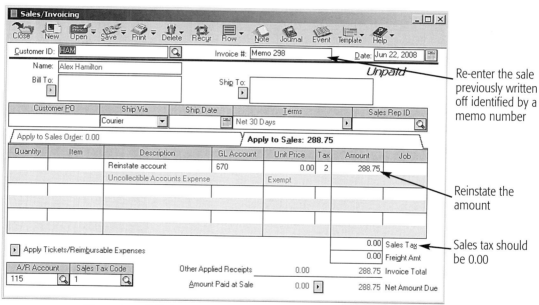

Figure 24-4C *Reinstated Account (June 22)*

Step 8   Print the following reports: Cash Receipts Journal, Sales Journal, Customer Ledgers, and General Ledger.

Step 9   Proof your work.

Step 10   Answer the Analyze question.

Ending the Session

Step 11   Click the **Close Problem** button in the Glencoe Smart Guide window to save your work.

Mastering Peachtree

Does Peachtree include any reports that show you which accounts are overdue and by how long? On a separate sheet of paper, describe the information provided on the report.

Peachtree Guide

Problem 24-5   Calculating and Recording Estimated Uncollectible Accounts Expense

INSTRUCTIONS

Beginning a Session

Step 1   Select the problem set: InBeat CD Shop (Prob. 24-5).

Step 2   Rename the company, and set the system date to June 30, 2008.

Completing the Accounting Problem

Step 3   Review the information provided in your textbook.

Step 4   Record the uncollectible accounts adjustment using the **General Journal Entry** option.

Step 5   Print a General Journal report and a General Ledger report to proof your work.

Notes

Peachtree automatically posts entries to the general ledger.

TIP: You can access the reports using the General Ledger navigation aid.

Step 6   Answer the Analyze question.

Checking Your Work and Ending the Session

Step 7   Click the **Close Problem** button in the Glencoe Smart Guide window.

Step 8   If your teacher has asked you to check your solution, select *Check my answer to this problem.* Review, print, and close the report.

Step 9   Click the **Close Problem** button and select a save option.

Problem 24-6   Writing Off Accounts Under the Allowance Method

INSTRUCTIONS

Beginning a Session

    Step 1   Select the problem set: Shutterbug Cameras (Prob. 24-6).

    Step 2   Rename the company, and set the system date to June 30, 2008.

Completing the Accounting Problem

    Step 3   Review the information provided in your textbook.

    Step 4   Record the transactions to write off the bad debts using the **Receipts** option.

> **TIP:** Remember to change the Cash Account to **Allowance for Uncollectible Accounts** when you write off an account using the allowance method. Also, check the Accounting Behind the Screens window to verify your work.

    Step 5   Enter the transactions to reinstate Jimmy Thompson's account and record the receipt on account.

        IMPORTANT: Use the **Sales/Invoicing** option to reinstate an account that was previously written off. For the allowance method, record the **Allowance for Uncollectible Accounts** general ledger on the Apply to Sales tab. Also, make sure the sales tax is 0.00.

    Step 6   Record the adjustment for estimated uncollectible accounts.

    Step 7   Record the closing entry for **Uncollectible Accounts Expense** using the **General Journal Entry. (Note:** Don't use the **Year-End Wizard** option. Manually enter the closing entry.)

    Step 8   Print the following reports to proof your work: Cash Receipts Journal, Sales Journal, Customer Ledgers, General Journal, and General Ledger.

    Step 9   Print a Balance Sheet.

    Step 10 Answer the Analyze question.

Ending the Session

    Step 11 Click the **Close Problem** button in the Glencoe Smart Guide window to save your work.

Problem 24-7   Estimating Uncollectible Accounts Expense

INSTRUCTIONS

Beginning a Session

Step 1   Select the problem set: Cycle Tech Bicycles (Prob. 24-7).

Step 2   Rename the company, and set the system date to June 30, 2008.

Completing the Accounting Problem

Step 3   Review the information provided in your textbook.

Step 4   Complete the analysis of the accounts receivable provided in your working papers.

Step 5   Based on your analysis, record the uncollectible accounts adjustment using the **General Journal Entry** option.

Step 6   Print a General Journal report and a General Ledger report to proof your work.

Step 7   Answer the Analyze question.

Checking Your Work and Ending the Session

Step 8   Click the **Close Problem** button in the Glencoe Smart Guide window.

Step 9   If your teacher has asked you to check your solution, select *Check my answer to this problem*. Review, print, and close the report.

Step 10 Click the **Close Problem** button and select a save option.

Peachtree Guide

Problem 24-8 Reporting Uncollectible Amounts on the Financial Statements

INSTRUCTIONS

Beginning a Session

Step 1 Select the problem set: River's Edge Canoe & Kayak (Prob. 24-8).

Step 2 Rename the company, and set the system date to December 31, 2008.

Completing the Accounting Problem

Step 3 Review the information provided in your textbook.

Step 4 Record the adjustment for the uncollectible accounts expense.

Step 5 Print a General Journal report and proof your work.

Step 6 Print a Balance Sheet, a Statement of Retained Earnings, and an Income Statement.

Step 7 Click the **Save Pre-closing Balances** button in the Glencoe Smart Guide window.

Step 8 Close the fiscal year.

Notes

Peachtree does not include an option to print a ten-column work sheet.

TIP: To close the fiscal year, choose **System** from the *Tasks* menu and then choose **Year-End Wizard.**

Step 9 Print a Post-Closing Trial Balance.

Step 10 Answer the Analyze question.

Ending the Session

Step 11 Click the **Close Problem** button in the Glencoe Smart Guide window to save your work.

Problem 24-9   Using the Allowance Method for Write-Offs

INSTRUCTIONS

Beginning a Session

Step 1   Select the problem set: Buzz Newsstand (Prob. 24-9).

Step 2   Rename the company, and set the system date to December 31, 2008.

Completing the Accounting Problem

Step 3   Review the information provided in your textbook.

Step 4   Record the transactions. Remember to enter each transaction in the proper accounting period. Use the **System** option from the *Tasks* menu to change the accounting period before entering each transaction.

Step 5   Print the following reports to proof your work: Cash Receipts Journal, Sales Journal, and Customer Ledgers.

> **TIP:** When you print the reports, change the date range options to include the transactions for the entire year, not just the current period.

Step 6   Answer the Analyze question.

Checking Your Work and Ending the Session

Step 7   Verify that the accounting period displayed is Period 12—12/1/08 to 12/31/08. Click the **Close Problem** button in the Glencoe Smart Guide window.

Step 8   If your teacher has asked you to check your solution, select *Check my answer to this problem*. Review, print, and close the report.

Step 9   Click the **Close Problem** button and select a save option.

FAQs

Why don't some Peachtree reports show all of the transactions?

By default, most Peachtree reports show only those transactions entered in the current period. If you entered transactions over multiple periods, you must change the report date option. For example, change the date range to include the entire year.

Working Papers *for End-of-Chapter Problems*

Problem 24-4   Using the Direct Write-Off Method

(1)

GENERAL JOURNAL   PAGE _____

	DATE	DESCRIPTION	POST. REF.	DEBIT	CREDIT	
1						1
2						2
3						3
4						4
5						5
6						6
7						7
8						8
9						9
10						10
11						11
12						12
13						13
14						14
15						15
16						16
17						17
18						18
19						19
20						20
21						21
22						22
23						23
24						24
25						25
26						26
27						27
28						28
29						29
30						30
31						31
32						32
33						33
34						34

Problem 24-4   (continued)

(2)

ACCOUNTS RECEIVABLE SUBSIDIARY LEDGER

Name *Martha Adams* _____

Address _____

DATE		DESCRIPTION	POST. REF.	DEBIT	CREDIT	BALANCE
20--						
June	1	Balance	✓			100 80

Name *Alex Hamilton* _____

Address _____

DATE		DESCRIPTION	POST. REF.	DEBIT	CREDIT	BALANCE
20--						
June	1	Balance	✓			288 75

Name *Helen Jun* _____

Address _____

DATE		DESCRIPTION	POST. REF.	DEBIT	CREDIT	BALANCE
20--						
June	1	Balance	✓			243 60

Name *Nate Moulder* _____

Address _____

DATE		DESCRIPTION	POST. REF.	DEBIT	CREDIT	BALANCE
20--						
June	1	Balance	✓			57 75

Problem 24-4   (concluded)

GENERAL LEDGER

ACCOUNT *Cash in Bank* _____ ACCOUNT NO. ___101___

DATE		DESCRIPTION	POST. REF.	DEBIT	CREDIT	BALANCE	
						DEBIT	CREDIT
20--							
June	1	Balance	✓			10 650 16	

ACCOUNT *Accounts Receivable* _____ ACCOUNT NO. ___115___

DATE		DESCRIPTION	POST. REF.	DEBIT	CREDIT	BALANCE	
						DEBIT	CREDIT
20--							
June	1	Balance	✓			8 016 50	

ACCOUNT *Uncollectible Accounts Expense* _____ ACCOUNT NO. ___670___

DATE		DESCRIPTION	POST. REF.	DEBIT	CREDIT	BALANCE	
						DEBIT	CREDIT

Analyze: _____

_____

Problem 24-5   Calculating and Recording Estimated Uncollectible Accounts Expense

(1) _____

(2)

GENERAL JOURNAL

PAGE _____

	DATE	DESCRIPTION	POST. REF.	DEBIT	CREDIT	
1						1
2						2
3						3
4						4
5						5
6						6

(3)

GENERAL LEDGER (PARTIAL)

ACCOUNT _Allowance for Uncollectible Accounts_   ACCOUNT NO. _117_

DATE		DESCRIPTION	POST. REF.	DEBIT	CREDIT	BALANCE DEBIT	BALANCE CREDIT
20--							
June	1	Balance	✓				4000 00

ACCOUNT _Uncollectible Accounts Expense_   ACCOUNT NO. _670_

DATE	DESCRIPTION	POST. REF.	DEBIT	CREDIT	BALANCE DEBIT	BALANCE CREDIT

Analyze: _____

_____

_____

Problem 24-6   Source Documents

Instructions: *Use the following source documents to record the transactions for this problem.*

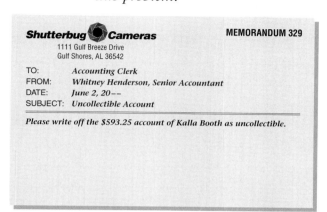

Shutterbug Cameras   MEMORANDUM 329
1111 Gulf Breeze Drive
Gulf Shores, AL 36542

TO: *Accounting Clerk*
FROM: *Whitney Henderson, Senior Accountant*
DATE: *June 2, 20--*
SUBJECT: *Uncollectible Account*

Please write off the $593.25 account of Kalla Booth as uncollectible.

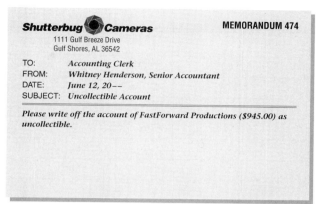

Shutterbug Cameras   MEMORANDUM 474
1111 Gulf Breeze Drive
Gulf Shores, AL 36542

TO: *Accounting Clerk*
FROM: *Whitney Henderson, Senior Accountant*
DATE: *June 12, 20--*
SUBJECT: *Uncollectible Account*

Please write off the account of FastForward Productions ($945.00) as uncollectible.

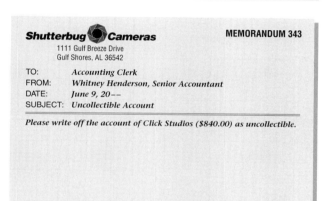

Shutterbug Cameras   MEMORANDUM 343
1111 Gulf Breeze Drive
Gulf Shores, AL 36542

TO: *Accounting Clerk*
FROM: *Whitney Henderson, Senior Accountant*
DATE: *June 9, 20--*
SUBJECT: *Uncollectible Account*

Please write off the account of Click Studios ($840.00) as uncollectible.

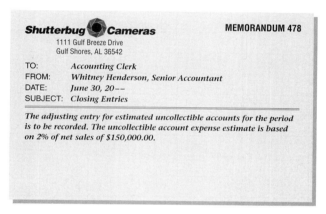

Shutterbug Cameras   MEMORANDUM 478
1111 Gulf Breeze Drive
Gulf Shores, AL 36542

TO: *Accounting Clerk*
FROM: *Whitney Henderson, Senior Accountant*
DATE: *June 30, 20--*
SUBJECT: *Closing Entries*

The adjusting entry for estimated uncollectible accounts for the period is to be recorded. The uncollectible account expense estimate is based on 2% of net sales of $150,000.00.

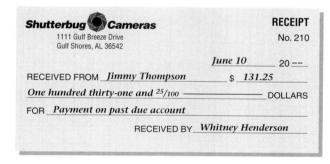

Shutterbug Cameras   RECEIPT
1111 Gulf Breeze Drive   No. 210
Gulf Shores, AL 36542

June 10   20 --

RECEIVED FROM *Jimmy Thompson* $ *131.25*
One hundred thirty-one and $^{25}/_{100}$ ———— DOLLARS
FOR *Payment on past due account*

RECEIVED BY *Whitney Henderson*

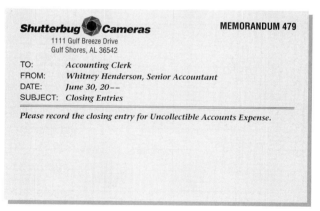

Shutterbug Cameras   MEMORANDUM 479
1111 Gulf Breeze Drive
Gulf Shores, AL 36542

TO: *Accounting Clerk*
FROM: *Whitney Henderson, Senior Accountant*
DATE: *June 30, 20--*
SUBJECT: *Closing Entries*

Please record the closing entry for Uncollectible Accounts Expense.

Problem 24-6 Writing Off Accounts Under the Allowance Method

(1)

GENERAL JOURNAL

PAGE _____

	DATE	DESCRIPTION	POST. REF.	DEBIT	CREDIT	
1						1
2						2
3						3
4						4
5						5
6						6
7						7
8						8
9						9
10						10
11						11
12						12
13						13
14						14
15						15
16						16
17						17
18						18
19						19
20						20
21						21
22						22
23						23
24						24
25						25
26						26
27						27
28						28
29						29
30						30
31						31
32						32
33						33
34						34

Problem 24-6 (continued)

(2)

GENERAL LEDGER

ACCOUNT **Cash in Bank**   ACCOUNT NO. **101**

DATE		DESCRIPTION	POST. REF.	DEBIT	CREDIT	BALANCE DEBIT	BALANCE CREDIT
20--							
June	1	Balance	✓			9306 54	

ACCOUNT **Accounts Receivable**   ACCOUNT NO. **115**

DATE		DESCRIPTION	POST. REF.	DEBIT	CREDIT	BALANCE DEBIT	BALANCE CREDIT
20--							
June	1	Balance	✓			23102 00	

ACCOUNT **Allowance for Uncollectible Accounts**   ACCOUNT NO. **117**

DATE		DESCRIPTION	POST. REF.	DEBIT	CREDIT	BALANCE DEBIT	BALANCE CREDIT
20--							
June	1	Balance	✓				3164 00

ACCOUNT **Income Summary**   ACCOUNT NO. **310**

DATE	DESCRIPTION	POST. REF.	DEBIT	CREDIT	BALANCE DEBIT	BALANCE CREDIT

ACCOUNT **Uncollectible Accounts Expense**   ACCOUNT NO. **665**

DATE	DESCRIPTION	POST. REF.	DEBIT	CREDIT	BALANCE DEBIT	BALANCE CREDIT

Problem 24-6   (continued)

(2)

ACCOUNTS RECEIVABLE SUBSIDIARY LEDGER

Name *Kalla Booth*

Address *1416 Halprin Avenue, Mobile, AL 36604*

DATE		DESCRIPTION	POST. REF.	DEBIT	CREDIT	BALANCE
20--						
June	1	Balance	✓			5 9 3 25

Name *Click Studios*

Address *1300 Nice Avenue, Mobile, AL 36610*

DATE		DESCRIPTION	POST. REF.	DEBIT	CREDIT	BALANCE
20--						
June	1	Balance	✓			8 4 0 00

Name *FastForward Productions*

Address *3937 Channel Drive, Mobile, AL 36617*

DATE		DESCRIPTION	POST. REF.	DEBIT	CREDIT	BALANCE
20--						
June	1	Balance	✓			9 4 5 00

Name *Jimmy Thompson*

Address *1616 Parkway Drive, Mobile, AL 36609*

DATE		DESCRIPTION	POST. REF.	DEBIT	CREDIT	BALANCE
20--						
June	1	Balance	✓			—

Problem 24-6 (concluded)

(3)

Analyze: _____

_____

Problem 24-7   Estimating Uncollectible Accounts Expense

(1)

Customer Name	Total Amount Owed	Not Yet Due	Days Past Due				
			1–30 Days	31–60 Days	61–90 Days	91–180 Days	Over 180
N. Bellis	$ 722	$ 722					
G. Buresh	1,362		$ 761		$ 601		
Rachel D'Souza	209	209					
S. Garfield	449		132	$ 317			
Greg Kellogg	271					$ 271	
Rishi Nadal	1,066	640		426			
Megan O'Hara	48					48	
ProTeam Sponsors Inc.	1,998	1,998					
Heidi Spencer	790	428	362				
Ed Young	296						$ 296
Totals							

(2)

Age Group	Amount	Estimated Percentage Uncollectible	Estimated Uncollectible Amount
Not yet due		2%	
1–30 days past due		4%	
31–60 days past due		20%	
61–90 days past due		30%	
91–180 days past due		45%	
Over 180 days past due		60%	
Totals			

Problem 24-7   (concluded)

(3)

GENERAL JOURNAL

PAGE _____

	DATE	DESCRIPTION	POST. REF.	DEBIT	CREDIT	
1						1
2						2
3						3
4						4
5						5
6						6

(4)

GENERAL LEDGER (PARTIAL)

ACCOUNT   *Accounts Receivable*   ACCOUNT NO. ___ *115*

DATE		DESCRIPTION	POST. REF.	DEBIT	CREDIT	BALANCE DEBIT	BALANCE CREDIT
20--							
June	*30*	*Balance*	✓			*7211 00*	

ACCOUNT   *Allowance for Uncollectible Accounts*   ACCOUNT NO. ___ *117*

DATE		DESCRIPTION	POST. REF.	DEBIT	CREDIT	BALANCE DEBIT	BALANCE CREDIT
20--							
June	*30*	*Balance*	✓				*142 00*

ACCOUNT   *Uncollectible Accounts Expense*   ACCOUNT NO. ___ *670*

DATE		DESCRIPTION	POST. REF.	DEBIT	CREDIT	BALANCE DEBIT	BALANCE CREDIT

Analyze: Book value of accounts receivable: _____

Problem 24-8 Reporting Uncollectible Amounts on the Financial Statements

(1), (2)

River's Edge

Work

For the Year Ended

#	ACCT. NO.	ACCOUNT NAME	TRIAL BALANCE DEBIT	TRIAL BALANCE CREDIT	ADJUSTMENTS DEBIT	ADJUSTMENTS CREDIT
1	101	Cash in Bank	21 633 50			
2	115	Accounts Receivable	10 168 45			
3	117	Allow. for Uncollectible Accounts		400 00		
4	130	Merchandise Inventory	39 391 75			(b) 3 630 25
5	135	Supplies	2 875 00			(c) 1 900 00
6	140	Prepaid Insurance	3 100 00			(d) 2 000 00
7	150	Store Equipment	30 000 00			
8	152	Accum. Depr.—Store Equip.		3 610 00		(e) 500 00
9	201	Accounts Payable		3 960 00		
10	210	Fed. Corporate Income Tax Pay.	—	—		(f) 1 806 00
11	215	Sales Tax Payable		613 10		
12	301	Capital Stock		40 000 00		
13	305	Retained Earnings		9 764 15		
14	310	Income Summary	—	—	(b) 3 630 25	
15	401	Sales		152 875 20		
16	410	Sales Returns and Allowances	3 585 00			
17	501	Purchases	81 860 00			
18	505	Transportation In	3 956 00			
19	510	Purchases Discounts		740 00		
20	515	Purchases Returns and Allow.		1 221 00		
21	605	Bankcard Fees Expense	452 75			
22	615	Depr. Exp.—Store Equip.	—		(e) 500 00	
23	625	Fed. Corporate Income Tax Exp.	3 750 00		(f) 1 806 00	
24	635	Insurance Expense	—		(d) 2 000 00	
25	655	Miscellaneous Expense	730 00			
26	660	Rent Expense	9 600 00			
27	670	Supplies Expense	—		(c) 1 900 00	
28	675	Uncollectible Accounts Expense	—			
29	680	Utilities Expense	2 081 00			
30			213 183 45	213 183 45		
31						
32						
33						

Canoe & Kayak

Sheet

June 30, 20--

	ADJUSTED TRIAL BALANCE		INCOME STATEMENT		BALANCE SHEET		
	DEBIT	CREDIT	DEBIT	CREDIT	DEBIT	CREDIT	
							1
							2
							3
							4
							5
							6
							7
							8
							9
							10
							11
							12
							13
							14
							15
							16
							17
							18
							19
							20
							21
							22
							23
							24
							25
							26
							27
							28
							29
							30
							31
							32
							33

Problem 24-8   (continued)
(3)

Problem 24-8   (continued)

(3)

(3)

Name _____ Date _____ Class _____

Problem 24-8 (continued)
(4), (5)

GENERAL JOURNAL   PAGE _____

DATE	DESCRIPTION	POST. REF.	DEBIT	CREDIT	
1					1
2					2
3					3
4					4
5					5
6					6
7					7
8					8
9					9
10					10
11					11
12					12
13					13
14					14
15					15
16					16
17					17
18					18
19					19
20					20
21					21
22					22
23					23
24					24
25					25
26					26
27					27
28					28
29					29
30					30
31					31
32					32
33					33
34					34
35					35
36					36

Problem 24-8   (continued)
(5)

GENERAL LEDGER

ACCOUNT *Cash in Bank* _____   ACCOUNT NO. ___101___

DATE	DESCRIPTION	POST. REF.	DEBIT	CREDIT	BALANCE DEBIT	BALANCE CREDIT
20--						
June 30	Balance	✓			2 1 6 3 3 50	

ACCOUNT *Accounts Receivable* _____   ACCOUNT NO. ___115___

DATE	DESCRIPTION	POST. REF.	DEBIT	CREDIT	BALANCE DEBIT	BALANCE CREDIT
20--						
June 30	Balance	✓			10 1 6 8 45	

ACCOUNT *Allowance for Uncollectible Accounts* _____   ACCOUNT NO. ___117___

DATE	DESCRIPTION	POST. REF.	DEBIT	CREDIT	BALANCE DEBIT	BALANCE CREDIT
20--						
June 30	Balance	✓				4 0 0 00

ACCOUNT *Merchandise Inventory* _____   ACCOUNT NO. ___130___

DATE	DESCRIPTION	POST. REF.	DEBIT	CREDIT	BALANCE DEBIT	BALANCE CREDIT
20--						
June 30	Balance	✓			3 9 3 9 1 75	

ACCOUNT *Supplies* _____   ACCOUNT NO. ___135___

DATE	DESCRIPTION	POST. REF.	DEBIT	CREDIT	BALANCE DEBIT	BALANCE CREDIT
20--						
June 30	Balance	✓			2 8 7 5 00	

ACCOUNT *Prepaid Insurance* _____   ACCOUNT NO. ___140___

DATE	DESCRIPTION	POST. REF.	DEBIT	CREDIT	BALANCE DEBIT	BALANCE CREDIT
20--						
June 30	Balance	✓			3 1 0 0 00	

Problem 24-8   (continued)

ACCOUNT *Store Equipment* _____ ACCOUNT NO. ___150___

DATE	DESCRIPTION	POST. REF.	DEBIT	CREDIT	BALANCE DEBIT	BALANCE CREDIT
20-- June 30	Balance	✓			3000000	

ACCOUNT *Accumulated Depreciation—Store Equipment* _____ ACCOUNT NO. ___152___

DATE	DESCRIPTION	POST. REF.	DEBIT	CREDIT	BALANCE DEBIT	BALANCE CREDIT
20-- June 30	Balance	✓				361000

ACCOUNT *Accounts Payable* _____ ACCOUNT NO. ___201___

DATE	DESCRIPTION	POST. REF.	DEBIT	CREDIT	BALANCE DEBIT	BALANCE CREDIT
20-- June 30	Balance	✓				396000

ACCOUNT *Federal Corporate Income Tax Payable* _____ ACCOUNT NO. ___204___

DATE	DESCRIPTION	POST. REF.	DEBIT	CREDIT	BALANCE DEBIT	BALANCE CREDIT

ACCOUNT *Sales Tax Payable* _____ ACCOUNT NO. ___215___

DATE	DESCRIPTION	POST. REF.	DEBIT	CREDIT	BALANCE DEBIT	BALANCE CREDIT
20-- June 30	Balance	✓				61310

ACCOUNT *Capital Stock* _____ ACCOUNT NO. ___301___

DATE	DESCRIPTION	POST. REF.	DEBIT	CREDIT	BALANCE DEBIT	BALANCE CREDIT
20-- June 30	Balance	✓				4000000

Problem 24-8 (continued)

ACCOUNT __Retained Earnings__ _____ ACCOUNT NO. ___305___

DATE		DESCRIPTION	POST. REF.	DEBIT	CREDIT	BALANCE	
						DEBIT	CREDIT
20--							
June	30	Balance	✓				9 7 6 4 15

ACCOUNT __Income Summary__ _____ ACCOUNT NO. ___310___

DATE	DESCRIPTION	POST. REF.	DEBIT	CREDIT	BALANCE	
					DEBIT	CREDIT

ACCOUNT __Sales__ _____ ACCOUNT NO. ___401___

DATE		DESCRIPTION	POST. REF.	DEBIT	CREDIT	BALANCE	
						DEBIT	CREDIT
20--							
June	30	Balance	✓				1 5 2 8 7 5 20

ACCOUNT __Sales Returns and Allowances__ _____ ACCOUNT NO. ___410___

DATE		DESCRIPTION	POST. REF.	DEBIT	CREDIT	BALANCE	
						DEBIT	CREDIT
20--							
June	30	Balance	✓			3 5 8 5 00	

ACCOUNT __Purchases__ _____ ACCOUNT NO. ___501___

DATE		DESCRIPTION	POST. REF.	DEBIT	CREDIT	BALANCE	
						DEBIT	CREDIT
20--							
June	30	Balance	✓			8 1 8 6 0 00	

ACCOUNT __Transportation In__ _____ ACCOUNT NO. ___505___

DATE		DESCRIPTION	POST. REF.	DEBIT	CREDIT	BALANCE	
						DEBIT	CREDIT
20--							
June	30	Balance	✓			3 9 5 6 00	

Problem 24-8 (continued)

ACCOUNT __Purchases Discounts__   ACCOUNT NO. ___510___

DATE		DESCRIPTION	POST. REF.	DEBIT	CREDIT	BALANCE DEBIT	BALANCE CREDIT
20--							
June	30	Balance	✓				740 00

ACCOUNT __Purchases Returns and Allowances__   ACCOUNT NO. ___515___

DATE		DESCRIPTION	POST. REF.	DEBIT	CREDIT	BALANCE DEBIT	BALANCE CREDIT
20--							
June	30	Balance	✓				1221 00

ACCOUNT __Bankcard Fees Expense__   ACCOUNT NO. ___605___

DATE		DESCRIPTION	POST. REF.	DEBIT	CREDIT	BALANCE DEBIT	BALANCE CREDIT
20--							
June	30	Balance	✓			452 75	

ACCOUNT __Depreciation Expense—Store Equipment__   ACCOUNT NO. ___620___

DATE		DESCRIPTION	POST. REF.	DEBIT	CREDIT	BALANCE DEBIT	BALANCE CREDIT

ACCOUNT __Federal Corporate Income Tax Expense__   ACCOUNT NO. ___625___

DATE		DESCRIPTION	POST. REF.	DEBIT	CREDIT	BALANCE DEBIT	BALANCE CREDIT
20--							
June	30	Balance	✓			3750 00	

ACCOUNT __Insurance Expense__   ACCOUNT NO. ___635___

DATE		DESCRIPTION	POST. REF.	DEBIT	CREDIT	BALANCE DEBIT	BALANCE CREDIT

Problem 24-8   (concluded)

ACCOUNT ___Miscellaneous Expense___   ACCOUNT NO. ___655___

DATE		DESCRIPTION	POST. REF.	DEBIT	CREDIT	BALANCE DEBIT	BALANCE CREDIT
20--							
June	30	Balance	✓			730 00	

ACCOUNT ___Rent Expense___   ACCOUNT NO. ___660___

DATE		DESCRIPTION	POST. REF.	DEBIT	CREDIT	BALANCE DEBIT	BALANCE CREDIT
20--							
June	30	Balance	✓			9600 00	

ACCOUNT ___Supplies Expense___   ACCOUNT NO. ___670___

DATE		DESCRIPTION	POST. REF.	DEBIT	CREDIT	BALANCE DEBIT	BALANCE CREDIT

ACCOUNT ___Uncollectible Accounts Expense___   ACCOUNT NO. ___675___

DATE		DESCRIPTION	POST. REF.	DEBIT	CREDIT	BALANCE DEBIT	BALANCE CREDIT

ACCOUNT ___Utilities Expense___   ACCOUNT NO. ___680___

DATE		DESCRIPTION	POST. REF.	DEBIT	CREDIT	BALANCE DEBIT	BALANCE CREDIT
20--							
June	30	Balance	✓			2081 00	

Analyze: _____

_____

Problem 24-9   Using the Allowance Method for Write-Offs

GENERAL JOURNAL

PAGE _____

	DATE	DESCRIPTION	POST. REF.	DEBIT	CREDIT	
1						1
2						2
3						3
4						4
5						5
6						6
7						7
8						8
9						9
10						10
11						11
12						12
13						13
14						14
15						15
16						16
17						17
18						18
19						19

ACCOUNTS RECEIVABLE SUBSIDIARY LEDGER

Name _Lee Adkins_

Address _____

DATE		DESCRIPTION	POST. REF.	DEBIT	CREDIT	BALANCE
20--						
Jan.	1	Balance	✓			1945 0

Analyze: _____

_____

CHAPTER 24   Uncollectible Accounts Receivable

Self-Test

Part A   True or False

Directions: *Circle the letter* T *in the Answer column if the statement is true; circle the letter* F *if the statement is false.*

Answer

T   F   **1.** When a business uses the allowance method, an adjusting entry must be made.

T   F   **2.** The two general ledger accounts affected by the direct write-off method of accounting for uncollectible accounts are Uncollectible Accounts Expense and Accounts Payable.

T   F   **3.** Another term for uncollectible accounts is bad debts.

T   F   **4.** The Allowance for Uncollectible Accounts account is classified as a contra asset account.

T   F   **5.** When a business sells goods or services on account, it knows which charge customers' accounts will be uncollectible.

T   F   **6.** When an account is written off as uncollectible, an explanation should be written on the account.

T   F   **7.** Large businesses with many charge customers normally use the direct write-off method of accounting for uncollectible accounts.

T   F   **8.** The normal balance of Allowance for Uncollectible Accounts is a credit.

T   F   **9.** Under the allowance method, Allowance for Uncollectible Accounts is debited when a charge customer's account is written off as a bad debt.

T   F   **10.** Businesses that sell on credit usually expect to sell more than if they accepted only cash.

T   F   **11.** When the direct write-off method of accounting is used, Accounts Receivable and the customer's account in the subsidiary ledger are credited when it is determined that the charge customer is not going to pay.

Part B   Multiple Choice

Directions: *Only one of the choices given with each of the following statements is correct. Write the letter of the correct answer in the Answer column.*

Answer

_____ 1. Allowance for Uncollectible Accounts is all of the following except
   (A) a contra asset account.
   (B) listed on the balance sheet just below Accounts Receivable.
   (C) increased on the debit side.
   (D) a valuation account.

_____ 2. If a company had net sales of $900,000 and estimates that its uncollectible accounts will be 3% of net sales, what is the amount of the adjustment?
   (A) $900,000
   (B) $27,000
   (C) $2,700
   (D) $270

_____ 3. Under the direct write-off method, the journal entry used to write off the account as uncollectible affects which two general ledger accounts?
   (A) Uncollectible Accounts Expense/Allowance for Uncollectible Accounts
   (B) Uncollectible Accounts Expense/Cash in Bank
   (C) Allowance for Uncollectible Accounts/Accounts Receivable
   (D) Uncollectible Accounts Expense/Accounts Receivable

_____ 4. With an Accounts Receivable account balance of $15,000 and an Allowance for Uncollectible Accounts balance of $2,500, what is the book value of accounts receivable?
   (A) $15,000
   (B) $2,500
   (C) $17,500
   (D) $12,500

CHAPTER 25   Inventories

Study Plan

Check Your Understanding

Section 1   *Read Section 1 on pages 680–682 and complete the following exercises on page 683.*
- ❏ Thinking Critically
- ❏ Communicating Accounting
- ❏ Problem 25-1   *Preparing Inventory Reports*

Section 2   *Read Section 2 on pages 684–687 and complete the following exercises on page 688.*
- ❏ Thinking Critically
- ❏ Computing in the Business World
- ❏ Problem 25-2   *Determining Inventory Costs*

Section 3   *Read Section 3 on pages 690–691 and complete the following exercises on page 692.*
- ❏ Thinking Critically
- ❏ Analyzing Accounting
- ❏ Problem 25-3   *Analyzing a Source Document*

Summary   *Review the Chapter 25 Summary on page 693 in your textbook.*
- ❏ Key Concepts

Review and Activities   *Complete the following questions and exercises on pages 694–695 in your textbook.*
- ❏ Using Key Terms
- ❏ Understanding Accounting Concepts and Procedures
- ❏ Case Study
- ❏ Conducting an Audit with Alex
- ❏ Internet Connection
- ❏ Workplace Skills

Computerized Accounting   *Read the Computerized Accounting information on page 696 in your textbook.*
- ❏ *Making the Transition from a Manual to a Computerized System*
- ❏ *Maintaining and Closing Inventories in Peachtree*

Problems   *Complete the following end-of-chapter problems for Chapter 25 in your textbook.*
- ❏ Problem 25-4   *Calculating the Cost of Ending Inventory*
- ❏ Problem 25-5   *Completing an Inventory Sheet*
- ❏ Problem 25-6   *Calculating Gross Profit on Sales*
- ❏ Problem 25-7   *Reporting Ending Inventory on the Income Statement*

Challenge Problem
- ❏ Problem 25-8   *Calculating Cost of Merchandise Sold and Gross Profit on Sales*

Chapter Reviews and Working Papers   *Complete the following exercises for Chapter 25 in your Chapter Reviews and Working Papers.*
- ❏ Chapter Review
- ❏ Self-Test

CHAPTER 25 REVIEW   Inventories

Part 1   Accounting Vocabulary (9 points)

Total Points	33
Student's Score	

Directions: *Using terms from the following list, complete the sentences below. Write the letter of the term you have chosen in the space provided.*

A. conservatism principle	**E.** market value	**H.** point-of-sale terminal
B. consistency principle	**F.** periodic inventory system	**I.** specific identification method
C. first-in, first-out method	**G.** perpetual inventory system	**J.** weighted average cost method
D. last-in, first-out method		

___E___ **0.** _____ is the current price that is being charged for similar items of merchandise in the market.

_____ **1.** The inventory costing method under which the cost of the items on hand is determined by the average cost of all identical items purchased during the period is the _____.

_____ **2.** The _____ is the inventory costing method that assumes the last items purchased are the first items sold.

_____ **3.** The _____ is the accounting guideline that states a business should report its financial position in amounts that are least likely to result in an overstatement of income or property values.

_____ **4.** The inventory costing method that assumes the first items purchased were the first items sold is the _____.

_____ **5.** The inventory costing method under which the exact cost of each item on the inventory sheet must be determined and assigned to that item is the _____.

_____ **6.** The _____ requires a constant, up-to-date record of merchandise on hand.

_____ **7.** The _____ requires a physical count of all merchandise on hand to determine the quantity of merchandise on hand.

_____ **8.** A _____ reads bar codes and enters the information into a computer.

_____ **9.** The _____ requires that businesses not normally change their chosen method of inventory costing.

Part 2   Comparing Inventory Costing Methods (6 points)

Directions: *Read each of the statements below and determine the inventory method that completes the statement. Write the identifying letter of your choice in the space provided.*

A. first-in, first-out method	**C.** specific identification method
B. last-in, first-out method	**D.** weighted average cost method

___C___ **0.** The _____ is a time-consuming process.

_____ **1.** The _____ is used by businesses that have a low unit volume of merchandise with high unit prices.

_____ **2.** According to the _____, the items still on hand at the end of the fiscal period are assumed to be the last items purchased.

_____ **3.** The _____ assumes that the items still on hand at the end of the period are the first ones purchased.

_____ **4.** The _____ takes into account the costs of all the merchandise available for sale during the period.

_____ **5.** The _____ is the most realistic costing method.

_____ **6.** The last-in, first-out method and the _____ are based on certain assumptions about the items remaining in inventory.

Part 3   Calculating Inventory Costs (6 points)

Directions: *The Lindborg Craft Shop has the following record of crewel kits for the month of April:*

Beginning inventory	13 units @ $3.48	=	$ 45.24
Purchased April 4	15 units @ $3.51	=	52.65
Purchased April 9	20 units @ $3.67	=	73.40
Purchased April 14	10 units @ $3.71	=	37.10
Purchased April 26	15 units @ $3.74	=	56.10
	73		$264.49

At the end of April, there were 22 units on hand. Based on the above information, complete the following statements.

_____$82.07_____   **0.** The value of the ending inventory using the FIFO method is _____.

_____   **1.** The value of the ending inventory using the LIFO method is _____.

_____   **2.** The value of the ending inventory using the weighted average cost method is _____.

_____   **3.** The purchase that will have the greatest impact on the weighted average cost method was made on _____.

_____   **4.** If the LIFO method is used and if the current market value of its inventory is $80.37, the Lindborg Craft Shop would report the value of its inventory at _____.

_____   **5.** The _____ method produces the highest value for the ending inventory.

_____   **6.** The _____ method produces the lowest value for the ending inventory.

Part 4   Choosing an Inventory Costing Method (12 points)

Directions: *Read each of the following statements to determine whether the statement is true or false. Write your answer in the space provided.*

____True____   **0.** A perpetual inventory system can be established without a computer.

_____   **1.** When a cash register is linked to a computer, it is said to be online.

_____   **2.** The costing method used to determine the value of the ending inventory will not affect a company's gross profit on sales or net income.

_____   **3.** A business may change inventory costing methods without obtaining permission from the Internal Revenue Service.

_____   **4.** The market value of the merchandise on hand is always lower than the original cost.

_____   **5.** The most commonly used method of determining the quantity of merchandise on hand is the perpetual inventory system.

_____   **6.** A physical inventory should be conducted at least once a year.

_____   **7.** The weighted average cost method is the most accurate inventory costing method.

_____   **8.** A periodic inventory system provides management with continuous merchandise inventory information.

_____   **9.** The specific identification method can often be used by businesses that sell large items such as automobiles.

_____   **10.** When choosing an inventory costing method, a business's owner or manager should consider only the present economic conditions.

_____   **11.** An inventory control system includes the quantity of merchandise on hand at a given time and the selling price of that merchandise.

_____   **12.** The lower-of-cost-or-market rule for reporting inventory value allows a business to follow a conservative approach.

Working Papers *for Section Problems*

Problem 25-1   Preparing Inventory Reports

INVENTORY SHEET

Date _____ Clerk _____ Page _____

STOCK NO.	ITEM	UNIT	QUANTITY	UNIT COST	TOTAL VALUE
					TOTAL

Problem 25-2   Determining Inventory Costs

(a) Specific Identification Method   _____

(b) First-In, First-Out Method   _____

(c) Last-In, First-Out Method   _____

(d) Weighted Average Cost Method   _____

Problem 25-3   Analyzing a Source Document

1. _____

_____

_____

2. _____

_____

_____

_____

_____

Computerized Accounting Using Peachtree

Software Objectives

When you have completed this chapter, you will be able to use Peachtree to:

1. Update an inventory record.
2. Print an Inventory Valuation Report, a Cost of Goods Sold Journal, and an Item Costing Report.
3. Record the purchase of inventory.
4. Record the sale of merchandise.
5. Print an Income Statement and Balance Sheet that shows the cost of goods sold.

Problem 25-4 **Calculating the Cost of Ending Inventory**

INSTRUCTIONS

Beginning a Session

Step 1 Select the problem set: Sunset Surfwear (Prob. 25-4).
Step 2 Rename the company, and set the system date to December 31, 2008.

Completing the Accounting Problem

Step 3 Review the information provided in your textbook.

Step 4 Print an Inventory Valuation Report.

The Inventory Valuation Report shows costing method, quantity on hand, and item value for each inventory item. Peachtree automatically calculates the inventory valuation based on the method set up for each item.

> **DO YOU HAVE A QUESTION**
>
> **Q.** *What inventory valuation methods does Peachtree support?*
>
> **A.** Peachtree supports the following inventory valuation methods: FIFO, LIFO, and average cost. However, once you set up an item and specify the valuation method, you cannot change this information. For that reason, you can print an Inventory Valuation Report based on only one valuation method.

 TIP: Choose **Inventory** from the *Reports* menu to access inventory reports, such as Inventory Valuation Report, Cost of Goods Sold Journal, and Item Costing Report.

Step 5 Print a Cost of Goods Sold Journal.

By default, Peachtree uses the current period. To view the Cost of Goods Sold Journal, you must change the report options date to **This Year**. The report shows the cost of the items at the time they are sold.

Notes

When you record purchases and sales, Peachtree automatically updates the merchandise inventory account.

Peachtree Guide

Step 6   Print an Item Costing Report.

The Item Costing Report gives a detailed listing of all transactions that involved inventory items and their effect on inventory valuation. Compare the information shown on the report to the purchases shown in your textbook.

TIP: When you choose to print the Item Costing Report, change the printer setup to print in landscape mode, then change the report date range to **This Year.**

Ending the Session

Step 7   Click the **Close Problem** button in the Glencoe Smart Guide window to save your work.

Mastering Peachtree

On a separate sheet of paper, explain how you can record minimum quantity and reorder quantity values for an inventory item. How would a company use this information?

Problem 25-7   Reporting Ending Inventory on the Income Statement

INSTRUCTIONS

Beginning a Session

Step 1   Select the problem set: Cycle Tech Bicycles (Prob. 25-7).
Step 2   Rename the company, and set the system date to December 31, 2008.

Completing the Accounting Problem

Step 3   Review the information provided in your textbook.
Step 4   Print the following reports: Inventory Valuation Report, Cost of Goods Sold Journal, and Item Costing Report.

TIP: Remember to change the date range for the Cost of Goods Sold Journal to **This Year.**

Step 5   Print an Income Statement and a Balance Sheet.

Ending the Session

Step 6   Click the **Close Problem** button in the Glencoe Smart Guide window.

Mastering Peachtree

On a separate sheet of paper, record what inventory valuation method is set for the Model #8274 ten-speed bicycle? How do you set the inventory valuation method?

Problem 25-8   Calculating Cost of Merchandise Sold and Gross Profit on Sales

INSTRUCTIONS

Beginning a Session

Step 1   Select the problem set: Buzz Newsstand (Prob. 25-8).

Step 2   Rename the company, and set the system date to May 31, 2008.

Completing the Accounting Problem

Step 3   Review the information provided in your textbook.

Step 4   Record the merchandise purchases transactions using the **Purchases/ Receive Inventory** option.

> **IMPORTANT:** The invoices are provided in your working papers. Be sure to complete the *Quantity, Item,* and *Unit Price* fields on the Apply to Purchases tab when you record the purchases using the **Purchases/ Receive Inventory** option. See the completed Purchases/Receive Inventory transaction shown in Figure 25-8A.

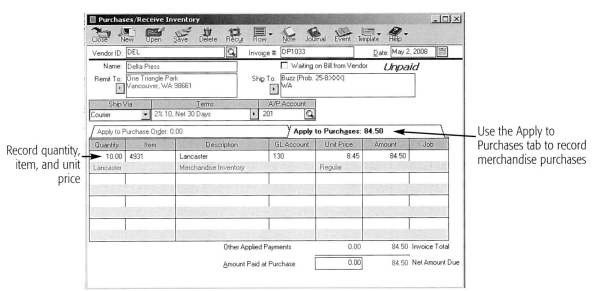

Figure 25-8A *Completed Merchandise Purchase Transaction (May 2)*

Step 5   Record the transportation charges using the **Payments** option. All of the transportation charges are paid in cash to Wolfe Trucking. The May 9 payment is made with Check No. 123, and the May 19 payment is made with Check No. 124.

Step 6 Record the camera sales using the **Receipts** option.

To record the camera sales:

- Determine the number of cameras sold.
- Choose the **Receipts** option.
- Enter **5/31** for the date, type **Cash Sales** in the *Name* field, and input **T290** for the reference.
- Enter **5/31/08** in the *Date* field.
- On the Apply to Revenues tab enter the information (quantity sold, item ID, and price) for each camera brand as a separate line item. Override the unit price field as necessary to enter the correct unit cost.
- Select **1** for the sales tax code.
- Verify the total receipt amount ($857.37) and post.

Step 7 Print a Purchases Journal, Cash Disbursements Journal, and a Cash Receipts Journal to proof your work.

Step 8 Print the following reports: Inventory Valuation Report, Cost of Goods Sold Journal and Item Costing Report.

Step 9 Print an Income Statement.

Step 10 Answer the Analyze question.

Ending the Session

Step 11 Click the **Close Problem** button in the Glencoe Smart Guide window to save your work.

FAQs

Can you change the inventory valuation method for an inventory item?

No, once you set up an inventory item, you cannot change the inventory valuation method. You could delete the item and re-enter it if you have not entered any transactions that affect the item.

Computerized Accounting Using Spreadsheets

Problem 25-6   Calculating Gross Profit on Sales

Completing the Spreadsheet

Step 1 Read the instructions for Problem 25-6 in your textbook. This problem involves calculating gross profit on sales using each of the four inventory costing methods.

Step 2 Open the Glencoe Accounting: Electronic Learning Center software.

Step 3 From the Program Menu, click on the **Peachtree Complete®** **Accounting Software and Spreadsheet Applications** icon.

Step 4 Log onto the Management System by typing your user name and password.

Step 5 Under the **Problems & Tutorials** tab, select template 25-6 from the Chapter 25 drop-down menu. The template should look like the one shown below.

```
PROBLEM 25-6
CALCULATING GROSS PROFIT ON SALES

(name)
(date)


                            SPECIFIC                             WEIGHTED
                         IDENTIFICATION     FIFO        LIFO    AVERAGE COST
                            METHOD         METHOD      METHOD      METHOD

Net Sales
  Purchases Available for Sale
  Cost of Ending Inventory
Cost of Merchandise Sold         $0.00       $0.00       $0.00       $0.00
Gross Profit on Sales            $0.00       $0.00       $0.00       $0.00
```

Step 6 Key your name and today's date in the cells containing the *(name)* and *(date)* placeholders.

Step 7 For each of the four inventory costing methods, enter the following data in the appropriate cells of the spreadsheet template: net sales, purchases available for sale, and cost of ending inventory. The spreadsheet will automatically calculate the cost of merchandise sold and the gross profit on sales for each inventory costing method.

Step 8 Save the spreadsheet using the **Save** option from the *File* menu. You should accept the default location for the save as this is handled by the management system.

Step 9 Print the completed spreadsheet.

Step 10 Exit the spreadsheet program.

Step 11 In the Close Options box, select the location where you would like to save your work.

Step 12 Answer the Analyze question from your textbook for this problem.

What-If Analysis

If the cost of ending inventory using the last-in, first-out method were $21,399.13, what would gross profit on sales be?

Working Papers *for End-of-Chapter Problems*

Problem 25-4   Calculating the Cost of Ending Inventory

a. Specific Identification Method _____

b. FIFO _____

c. LIFO _____

d. Weighted Average Method _____

Analyze: _____

Problem 25-5   Completing an Inventory Sheet

INVENTORY RECORD						
ITEM NO.	ITEM	ENDING INVENTORY	COST PER UNIT	CURRENT MARKET VALUE	PRICE TO BE USED	TOTAL COST
0247	Blank CDs	24	2.67	2.88	2.67	64.08
0391	Blank CDs	36	2.80	2.74		
0388	Cable #4	21	2.91	3.05		
0379	CD Cleaner	6	6.36	8.33		
0380	CD Cleaner	19	7.49	7.51		
0274	Audio Plug	23	6.90	6.95		
0276	Dust Cover	12	8.13	7.95		
0277	Headset	14	9.25	9.57		
0181	Cable #9	18	2.06	2.52		
0193	Cable #5	9	2.29	2.74		
0419	Headset	8	8.42	8.73		
0420	Headset	14	8.98	9.19		
					TOTAL COST	

Analyze: _____

Problem 25-6   Calculating Gross Profit on Sales

	Specific Identification Method	First-In, First-Out Method	Last-In, First-Out Method	Weighted Average Cost Method
Net Sales	$	$	$	$
Cost of Merchandise Sold	$	$	$	$
Gross Profit on Sales	$	$	$	$

Analyze: _____

Problem 25-7   Reporting Ending Inventory on the Income Statement

	FIFO Method	LIFO Method	Weighted Average Cost Method
Ending Inventory	$	$	$
Cost of Merchandise Sold			

FIFO Method

Problem 25-7   (concluded)
LIFO Method

Weighted Average Cost Method

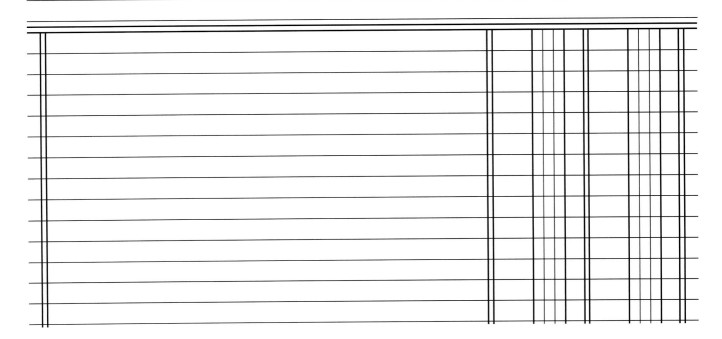

Analyze: _____

Problem 25-8   Source Documents

Instructions: *Use the following source documents to record the transactions for this problem.*

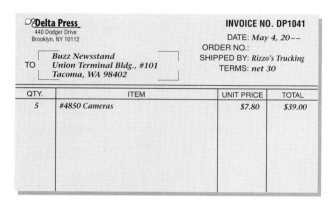

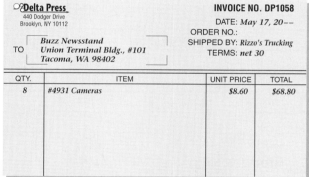

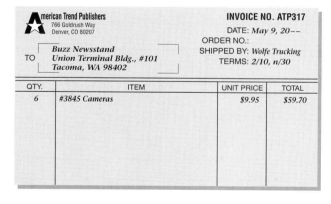

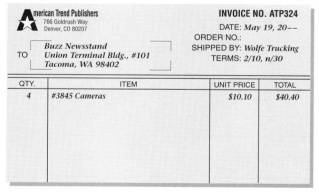

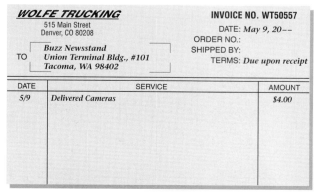

Problem 25-8   (continued)

Delta Press
440 Dodger Drive
Brooklyn, NY 10112

INVOICE NO. DP1067

DATE: *May 27, 20--*
ORDER NO.:
SHIPPED BY: *Rizzo's Trucking*
TERMS: *net 30*

TO   *Buzz Newsstand*
Union Terminal Bldg., #101
Tacoma, WA 98402

QTY.	ITEM	UNIT PRICE	TOTAL
8	#9265 Cameras	$8.30	$66.40

Delta Press
440 Dodger Drive
Brooklyn, NY 10112

INVOICE NO. DP1071

DATE: *May 29, 20--*
ORDER NO.:
SHIPPED BY: *Rizzo's Trucking*
TERMS: *net 30*

TO   *Buzz Newsstand*
Union Terminal Bldg., #101
Tacoma, WA 98402

QTY.	ITEM	UNIT PRICE	TOTAL
4	#4931 Cameras	$8.85	$35.40

Problem 25-8   Calculating Cost of Merchandise Sold and Gross Profit on Sales

(1)

#3845 _____   #4931 _____

#9265 _____   #4850 _____

(2)

	Item			
	3845	**4931**	**9265**	**4850**
Sales for Month				
Value of Beginning Inventory				
Purchases for May				
Transportation Costs				
Net Purchases for May				
Goods Available for Sale				
Value of Ending Inventory				
Cost of Merchandise Sold				
Gross Profit on Sales				

Analyze: _____

CHAPTER 25 Inventories

Self-Test

Part A True or False

Directions: *Circle the letter* T *in the Answer column if the statement is true; circle the letter* F *if the statement is false.*

Answer

T	F	**1.** The specific identification method is used by most businesses.
T	F	**2.** The LIFO method assumes that the first items purchased are the ones still remaining.
T	F	**3.** A perpetual inventory system can be established without using a computer.
T	F	**4.** When the inventory is valued using the weighted average cost method, the value is usually between the values determined by the LIFO and FIFO methods.
T	F	**5.** The market value of the merchandise on hand is always higher than the original cost.
T	F	**6.** The consistency principle does not permit businesses to change inventory costing methods so the financial statements can be compared.
T	F	**7.** The weighted average cost method is the most accurate of the inventory costing methods.
T	F	**8.** The costing method used to determine the value of the ending inventory will affect the gross profit on sales.
T	F	**9.** An automobile dealership could use the specific identification costing method without much difficulty.
T	F	**10.** If a business uses a perpetual inventory system, a periodic inventory is never needed.
T	F	**11.** The FIFO method assumes that the items still in inventory were the last ones purchased.
T	F	**12.** A business must take the higher of the determined cost of the existing inventory or the current market replacement value.

Part B   Fill in the Missing Term

Directions: *In the Answer column, write the letter of the word or phrase that best completes the sentence. Some answers may be used more than once.*

A. conservatism principle	**F.** periodic inventory system
B. consistency principle	**G.** perpetual inventory system
C. FIFO method	**H.** point-of-sale terminal
D. LIFO method	**I.** specific identification method
E. market value	**J.** weighted average cost method

Answer

_____   **1.** A cash register that inputs data into a computer is called a _____.

_____   **2.** The current price being charged for an item of merchandise in the market is called _____.

_____   **3.** The _____ is usually used by businesses that input all purchases and sales into a computer system when the transactions occur.

_____   **4.** _____ is an inventory method that assumes that the items purchased at the beginning of the period are the first items sold.

_____   **5.** The _____ presents amounts that are least likely to result in an overstatement of income or property values when reporting a business's financial position.

_____   **6.** An inventory method in which the actual cost of each item must be determined is the _____.

_____   **7.** A _____ is used when items on hand are counted to update the inventory records.

_____   **8.** _____ is an inventory method in which the cost of items on hand is determined by averaging the costs of all similar items purchased during the period.

_____   **9.** Businesses do not normally change inventory costing methods because of the _____.

_____   **10.** _____ is an inventory method that assumes that the items purchased at the end of the period are the first items sold.

CHAPTER  26 Notes Payable and Receivable

Study Plan

Check Your Understanding

Section 1
Read Section 1 on pages 702–706 and complete the exercises on page 707.
- ❏ Thinking Critically
- ❏ Computing in the Business World
- ❏ Problem 26-1 *Calculating Interest and Finding Maturity Values*
- ❏ Problem 26-2 *Calculating Interest*

Section 2
Read Section 2 on pages 708–714 and complete the exercises on page 715.
- ❏ Thinking Critically
- ❏ Communicating Accounting
- ❏ Problem 26-3 *Recording the Issuance of an Interest-Bearing Note Payable*
- ❏ Problem 26-4 *Recording the Issuance of a Noninterest-Bearing Note Payable*

Section 3
Read Section 3 on pages 716–717 and complete the exercises on page 718.
- ❏ Thinking Critically
- ❏ Analyzing Accounting
- ❏ Problem 26-5 *Analyzing a Source Document*

Summary
Review the Chapter 26 Summary on page 719 in your textbook.
- ❏ Key Concepts

Review and Activities
Complete the following questions and exercises on pages 720–721.
- ❏ Using Key Terms
- ❏ Understanding Accounting Concepts and Procedures
- ❏ Case Study
- ❏ Conducting an Audit with Alex
- ❏ Internet Connection
- ❏ Workplace Skills

Computerized Accounting
Read the Computerized Accounting information on page 722.
- ❏ *Making the Transition from a Manual to a Computerized System*
- ❏ *Recording Notes Receivable and Payable in Peachtree*

Problems
Complete the following end-of-chapter problems for Chapter 26.
- ❏ Problem 26-6 *Recording Transactions for Interest-Bearing Notes Payable*
- ❏ Problem 26-7 *Recording Transactions for Noninterest-Bearing Notes Payable*
- ❏ Problem 26-8 *Recording Notes Payable and Notes Receivable*
- ❏ Problem 26-9 *Recording Notes Payable and Notes Receivable*

Challenge Problem
- ❏ Problem 26-10 *Renewing a Note Receivable*

Chapter Reviews and Working Papers
Complete the following exercises for Chapter 26 in your Chapter Reviews and Working Papers.
- ❏ Chapter Review
- ❏ Self-Test

CHAPTER 26 REVIEW Notes Payable and Receivable

Part 1 Accounting Vocabulary (18 points)

Directions: *Using terms from the following list, complete the sentences below. Write the letter of the term you have chosen in the space provided.*

A. bank discount	**F.** issue date	**K.** note payable	**P.** principal
B. face value	**G.** maker	**L.** note receivable	**Q.** proceeds
C. interest	**H.** maturity date	**M.** other expense	**R.** promissory note
D. interest-bearing note payable	**I.** maturity value	**N.** other revenue	**S.** term
E. interest rate	**J.** noninterest-bearing note payable	**O.** payee	

_____**B**_____   **0.** The _____ of a promissory note is the amount of money written on the face of the note.

_____   **1.** The date on which a note is written is called its _____.

_____   **2.** The _____ of the note is the person or business promising to repay the principal and interest.

_____   **3.** A(n) _____ is a promissory note issued to a creditor or by a business to borrow money from a bank.

_____   **4.** The _____ is the date on which the note must be paid.

_____   **5.** The amount of interest to be charged stated as a percentage of the principal is called the _____.

_____   **6.** The amount being borrowed is the _____ of the note.

_____   **7.** A promissory note that a business accepts from a customer who needs additional time to pay a debt is called a(n) _____.

_____   **8.** _____ is the fee charged for the use of money.

_____   **9.** A note that requires the face value plus interest be paid on the maturity date is called a(n) _____.

_____   **10.** The _____ is the amount of time that the borrower has to repay a promissory note.

_____   **11.** The person or business to whom a promissory note is made payable is the _____.

_____   **12.** A(n) _____ is a written promise to pay a business or a person a certain amount of money at a specific time.

_____   **13.** The interest deducted in advance from a non-interest-bearing note payable is called the _____.

_____   **14.** An expense that does not result from the normal operations of the business is called a(n) _____.

_____   **15.** A promissory note from which the interest has been deducted in advance and which therefore has no interest rate stated on the note itself is called a(n) _____.

_____   **16.** The amount of cash actually received by the borrower of a non-interest-bearing note payable is called the _____.

_____   **17.** The _____ of a note is the principal plus the interest.

_____   **18.** _____ is revenue that a business receives or earns from activities outside the normal operations of the business.

Part 2 Examining Notes Payable and Receivable (15 points)

Directions: *Read each of the following statements to determine whether the statement is true or false. Write your answer in the space provided.*

___**True**___   **0.** Promissory notes are formal documents that provide evidence that credit was granted or received.

_____   **1.** When a noninterest-bearing note payable is paid, the interest charge is transferred from the Notes Payable account to the Interest Expense account.

_____   **2.** Interest = Principal × Interest Rate × Time is the equation for calculating interest on a promissory note.

    3. A note receivable is an asset to the business receiving the note.

    4. Discount on Notes Payable is a liability account.

    5. Interest rates are usually stated on an annual basis.

    6. The maturity value of a noninterest-bearing note payable is the same as its face value.

    7. Notes Receivable is classified as a contra asset account, and its normal balance is a credit.

    8. Both the term and the issue date are needed to determine the maturity date of a note.

    9. Borrowing periods of less than one year cannot be used in calculating interest.

    10. The interest on a 12.5%, $3,500 promissory note for two years is $437.50.

    11. On interest-bearing notes, the face value and the principal are the same.

    12. A business may not issue a note payable to borrow money from a bank.

    13. A noninterest-bearing note payable is the same as an interest-free note.

    14. Interest income is an example of other revenue and is reported separately on the income statement.

    15. The payment of a noninterest-bearing note payable and the recognition of the interest expense are always recorded with two separate journal entries in the cash payments journal and in the general journal.

Part 3 Analyzing Transactions Affecting Notes Payable and Receivable
(14 points)

Directions: *Using the following account names, analyze the transactions below. Determine the account(s) to be debited and credited. Write your answers in the space provided.*

A. Cash in Bank	**D.** Notes Payable	**G.** Sales
B. Accounts Receivable	**E.** Discount on Notes Payable	**H.** Interest Expense
C. Notes Receivable	**F.** Interest Income	

Debit	Credit	
B	*G*	**0.** Sold merchandise on account to a customer.
		1. Received an interest-bearing note from a customer for payment on account.
		2. Issued an interest-bearing note payable to the bank for cash.
		3. Received a check for the payment of the interest-bearing note from the customer.
		4. Issued a noninterest-bearing note to the bank for cash.
		5. Wrote a check to the bank for payment of the interest-bearing note.
		6. Wrote a check to the bank for payment on the noninterest-bearing note.
		7. Received the interest due on a note receivable and renewed the note for 90 days.

Working Papers *for Section Problems*

Problem 26-1   Calculating Interest and Finding Maturity Values

	Principal	Interest Rate	Term (in days)	Interest	Maturity Value
1	$ 4,000.00	11.50%	60	$	$
2	10,000.00	11.75%	90		
3	6,500.00	12.75%	60		
4	900.00	12.25%	120		
5	2,400.00	12.00%	180		

Problem 26-2   Calculating Interest

	Principal	Interest Rate	Term	Interest
1	$ 600.00	15.00%	90 days	$
2	3,500.00	12.00%	60 days	
3	9,600.00	9.00%	4 months	
4	2,500.00	10.00%	180 days	
5	1,500.00	11.50%	6 months	

Problem 26-3   Recording the Issuance of an Interest-Bearing Note Payable

1. _____

2. _____

3. _____

4. _____

Problem 26-4　Recording the Issuance of a Noninterest-Bearing Note Payable

1. _____

_____

2. _____

_____

Problem 26-5　Analyzing a Source Document

GENERAL JOURNAL

PAGE _____

	DATE	DESCRIPTION	POST. REF.	DEBIT	CREDIT	
1						1
2						2
3						3
4						4
5						5
6						6
7						7
8						8
9						9
10						10

Computerized Accounting Using Peachtree

Software Objectives

When you have completed this chapter, you will be able to use Peachtree to:

1. Record transactions involving notes payable.
2. Record notes receivable transactions.

Problem 26-6   Recording Transactions for Interest-Bearing Notes Payable

INSTRUCTIONS

Beginning a Session

Step 1   Select the problem set: Sunset Surfwear (Prob. 26-6).
Step 2   Rename the company, and set the system date to December 31, 2008.

Completing the Accounting Problem

Step 3   Review the information provided in your textbook.
Step 4   Record the transactions using the **Receipts** and **Payments** options. Remember to enter each transaction in the appropriate accounting period. Use the **System** option from the **Tasks** menu to change the accounting period before entering each transaction.

DO YOU HAVE A QUESTION

Q. *Can you use Peachtree to record a multi-part cash payment entry?*

A. Using Peachtree you can record a multi-part cash payment entry. For example, you can record the payment of a notes payable with interest using the **Payments** option. Simply record each part of the entry in the space provided on the Apply to Expenses tab. Be sure to change the *GL Account* field for each part.

TIP: You can enter a multi-part transaction using the **Payments** option to record the payment of a note and interest. (See Figure 26-6A.)

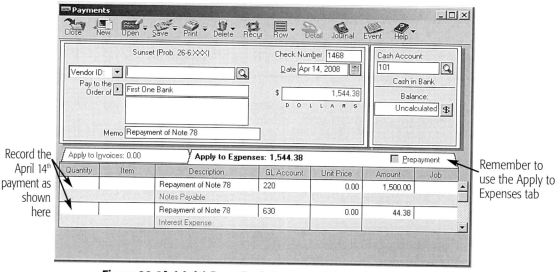

Record the April 14th payment as shown here

Remember to use the Apply to Expenses tab

Figure 26-6A *Multi-Part Cash Payment Entry (April 14)*

Step 5   Print a Cash Receipts Journal and a Cash Disbursements Journal to proof your work.

TIP: Be sure to set the date range on the journal reports to print the transactions for the entire year.

Step 6   Answer the Analyze question.

Ending the Session

Step 7   Verify that Period 8—8/1/08 to 8/31/08 is displayed in the lower right hand corner. Click the **Close Problem** button in the Glencoe Smart Guide window.

Mastering Peachtree

Answer the following on a separate sheet of paper. Does Peachtree provide any features that allow you to set up a reminder to pay a note? Explain your answer.

Problem 26-7   Recording Transactions for Noninterest-Bearing Notes Payable

INSTRUCTIONS

Beginning a Session

Step 1   Select the problem set: InBeat CD Shop (Prob. 26-7).
Step 2   Rename the company, and set the system date to December 31, 2008.

Completing the Accounting Problem

Step 3   Review the information provided in your textbook.
Step 4   Record the notes payable transactions. Remember to enter each transaction in the appropriate accounting period. Use the **System** option from the *Tasks* menu to change the accounting period.

TIP: When you record the cash receipt for a discounted note, enter the discount as a negative amount on the Apply to Revenues tab.

Step 5   Print a Cash Receipts Journal and a Cash Disbursements Journal to proof your work. Set the date range to include the entire year.
Step 6   Answer the Analyze question.

Checking Your Work and Ending the Session

Step 7   Verify that the accounting period displayed is Period 12—12/1/08 to 12/31/08. Click the **Close Problem** button in the Glencoe Smart Guide window.
Step 8   If your teacher has asked you to check your solution, select *Check my answer to this problem.* Review, print, and close the report.
Step 9   Click the **Close Problem** button and select a save option.

Problem 26-8   Recording Notes Payable and Notes Receivable

INSTRUCTIONS

Beginning a Session

Step 1   Select the problem set: Cycle Tech Bicycles (Prob. 26-8).
Step 2   Rename the company, and set the system date to December 31, 2008.

Completing the Accounting Problem

Step 3   Review the information provided in your textbook.
Step 4   Record the transactions. Be sure to enter each transaction in the proper accounting period (month). Use the **System** option from the *Tasks* menu to set the accounting period before entering each transaction.

> **IMPORTANT:** Use the **Receipts** option to record a transaction where a customer replaces an accounts receivable balance with a promissory note. Change the *Cash Account* to **Notes Receivable**. Mark the invoice(s) as "Paid" and post the transaction. Be sure to reset the *Cash Account* before you record any other cash receipts.

Step 5   Print a Cash Receipts Journal and a Cash Disbursements Journal to proof your work.
Step 6   Answer the Analyze question.

Ending the Session

Step 7   Verify that the accounting period displayed in the lower right hand corner of your screen is Period 12—12/1/08 to 12/31/08. Click the **Close Problem** button in the Glencoe Smart Guide window to save your work.

Problem 26-9   Recording Notes Payable and Notes Receivable

INSTRUCTIONS

Beginning a Session

Step 1   Select the problem set: River's Edge Canoe & Kayak (Prob. 26-9).
Step 2   Rename the company, and set the system date to September 30, 2008.

Completing the Accounting Problem

Step 3   Review the information provided in your textbook.
Step 4   Record the transactions. Be sure to enter each transaction in the proper accounting period. Use the **System** option from the *Tasks* menu to set the accounting period before entering each transaction.

> **IMPORTANT:** Use the **Receipts** option to record a transaction where a customer replaces an accounts receivable balance with a promissory note. Change the *Cash Account* to **Notes Receivable**. Mark the invoice(s) as "Paid" and post the transaction. Be sure to reset the *Cash Account* before you record any other cash receipts.
>
> Use the **Payments** option to record a transaction when a company issues a note to a vendor in place of an amount owed on account. Change the *Cash Account* to **Notes Payable**. Mark the invoice(s) as "Paid" and post the transaction. Be sure to reset the *Cash Account* before you record any other cash payments.

Step 5 Print a Cash Receipts Journal and a Cash Disbursements Journal to proof your work.

TIP: Be sure to set the date range on the journal reports to print the transactions for the entire year.

Step 6 Answer the Analyze question.

Checking Your Work and Ending the Session

Step 7 Verify that the accounting period displayed in the lower right hand corner of your screen is Period 9—9/1/08 to 9/30/08. Click the **Close Problem** button in the Glencoe Smart Guide window.

Step 8 If your teacher has asked you to check your solution, select *Check my answer to this problem.* Review, print, and close the report.

Step 9 Click the **Close Problem** button and select a save option.

Problem 26-10   Renewing a Note Receivable

INSTRUCTIONS

Beginning a Session

Step 1 Select the problem set: Buzz Newsstand (Prob. 26-10).

Step 2 Rename the company, and set the system date to September 30, 2008.

Completing the Accounting Problem

Step 3 Review the information provided in your textbook.

Step 4 Record the transactions. Be sure to record each transaction in the proper accounting period.

Step 5 Print a Sales Journal, Cash Receipts Journal and a General Journal to proof your work.

Step 6 Answer the Analyze question.

Ending the Session

Step 7 Verify that the accounting period displayed on your screen is Period 9—9/1/08 to 9/30/08. Click the **Close Problem** button in the Glencoe Smart Guide window to save your work.

FAQs

Why don't some Peachtree reports show all of the transactions?

By default, most Peachtree reports show only those transactions entered in the current period. If you entered transactions over multiple periods, you must change the report date option. For example, change the date range to include the entire year.

Problem 26-6 Recording Transactions for Interest-Bearing Notes Payable

CASH RECEIPTS JOURNAL

PAGE _____

DATE	DOC. NO.	ACCOUNT NAME	POST. REF.	GENERAL DEBIT	GENERAL CREDIT	SALES CREDIT	SALES TAX PAYABLE CREDIT	ACCOUNTS RECEIVABLE CREDIT	CASH IN BANK DEBIT
1									
2									
3									
4									
5									
6									
7									
8									
9									
10									

CASH PAYMENTS JOURNAL

PAGE _____

DATE	DOC. NO.	ACCOUNT NAME	POST. REF.	GENERAL DEBIT	GENERAL CREDIT	ACCOUNTS PAYABLE DEBIT	PURCHASES DISCOUNTS CREDIT	CASH IN BANK CREDIT
1								
2								
3								
4								
5								
6								
7								
8								
9								
10								

Analyze:

Problem 26-7 Recording Transactions for Noninterest-Bearing Notes Payable

CASH RECEIPTS JOURNAL

DATE	DOC. NO.	ACCOUNT NAME	POST. REF.	GENERAL DEBIT	GENERAL CREDIT	SALES CREDIT	SALES TAX PAYABLE CREDIT	ACCOUNTS RECEIVABLE CREDIT	CASH IN BANK DEBIT	
										1
										2
										3
										4
										5
										6
										7
										8
										9
										10

CASH PAYMENTS JOURNAL

DATE	DOC. NO.	ACCOUNT NAME	POST. REF.	GENERAL DEBIT	GENERAL CREDIT	ACCOUNTS PAYABLE DEBIT	PURCHASES DISCOUNTS CREDIT	CASH IN BANK CREDIT	
									1
									2
									3
									4
									5
									6
									7
									8
									9
									10

Analyze:

Problem 26-8   Recording Notes Payable and Notes Receivable

CASH RECEIPTS JOURNAL

PAGE _____

DATE	DOC. NO.	ACCOUNT NAME	POST. REF.	GENERAL DEBIT	GENERAL CREDIT	SALES CREDIT	SALES TAX PAYABLE CREDIT	ACCOUNTS RECEIVABLE CREDIT	CASH IN BANK DEBIT	
										1
										2
										3
										4
										5
										6
										7
										8
										9
										10

CASH PAYMENTS JOURNAL

PAGE _____

DATE	DOC. NO.	ACCOUNT NAME	POST. REF.	GENERAL DEBIT	GENERAL CREDIT	ACCOUNTS PAYABLE DEBIT	PURCHASES DISCOUNTS CREDIT	CASH IN BANK CREDIT	
									1
									2
									3
									4
									5
									6
									7
									8
									9
									10

Problem 26-8   (concluded)

GENERAL JOURNAL   PAGE _____

	DATE	DESCRIPTION	POST. REF.	DEBIT	CREDIT	
1						1
2						2
3						3
4						4
5						5
6						6
7						7
8						8
9						9
10						10
11						11
12						12
13						13
14						14
15						15
16						16
17						17
18						18
19						19
20						20
21						21
22						22
23						23
24						24
25						25
26						26

Analyze: _____

Problem 26-9   Recording Notes Payable and Notes Receivable

CASH RECEIPTS JOURNAL

PAGE _____

DATE	DOC. NO.	ACCOUNT NAME	POST. REF.	GENERAL DEBIT	GENERAL CREDIT	SALES CREDIT	SALES TAX PAYABLE CREDIT	ACCOUNTS RECEIVABLE CREDIT	CASH IN BANK DEBIT	
										1
										2
										3
										4
										5
										6
										7
										8
										9
										10

CASH PAYMENTS JOURNAL

PAGE _____

DATE	DOC. NO.	ACCOUNT NAME	POST. REF.	GENERAL DEBIT	GENERAL CREDIT	ACCOUNTS PAYABLE DEBIT	PURCHASES DISCOUNTS CREDIT	CASH IN BANK CREDIT	
									1
									2
									3
									4
									5
									6
									7
									8
									9
									10

Problem 26-9   (concluded)

GENERAL JOURNAL   PAGE _____

	DATE	DESCRIPTION	POST. REF.	DEBIT	CREDIT	
1						1
2						2
3						3
4						4
5						5
6						6
7						7
8						8
9						9
10						10
11						11
12						12
13						13
14						14
15						15
16						16
17						17
18						18
19						19
20						20
21						21
22						22
23						23
24						24
25						25
26						26

Analyze: _____

Problem 26-10 Source Documents

Instructions: *Use the following source documents to record the transactions for this business.*

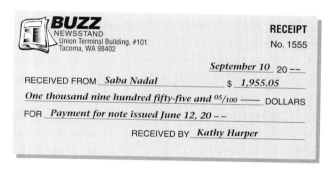

BUZZ NEWSSTAND
Union Terminal Building, #101
Tacoma, WA 98402

DATE: March 14, 20-- NO. 388

SOLD TO: Saba Nadal, 1306 Hampstead Ct., Seattle, WA 98134

CLERK B.A. | CASH | CHARGE ✓ | TERMS n/30

QTY.	DESCRIPTION	UNIT PRICE	AMOUNT	
30	Travel planning software	$40.00	$1,200	00
6 cs	Magazines	100/cs	600	00
		SUBTOTAL	$1,800	00
		SALES TAX	108	00
		TOTAL	$1,908	00

Thank You!

NOTE NO. 416

$ 1,908.00   Date April 13   20--

Sixty days —— after date I promise to pay to

Buzz Newsstand —— the sum of

One thousand nine hundred eight dollars —— with interest at the rate of 9% per year.

Due date June 12   20--

Saba Nadal

BUZZ NEWSSTAND
Union Terminal Building, #101
Tacoma, WA 98402

RECEIPT No. 1387

June 12   20--

RECEIVED FROM Saba Nadal $ 28.23

Twenty-eight and 23/100 —— DOLLARS

FOR Interest on note

RECEIVED BY Kathy Harper

NOTE NO. 417

$ 1,908.00   Date June 12   20--

Ninety days —— after date I promise to pay to

Buzz Newsstand —— the sum of

One thousand nine hundred eight dollars —— with interest at the rate of 10% per year.

Due date September 10   20--

Saba Nadal

BUZZ NEWSSTAND
Union Terminal Building, #101
Tacoma, WA 98402

RECEIPT No. 1555

September 10   20--

RECEIVED FROM Saba Nadal $ 1,955.05

One thousand nine hundred fifty-five and 05/100 —— DOLLARS

FOR Payment for note issued June 12, 20--

RECEIVED BY Kathy Harper

Problem 26-10   Renewing a Note Receivable

GENERAL JOURNAL   PAGE _____

	DATE	DESCRIPTION	POST. REF.	DEBIT	CREDIT	
1						1
2						2
3						3
4						4
5						5
6						6
7						7
8						8
9						9
10						10
11						11
12						12
13						13
14						14
15						15
16						16
17						17
18						18
19						19
20						20
21						21
22						22
23						23
24						24
25						25
26						26

Analyze: _____

CHAPTER 26 Notes Payable and Receivable

Self-Test

Part A Fill in the Missing Term

Directions: *In the Answer column, write the letter of the word or phrase that best completes the sentence. Not all of the terms will be used.*

A. bank discount	**G.** issue date	**M.** note receivable
B. Discount on Notes Payable	**H.** maker	**N.** payee
C. face value	**I.** maturity date	**O.** principal
D. interest	**J.** maturity value	**P.** proceeds
E. interest-bearing note	**K.** noninterest-bearing note payable	**Q.** promissory note
F. interest rate	**L.** note payable	**R.** term

Answer

_____ 1. The _____ is the percentage of the principal that is charged for the use of the money.

_____ 2. The amount being borrowed when a promissory note is issued or received is called the _____.

_____ 3. A contra liability account representing future interest charges that have already been paid is the _____ account.

_____ 4. The _____ is the due date of a note or the date on which the principal and interest must be paid.

_____ 5. The _____ of a promissory note is the amount of time the borrower has to repay it.

_____ 6. _____ is the principal plus the interest on a promissory note.

_____ 7. The _____ is the date on which a note is written.

_____ 8. The _____ on a promissory note is the person or business that promises to repay the principal and the interest.

_____ 9. A(n) _____ is a note from which the interest has been deducted in advance.

_____ 10. The amount of cash actually received by the borrower when a noninterest-bearing note payable is issued is the _____.

_____ 11. The charge for the use of the principal borrowed on a promissory note is the _____.

_____ 12. A(n) _____ is a note that requires the face value plus interest to be paid at maturity.

Part B   True or False

Directions: *Circle the letter* T *in the answer column if the statement is true;*
circle the letter F *if the statement is false.*

Answer

T   F   **1.** The due date for a 30-day note dated April 10 is May 9.

T   F   **2.** When a note is signed, the maker is agreeing to repay the
note within a certain period of time.

T   F   **3.** The Interest Expense account is increased by a credit.

T   F   **4.** The account Discount on Notes Payable has a normal
debit balance.

T   F   **5.** Notes Payable is credited when a company issues a
promise to repay a debt.

T   F   **6.** A person or business may issue a promissory note to obtain
a loan from a bank.

T   F   **7.** Interest expense for a noninterest-bearing note is recorded
on the maturity date.

MINI PRACTICE SET 5

Kite Loft Inc.

CHART OF ACCOUNTS

ASSETS
101   Cash in Bank
105   Accounts Receivable
110   Merchandise Inventory
115   Supplies
120   Prepaid Insurance
125   Office Equipment
130   Store Equipment

LIABILITIES
201   Accounts Payable
205   Federal Corp. Income
       Tax Payable
210   Sales Tax Payable

STOCKHOLDERS' EQUITY
301   Capital Stock
305   Retained Earnings
310   Income Summary

REVENUE
401   Sales
405   Sales Discounts
410   Sales Returns and Allowances

COST OF MERCHANDISE
501   Purchases
505   Transportation In
510   Purchases Discounts
515   Purchases Returns and Allowances

EXPENSES
605   Advertising Expense
610   Bankcard Fees Expense
615   Insurance Expense
620   Miscellaneous Expense
625   Rent Expense
630   Salaries Expense
635   Supplies Expense
640   Utilities Expense
650   Federal Corp. Income Tax Expense

Accounts Receivable Subsidiary Ledger
BES   Best Toys
LAR   Lars' Specialties
SER   Serendipity Shop
SMA   Small Town Toys
TOY   The Toy Store

Accounts Payable Subsidiary Ledger
BRA   Brad Kites, Ltd.
CRE   Creative Kites, Inc.
EAS   Easy Glide Co.
RED   Reddi-Bright Manufacturing
STA   Stars Kites Outlet
TAY   Taylor Office Supplies

Mini Practice Set 5   Source Documents

Instructions: *Use the following source documents to record the transactions for this practice set.*

Reddi-Bright MANUFACTURING 127 Hill Street, #5000 Druid Hills, GA 30333		INVOICE NO. 410

DATE: *Dec. 16, 20--*
ORDER NO.:
SHIPPED BY:
TERMS:

TO Kite Loft Inc.
112 Ashby Drive
Atlanta, GA 30308

QTY.	ITEM	UNIT PRICE	TOTAL
	General merchandise		$1,475.00

Taylor Office Supplies 212 Morningside Drive Atlanta, GA 30305		INVOICE NO. 830

DATE: *December 17, 20--*
ORDER NO.:
SHIPPED BY:
TERMS:

TO Kite Loft Inc.
112 Ashby Drive
Atlanta, GA 30308

QTY.	ITEM	UNIT PRICE	TOTAL
2	Calendar/Planner	$40.00	$80.00

Kite Loft Inc.
112 Ashby Drive
Atlanta, GA 30308

610
4-571
6212

DATE *December 16* 20--

PAY TO THE ORDER OF *United States Treasury* $ *1,050.00*

One thousand fifty and 00/100 ——————— DOLLARS

S *Sanwa Bank*

MEMO *Qtrly. fed. inc. tax*   *Michael Ramspart*

⑯621245７¦⑯ 2323 1112″ 0610

Kite Loft Inc.
112 Ashby Drive
Atlanta, GA 30308

RECEIPT
No. 358

December 17 20--

RECEIVED FROM *Best Toys* $ *1,965.60*

One thousand nine hundred sixty-five and 60/100 ——— DOLLARS

FOR *Sales slip #479 for $2,003.40, less $37.80 discount*

RECEIVED BY *Michael Ramspart*

Kite Loft Inc.
112 Ashby Drive
Atlanta, GA 30308

611
4-571
6212

DATE *December 16* 20--

PAY TO THE ORDER OF *Brad Kites, Ltd.* $ *2,548.00*

Two thousand five hundred forty-eight and 00/100 ——— DOLLARS

S *Sanwa Bank*

MEMO *#112 $2600 less disc.*   *Michael Ramspart*

⑯621245７¦⑯ 2323 1112″ 0611

Kite Loft Inc.
112 Ashby Drive
Atlanta, GA 30308

DATE: *December 19, 20--*   NO. *484*

SOLD TO *Best Toys*

CLERK	CASH	CHARGE	TERMS

QTY.	DESCRIPTION	UNIT PRICE	AMOUNT	
26	Kites	$100.00	$2,600	00
		SUBTOTAL	$2,600	00
		SALES TAX	156	00
Thank You!		TOTAL	$2,756	00

Kite Loft Inc.
112 Ashby Drive
Atlanta, GA 30308

612
4-571
6212

DATE *December 17* 20--

PAY TO THE ORDER OF *Sanwa Bank* $ *4,750.00*

Four thousand seven hundred fifty and 00/100 ——— DOLLARS

S *Sanwa Bank*

MEMO *Monthly payroll*   *Michael Ramspart*

⑯621245７¦⑯ 2323 1112″ 0612

Mini Practice Set 5   (continued)

Kite Loft Inc.
112 Ashby Drive
Atlanta, GA 30308

RECEIPT
No. 359

December 19 20--

RECEIVED FROM _Lars' Specialties_ $ 1,716.00

One thousand seven hundred sixteen and 00/100 —— DOLLARS

FOR _Payment of sales slip #480 for $1,749, less $33 discount_

RECEIVED BY _Michael Ramspart_

Brad Kites, Ltd.
633 Louise Street
Atlanta, GA 30303

INVOICE NO. 215
DATE: _Dec. 20, 20--_
ORDER NO.:
SHIPPED BY:
TERMS:

TO
Kite Loft Inc.
112 Ashby Drive
Atlanta, GA 30308

QTY.	ITEM	UNIT PRICE	TOTAL
50	Kites	$31.20	$1,560.00

Kite Loft Inc.
112 Ashby Drive
Atlanta, GA 30308

613
4-571
6212

DATE _December 20_ 20--

PAY TO THE ORDER OF _Creative Kites, Inc._ $ 375.00

Three hundred seventy-five and 00/100 —— DOLLARS

S Sanwa Bank

MEMO _on account_   _Michael Ramspart_

⑈621245７1⑈ 2323 1112⑈ 0613

Kite Loft Inc.
112 Ashby Drive
Atlanta, GA 30308

CREDIT MEMORANDUM
NO. 44

ORIGINAL SALES DATE	ORIGINAL SALES SLIP	APPROVAL	
Dec. 19, 20--	484	M.R.	☒ MDSE RET

DATE: _December 21, 20--_

NAME: _Best Toys_

ADDRESS:

QTY.	DESCRIPTION	AMOUNT
1	Kite	$ 100 00

REASON FOR RETURN _damaged_

THE TOTAL SHOWN AT THE RIGHT WILL BE CREDITED TO YOUR ACCOUNT.

SUB TOTAL	$	100 00
SALES TAX		6 00
TOTAL	$	106 00

Katie Sims

CUSTOMER SIGNATURE

Kite Loft Inc.
112 Ashby Drive
Atlanta, GA 30308

DATE: _December 23, 20--_   NO. _485_

SOLD TO _Lars' Specialties_

CLERK	CASH	CHARGE	TERMS

QTY.	DESCRIPTION	UNIT PRICE	AMOUNT
100	Kites	$15.80	$1,580 00
		SUBTOTAL	$1,580 00
		SALES TAX	94 80
		TOTAL	$1,674 80

Thank You!

Kite Loft Inc.
112 Ashby Drive
Atlanta, GA 30308

RECEIPT
No. 360

December 23 20--

RECEIVED FROM _Serendipity Shop_ $ 300.00

Three hundred and 00/100 —— DOLLARS

FOR _Payment on account_

RECEIVED BY _Michael Ramspart_

Kite Loft Inc.
112 Ashby Drive
Atlanta, GA 30308

614
4-571
6212

DATE _December 23_ 20--

PAY TO THE ORDER OF _Easy Glide Co._ $ 1,852.20

One thousand eight hundred fifty-two and 20/100 —— DOLLARS

S Sanwa Bank

MEMO _#326 $1890 less disc._   _Michael Ramspart_

⑈621245７1⑈ 2323 1112⑈ 0614

Kite Loft Inc.
112 Ashby Drive
Atlanta, GA 30308

RECEIPT
No. 361

December 26 20--

RECEIVED FROM _The Toy Store_ $ 1,102.40

One thousand one hundred two and 40/100 —— DOLLARS

FOR _Sales slip #483 for $1,123.60, less $21.20 discount_

RECEIVED BY _Michael Ramspart_

Mini Practice Set 5   (continued)

DEBIT MEMORANDUM   No. *28*

Kite Loft Inc.
112 Ashby Drive
Atlanta, GA 30308

Date: *December 26, 20--*
Invoice No.: *215*

To: *Brad Kites, Ltd.*
633 Louise Street
Atlanta, GA 30303

This day we have debited your account as follows:

Quantity	Item	Unit Price	Total
1	*misc. merchandise*	$150.00	$150.00

EASY GLIDE CO.
124 Merric Blvd., #2A
Atlanta, GA 30301

INVOICE NO. 335

DATE: *Dec. 26, 20--*
ORDER NO.:
SHIPPED BY:
TERMS:

TO   *Kite Loft Inc.*
112 Ashby Drive
Atlanta, GA 30308

QTY.	ITEM	UNIT PRICE	TOTAL
	Specialty kites		$1,630.00

Kite Loft Inc.   616
112 Ashby Drive   4-571
Atlanta, GA 30308   6212

DATE *December 29* 20--

PAY TO THE ORDER OF *Stars Kites Outlet* $ *1,625.00*

One thousand six hundred twenty-five and 00/100 ———— DOLLARS

S *Sanwa Bank*

MEMO *on account*   *Michael Ramspart*

⑆621245710⑆ 2323 1112⑈ 0616

Kite Loft Inc.
112 Ashby Drive
Atlanta, GA 30308

DATE: *December 29, 20--*   NO. *486*

SOLD TO   *The Toy Store*

CLERK	CASH	CHARGE	TERMS

QTY.	DESCRIPTION	UNIT PRICE	AMOUNT	
50	*Kites Variety Pack*	*$39.80*	*$1,990*	*00*
		SUBTOTAL	$1,990	00
		SALES TAX	119	40
		TOTAL	$2,109	40

Thank You!

Kite Loft Inc.   615
112 Ashby Drive   4-571
Atlanta, GA 30308   6212

DATE *December 28* 20--

PAY TO THE ORDER OF *Daily Examiner* $ *120.00*

One hundred twenty and 00/100 ———————— DOLLARS

S *Sanwa Bank*

MEMO *monthly advertising*   *Michael Ramspart*

⑆621245710⑆ 2323 1112⑈ 0615

Kite Loft Inc.   617
112 Ashby Drive   4-571
Atlanta, GA 30308   6212

DATE *December 30* 20--

PAY TO THE ORDER OF *Reddi-Bright Manufacturing* $ *700.00*

Seven hundred and 00/100 ———————————— DOLLARS

S *Sanwa Bank*

MEMO *on account*   *Michael Ramspart*

⑆621245710⑆ 2323 1112⑈ 0617

Kite Loft Inc.   RECEIPT
112 Ashby Drive
Atlanta, GA 30308   No. 362

December 28 20--

RECEIVED FROM *Small Town Toys* $ *450.00*

Four hundred fifty and 00/100 ———————— DOLLARS

FOR *Payment on account*

RECEIVED BY *Michael Ramspart*

Kite Loft Inc.
112 Ashby Drive
Atlanta, GA 30308

DATE: *December 30, 20--*   NO. *487*

SOLD TO   *Serendipity Shop*

CLERK	CASH	CHARGE	TERMS

QTY.	DESCRIPTION	UNIT PRICE	AMOUNT	
28	*Kites*	*$20.00*	*$560*	*00*
		SUBTOTAL	$560	00
		SALES TAX	33	60
		TOTAL	$593	60

Thank You!

Mini Practice Set 5   (continued)

	Dollars	Cents
$ _____ **No. 618**		
Date *December 31* 20——		
To _____		
For _____		
	Dollars	**Cents**
Balance brought forward		
Less Bank Svc. Chg.	*10*	*00*
Less Bankcard fee	*150*	*00*
Total		
Less this check		
Balance carried forward		

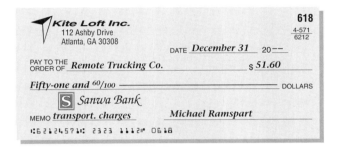

Kite Loft Inc.
112 Ashby Drive
Atlanta, GA 30308

618
4-571
6212

DATE *December 31* 20——

PAY TO THE
ORDER OF *Remote Trucking Co.* $ *51.60*

Fifty-one and ⁶⁰/₁₀₀ ———————————— DOLLARS

S *Sanwa Bank*

MEMO *transport. charges*   *Michael Ramspart*

⑆6212457⑆ 2323 1112⑈ 0618

	Dollars	Cents
$ _____ **No. 619**		
Date *December 31* 20——		
To _____		
For _____		
	Dollars	**Cents**
Balance brought forward		
Add deposits *12/31/—— (T41)*	*4,234*	*81*
12/31/—— (T41)	*1,840*	*45*
Total		
Less this check		
Balance carried forward		

```
Dec  31
Tape 41


     3,995.10   CA
       239.71   TX
     4,234.81   TTL

     1,736.27   CA
       104.18   TX
     1,840.45   TTL
```

Notes

Computerized Accounting Using Peachtree

Mini Practice Set 5

INSTRUCTIONS

Beginning a Session

Step 1 Select the problem set: Kite Loft Inc. (MP-5).

Step 2 Rename the company, and set the system date to December 31, 2008.

Completing the Accounting Problem

Step 3 Review the transactions provided in your textbook (December 16–December 31).

 TIP: To save time entering transactions, group them by type and then enter the transactions in batches.

Sales and Cash Receipts

Step 4 Record the sales on account using the **Sales/Invoicing** option.

 TIP: Remember that you need to change the GL Account in the Receipts window to enter a cash receipt for the sale of an asset (e.g., supplies, office equipment).

Step 5 Enter and apply any sales returns.

Step 6 Record all of the cash receipts using the **Receipts** option.

 TIP: Always verify the sales discount amount when you record a Cash Receipt. Peachtree may not always compute the discount correctly.

Purchases and Cash Payments

Step 7 Enter the purchases on account using the **Purchases/Receive Inventory** option.

 TIP: Always verify the *GL Account* field when you enter a purchases on account transaction.

Step 8 Record and apply any purchases returns.

Step 9 Process all of the cash payments with the **Payments** option.

General Journal

Step 10 Use the **General Journal Entry** option to record the adjusting entries.

Peachtree Guide

TIP: You can use the , and buttons to edit a multi-line general journal entry.

Printing Reports and Proofing Your Work

Step 11   Print the following reports: General Journal, Purchases Journal, Cash Disbursements Journal, Sales Journal, and Cash Receipts Journal.

TIP: Double-click a report title in the Select a Report window to go directly to that report.

Step 12   Proof your work. Print updated reports, if necessary.
Step 13   Print the following reports: General Ledger, Vendor Ledgers, and Customer Ledgers.
Step 14   Print a General Ledger Trial Balance.
Step 15   Print an Income Statement, a Statement of Retained Earnings, and a Balance Sheet.

Saving Your Work

Step 16   Click the **Save the Pre-closing Balances** button in the Glencoe Smart Guide window.

Closing the Fiscal Year

Step 17   Use the Year-End Wizard option in the *Tasks* menu to perform the year-end closing.
Step 18   Print a Post-Closing Trial Balance.

Analyzing Your Work

Step 19   Answer the Analyze questions.

Ending the Session

Step 20   Click the **Close Problem** button in the Glencoe Smart Guide window.

Mini Practice Set 5 (continued)
(1), (4)

SALES JOURNAL   PAGE ___22___

	DATE		SALES SLIP NO.	CUSTOMER'S ACCOUNT DEBITED	POST. REF.	SALES CREDIT	SALES TAX PAYABLE CREDIT	ACCOUNTS RECEIVABLE DEBIT	
1	20--								1
2	Dec.	7	479	Best Toys	✓	1 890 00	113 40	2 003 40	2
3		9	480	Lars' Specialties	✓	1 650 00	99 00	1 749 00	3
4		9	481	Serendipity Shop	✓	1 219 00	73 14	1 292 14	4
5		12	482	Small Town Toys	✓	875 00	52 50	927 50	5
6		15	483	The Toy Store	✓	1 060 00	63 60	1 123 60	6
7									7
8									8
9									9
10									10
11									11
12									12
13									13
14									14
15									15
16									16
17									17
18									18
19									19
20									20
21									21
22									22
23									23
24									24
25									25
26									26
27									27
28									28
29									29
30									30
31									31
32									32
33									33

Mini Practice Set 5 (continued)

CASH RECEIPTS JOURNAL

DATE	DOC. NO.	ACCOUNT NAME	POST. REF.	GENERAL CREDIT	SALES CREDIT	SALES TAX PAYABLE CREDIT	ACCOUNTS RECEIVABLE CREDIT	SALES DISCOUNTS DEBIT	CASH IN BANK DEBIT	
20--										1
Dec. 2	R351	Best Toys	✓				319590	6030	313560	2
3	R352	Store Equipment	130	20000					20000	3
6	R353	Lars' Specialties	✓				189740	3580	186160	4
8	R354	Serendipity Shop	✓				76360	1440	74920	5
10	R355	Small Town Toys	✓				128480		128480	6
12	R356	The Toy Store	✓				124020	2340	121680	7
13	R357	Supplies	115	3000					3000	8
15	T40	Cash Sales	—		365070	21904			386974	9
15	T40	Bankcard Sales	—		181240	10874			192114	10
										11
										12
										13
										14
										15
										16
										17
										18
										19
										20
										21
										22
										23
										24
										25
										26

Mini Practice Set 5   (continued)

PAGE 21

PURCHASES JOURNAL

	DATE	INVOICE NO.	CREDITOR'S ACCOUNT CREDITED	POST. REF.	ACCOUNTS PAYABLE CREDIT	PURCHASES DEBIT	GENERAL ACCOUNT DEBITED	GENERAL POST. REF.	GENERAL DEBIT
1	20-- Dec.								
2	3	CL213	Creative Kites, Inc.	✓	1500 00	1500 00			
3	4	803	Taylor Office Supplies	✓	125 00		Office Equipment	125	125 00
4	7	112	Brad Kites, Ltd.	✓	2600 00	2600 00			
5	11	514	Stars Kites Outlet	✓	3250 00	3250 00			
6	14	326	Easy Glide Co.	✓	1890 00	1890 00			
7									
8									
9									
10									
11									
12									
13									
14									
15									
16									
17									
18									
19									
20									
21									
22									
23									
24									
25									

Mini Practice Set 5   (continued)

CASH PAYMENTS JOURNAL

	DATE	DOC. NO.	ACCOUNT NAME	POST. REF.	GENERAL DEBIT	ACCOUNTS PAYABLE DEBIT	PURCHASES DISCOUNTS CREDIT	CASH IN BANK CREDIT	
1	20--								1
2	Dec. 2	601	Rent Expense	625	70000			70000	2
3	5	602	Brad Kites, Ltd.	✓		137500	2750	134750	3
4	6	603	Stars Kites Outlet	✓		147000	2940	144060	4
5	7	604	Transportation In	505	3720			3720	5
6	9	605	Creative Kites, Inc.	✓		109000	2180	106820	6
7	11	606	Easy Glide Co.	✓		123500	2470	121030	7
8	14	607	Reddi-Bright Mfg.	✓		228000		228000	8
9	15	608	Utilities Expense	640	16500			16500	9
10	15	609	Taylor Office Supplies	✓		12500		12500	10
11									11
12									12
13									13
14									14
15									15
16									16
17									17
18									18
19									19
20									20
21									21
22									22
23									23
24									24
25									25
26									26
27									27
28									28

Mini Practice Set 5 (continued)
(1), (13), (14)

GENERAL JOURNAL   PAGE ___12___

	DATE		DESCRIPTION	POST. REF.	DEBIT	CREDIT	
1	20--						1
2	Dec.	3	Purchases	501	1 50 00 0		2
3			Merchandise Inventory	110		1 50 00 0	3
4			Memo 30				4
5		5	Sales Returns and Allowances	410	1 20 00		5
6			Sales Tax Payable	210	7 20		6
7			Accts. Rec./Small Town Toys	105 ✓		1 27 20	7
8			Credit Memo 43				8
9		12	Accts. Pay./Reddi-Bright Mfg.	201 ✓	8 0 00		9
10			Purchases Returns and Allowances	515		8 0 00	10
11			Debit Memo 27				11
12							12
13							13
14							14
15							15
16							16
17							17
18							18
19							19
20							20
21							21
22							22
23							23
24							24
25							25
26							26
27							27
28							28
29							29
30							30
31							31
32							32
33							33
34							34
35							35
36							36

Mini Practice Set 5   (continued)
(1)

GENERAL JOURNAL   PAGE _____

	DATE	DESCRIPTION	POST. REF.	DEBIT	CREDIT	
1						1
2						2
3						3
4						4
5						5
6						6
7						7
8						8
9						9
10						10
11						11
12						12
13						13
14						14
15						15
16						16
17						17
18						18
19						19
20						20

(2)

ACCOUNTS RECEIVABLE SUBSIDIARY LEDGER

Name *Best Toys* _____

Address *13400 Midway Road, Dallas, TX 75244* _____

DATE		DESCRIPTION	POST. REF.	DEBIT	CREDIT	BALANCE
20--						
Dec.	1	Balance	✓			3 1 9 5 90
	2		CR23		3 1 9 5 90	——
	7		S22	2 0 0 3 40		2 0 0 3 40

Mini Practice Set 5   (continued)

ACCOUNTS RECEIVABLE SUBSIDIARY LEDGER

Name *Lars' Specialties*

Address *601 O'Hara Road, Arlington, TX 76010*

DATE		DESCRIPTION	POST. REF.	DEBIT	CREDIT	BALANCE
20--						
Dec.	1	Balance	✓			1897 40
	6		CR23		1897 40	—
	9		S22	1749 00		1749 00

Name *Serendipity Shop*

Address *835 Coronado Drive, Corpus Christi, TX 78403*

DATE		DESCRIPTION	POST. REF.	DEBIT	CREDIT	BALANCE
20--						
Dec.	1	Balance	✓			763 60
	8		CR23		763 60	—
	9		S22	1292 14		1292 14

Name *Small Town Toys*

Address *103 Cedar Park, Dallas, TX 75244*

DATE		DESCRIPTION	POST. REF.	DEBIT	CREDIT	BALANCE
20--						
Dec.	1	Balance	✓			1412 00
	5		G12		127 20	1284 80
	10		CR23		1284 80	—
	12		S22	927 50		927 50

Mini Practice Set 5   (continued)

ACCOUNTS RECEIVABLE SUBSIDIARY LEDGER

Name *The Toy Store*

Address *70 South Washington Street, Fort Worth, TX 76101*

DATE		DESCRIPTION	POST. REF.	DEBIT	CREDIT	BALANCE
20--						
Dec.	1	Balance	✓			1 2 4 0 20
	12		CR23		1 2 4 0 20	—
	15		S22	1 1 2 3 60		1 1 2 3 60

ACCOUNTS PAYABLE SUBSIDIARY LEDGER

Name *Brad Kites, Ltd.*

Address *633 Louise Street NW, Atlanta, GA 30303*

DATE		DESCRIPTION	POST. REF.	DEBIT	CREDIT	BALANCE
20--						
Dec.	1	Balance	✓			1 3 7 5 00
	5		CP24	1 3 7 5 00		—
	7		P21		2 6 0 0 00	2 6 0 0 00

Name *Creative Kites, Inc.*

Address *1900 Talman Avenue North, Chicago, IL 60647*

DATE		DESCRIPTION	POST. REF.	DEBIT	CREDIT	BALANCE
20--						
Dec.	1	Balance	✓			1 0 9 0 00
	3		P21		1 5 0 0 00	2 5 9 0 00
	9		CP24	1 0 9 0 00		1 5 0 0 00

Mini Practice Set 5   (continued)

ACCOUNTS PAYABLE SUBSIDIARY LEDGER

Name *Easy Glide Co.*

Address *124 Merric Blvd. #2A, Atlanta, GA 30301*

DATE		DESCRIPTION	POST. REF.	DEBIT	CREDIT	BALANCE
20--						
Dec.	1	Balance	✓			1235 00
	11		CP24	1235 00		————
	14		P21		1890 00	1890 00

Name *Reddi-Bright Manufacturing*

Address *127 Hill Street #5000, Druid Hills, GA 30333*

DATE		DESCRIPTION	POST. REF.	DEBIT	CREDIT	BALANCE
20--						
Dec.	1	Balance	✓			2360 00
	12		G12	80 00		2280 00
	14		CP24	2280 00		————

Name *Stars Kites Outlet*

Address *150 Vista Avenue, St. Louis, MO 63110*

DATE		DESCRIPTION	POST. REF.	DEBIT	CREDIT	BALANCE
20--						
Dec.	1	Balance	✓			1470 00
	6		CP24	1470 00		————
	11		P21		3250 00	3250 00

Mini Practice Set 5   (continued)

ACCOUNTS PAYABLE SUBSIDIARY LEDGER

Name *Taylor Office Supplies*

Address *212 Morningside Drive, Atlanta, GA 30305*

DATE		DESCRIPTION	POST. REF.	DEBIT	CREDIT	BALANCE
20--						
Dec.	4		P21		1 2 5 00	1 2 5 00
	15		CP24	1 2 5 00		—

(3)

GENERAL LEDGER

ACCOUNT   *Cash in Bank*   ACCOUNT NO.   101

DATE		DESCRIPTION	POST. REF.	DEBIT	CREDIT	BALANCE DEBIT	BALANCE CREDIT
20--							
Dec.	1	Balance	✓			1 8 4 8 0 29	

ACCOUNT   *Accounts Receivable*   ACCOUNT NO.   105

DATE		DESCRIPTION	POST. REF.	DEBIT	CREDIT	BALANCE DEBIT	BALANCE CREDIT
20--							
Dec.	1	Balance	✓			8 5 0 9 10	
	5		G12		1 2 7 20	8 3 8 1 90	

ACCOUNT   *Merchandise Inventory*   ACCOUNT NO.   110

DATE		DESCRIPTION	POST. REF.	DEBIT	CREDIT	BALANCE DEBIT	BALANCE CREDIT
20--							
Dec.	1	Balance	✓			3 1 7 6 6 98	
	3		G12		1 5 0 0 00	3 0 2 6 6 98	

Mini Practice Set 5 (continued)

ACCOUNT **Supplies**   ACCOUNT NO. **115**

DATE		DESCRIPTION	POST. REF.	DEBIT	CREDIT	BALANCE DEBIT	BALANCE CREDIT
20--							
Dec.	1	Balance	✓			1251 46	
	13		CR23		30 00	1221 46	

ACCOUNT **Prepaid Insurance**   ACCOUNT NO. **120**

DATE		DESCRIPTION	POST. REF.	DEBIT	CREDIT	BALANCE DEBIT	BALANCE CREDIT
20--							
Dec.	1	Balance	✓			2460 00	

ACCOUNT **Office Equipment**   ACCOUNT NO. **125**

DATE		DESCRIPTION	POST. REF.	DEBIT	CREDIT	BALANCE DEBIT	BALANCE CREDIT
20--							
Dec.	1	Balance	✓			6600 00	
	4		P21	125 00		6725 00	

ACCOUNT **Store Equipment**   ACCOUNT NO. **130**

DATE		DESCRIPTION	POST. REF.	DEBIT	CREDIT	BALANCE DEBIT	BALANCE CREDIT
20--							
Dec.	1	Balance	✓			10800 00	
	3		CR23		200 00	10600 00	

ACCOUNT **Accounts Payable**   ACCOUNT NO. **201**

DATE		DESCRIPTION	POST. REF.	DEBIT	CREDIT	BALANCE DEBIT	BALANCE CREDIT
20--							
Dec.	1	Balance	✓				7530 00
	12		G12	80 00			7450 00

Mini Practice Set 5   (continued)

ACCOUNT ___Federal Corporate Income Tax Payable___   ACCOUNT NO. ___205___

DATE	DESCRIPTION	POST. REF.	DEBIT	CREDIT	BALANCE DEBIT	BALANCE CREDIT

ACCOUNT ___Sales Tax Payable___   ACCOUNT NO. ___210___

DATE		DESCRIPTION	POST. REF.	DEBIT	CREDIT	BALANCE DEBIT	BALANCE CREDIT
20--							
Dec.	1	Balance	✓				895 80
	5		G12	7 20			888 60

ACCOUNT ___Capital Stock___   ACCOUNT NO. ___301___

DATE		DESCRIPTION	POST. REF.	DEBIT	CREDIT	BALANCE DEBIT	BALANCE CREDIT
20--							
Dec.	1	Balance	✓				25 000 00

ACCOUNT ___Retained Earnings___   ACCOUNT NO. ___305___

DATE		DESCRIPTION	POST. REF.	DEBIT	CREDIT	BALANCE DEBIT	BALANCE CREDIT
20--							
Dec.	1	Balance	✓				13 000 00

ACCOUNT ___Income Summary___   ACCOUNT NO. ___310___

DATE	DESCRIPTION	POST. REF.	DEBIT	CREDIT	BALANCE DEBIT	BALANCE CREDIT

Mini Practice Set 5   (continued)

ACCOUNT _Sales_ _____ ACCOUNT NO. __401__

DATE		DESCRIPTION	POST. REF.	DEBIT	CREDIT	BALANCE DEBIT	BALANCE CREDIT
20--							
Dec.	1	Balance	✓				10815139

ACCOUNT __Sales Discounts__ _____ ACCOUNT NO. __405__

DATE		DESCRIPTION	POST. REF.	DEBIT	CREDIT	BALANCE DEBIT	BALANCE CREDIT
20--							
Dec.	1	Balance	✓			21000	

ACCOUNT __Sales Returns and Allowances__ _____ ACCOUNT NO. __410__

DATE		DESCRIPTION	POST. REF.	DEBIT	CREDIT	BALANCE DEBIT	BALANCE CREDIT
20--							
Dec.	1	Balance	✓			17540	
	5		G12	12000		29540	

ACCOUNT __Purchases__ _____ ACCOUNT NO. __501__

DATE		DESCRIPTION	POST. REF.	DEBIT	CREDIT	BALANCE DEBIT	BALANCE CREDIT
20--							
Dec.	1	Balance	✓			2376113	
	3		G12	150000		2526113	

Mini Practice Set 5   (continued)

ACCOUNT ___Transportation In___   ACCOUNT NO. ___505___

DATE		DESCRIPTION	POST. REF.	DEBIT	CREDIT	BALANCE	
						DEBIT	CREDIT
20--							
Dec.	1	Balance	✓			1275 80	
	7		CP24	37 20		1313 00	

ACCOUNT ___Purchases Discounts___   ACCOUNT NO. ___510___

DATE		DESCRIPTION	POST. REF.	DEBIT	CREDIT	BALANCE	
						DEBIT	CREDIT
20--							
Dec.	1	Balance	✓				415 75

ACCOUNT ___Purchases Returns and Allowances___   ACCOUNT NO. ___515___

DATE		DESCRIPTION	POST. REF.	DEBIT	CREDIT	BALANCE	
						DEBIT	CREDIT
20--							
Dec.	1	Balance	✓				390 85
	12		G12		80 00		470 85

ACCOUNT ___Advertising Expense___   ACCOUNT NO. ___605___

DATE		DESCRIPTION	POST. REF.	DEBIT	CREDIT	BALANCE	
						DEBIT	CREDIT
20--							
Dec.	1	Balance	✓			430 00	

ACCOUNT ___Bankcard Fees Expense___   ACCOUNT NO. ___610___

DATE		DESCRIPTION	POST. REF.	DEBIT	CREDIT	BALANCE	
						DEBIT	CREDIT
20--							
Dec.	1	Balance	✓			1420 57	

Name _____ Date _____ Class _____

Mini Practice Set 5 (continued)

ACCOUNT __Insurance Expense__   ACCOUNT NO. __615__

DATE	DESCRIPTION	POST. REF.	DEBIT	CREDIT	BALANCE DEBIT	BALANCE CREDIT

ACCOUNT __Miscellaneous Expense__   ACCOUNT NO. __620__

DATE	DESCRIPTION	POST. REF.	DEBIT	CREDIT	BALANCE DEBIT	BALANCE CREDIT
20--						
Dec. 1	Balance	✓			247 52	

ACCOUNT __Rent Expense__   ACCOUNT NO. __625__

DATE	DESCRIPTION	POST. REF.	DEBIT	CREDIT	BALANCE DEBIT	BALANCE CREDIT
20--						
Dec. 1	Balance	✓			7700 00	
2		CP24	700 00		8400 00	

ACCOUNT __Salaries Expense__   ACCOUNT NO. __630__

DATE	DESCRIPTION	POST. REF.	DEBIT	CREDIT	BALANCE DEBIT	BALANCE CREDIT
20--						
Dec. 1	Balance	✓			34871 18	

ACCOUNT __Supplies Expense__   ACCOUNT NO. __635__

DATE	DESCRIPTION	POST. REF.	DEBIT	CREDIT	BALANCE DEBIT	BALANCE CREDIT

Mini Practice Set 5   (continued)

ACCOUNT _Utilities Expense_   ACCOUNT NO. _640_

DATE		DESCRIPTION	POST. REF.	DEBIT	CREDIT	BALANCE DEBIT	BALANCE CREDIT
20--							
Dec.	1	Balance	✓			2274 36	
	15		CP24	165 00		2439 36	

ACCOUNT _Federal Corporate Income Tax Expense_   ACCOUNT NO. _650_

DATE		DESCRIPTION	POST. REF.	DEBIT	CREDIT	BALANCE DEBIT	BALANCE CREDIT
20--							
Dec.	1	Balance	✓			3150 00	

(6)

Mini Practice Set 5 (continued)

(7)

(7)

Mini Practice Set 5   (continued)

(8), (9)

	ACCT. NO.	ACCOUNT NAME	TRIAL BALANCE		ADJUSTMENTS	
			DEBIT	CREDIT	DEBIT	CREDIT
1						
2						
3						
4						
5						
6						
7						
8						
9						
10						
11						
12						
13						
14						
15						
16						
17						
18						
19						
20						
21						
22						
23						
24						
25						
26						
27						
28						
29						
30						
31						
32						
33						

ADJUSTED TRIAL BALANCE		INCOME STATEMENT		BALANCE SHEET		
DEBIT	CREDIT	DEBIT	CREDIT	DEBIT	CREDIT	
						1
						2
						3
						4
						5
						6
						7
						8
						9
						10
						11
						12
						13
						14
						15
						16
						17
						18
						19
						20
						21
						22
						23
						24
						25
						26
						27
						28
						29
						30
						31
						32
						33

Mini Practice Set 5 (continued)

(10)

Mini Practice Set 5 (continued)

(11)

(12)

Mini Practice Set 5   (concluded)
(15)

Analyze:

1. _____

2. _____

3. _____

MINI PRACTICE SET 5

Kite Loft Inc.

Audit Test

Directions: *Use your completed solutions to answer the following questions. Write the answer in the space to the left of each question.*

Answer

_____ **1.** What is the total of debits and credits in the Sales journal at the end of December?

_____ **2.** What is the total of debits and credits in the Cash Receipts journal at the end of December?

_____ **3.** What is the total of debits and credits in the Purchases journal at the end of December?

_____ **4.** What is the total of debits and credits in the Cash Payments journal at the end of December?

_____ **5.** For the first transaction on December 26, which account was credited?

_____ **6.** What amount was credited to the Merchandise Inventory account as an adjustment to inventory?

_____ **7.** What amount was debited to the Income Summary account to close the expense accounts for the month?

_____ **8.** What is the ending balance of the Best Toys accounts receivable subsidiary ledger account?

_____ **9.** What is the ending balance for the Accounts Receivable general ledger account? Does the balance agree with the total of the subsidiary ledger accounts?

_____ **10.** What is the ending balance for the Reddi-Bright Manufacturing accounts payable subsidiary ledger account?

_____

11. What is the ending balance for the Accounts Payable general ledger account? Does the balance agree with the total of the subsidiary ledger accounts?

_____

12. How many transactions during the month of December affected the Cash in Bank account?

_____

13. Has the amount owed to Stars Kites Outlet been paid off by the end of December?

_____

14. How many checks were issued by the business in December?

_____

15. What was the total amount credited to Sales for the month?

_____

16. What was the date of the trial balance?

_____

17. How many accounts were affected by adjusting entries?

_____

18. What was the total of operating expenses for the period?

_____

19. What is the ending balance for Retained Earnings for the period?

_____

20. What is the amount of total assets for the business at December 31?

_____

21. How many accounts are listed on the trial balance?

_____

22. How many accounts on the trial balance have debit balances?

_____

23. Which account on the trial balance has the largest balance?

_____

24. What is the amount of total liabilities for the business at December 31?

_____

25. How many accounts are listed on the post-closing trial balance?

CHAPTER  Introduction to Partnerships

Study Plan

Check Your Understanding

Section 1

Read Section 1 on pages 734–737 and complete the following exercises on page 738.
- ❏ Thinking Critically
- ❏ Communicating Accounting
- ❏ Problem 27-1   *Recording Partners' Investments*

Section 2

Read Section 2 on pages 739–743 and complete the following exercises on page 744.
- ❏ Thinking Critically
- ❏ Computing in the Business World
- ❏ Problem 27-2   *Determining Partners' Fractional Shares*
- ❏ Problem 27-3   *Analyzing a Source Document*

Summary

Review the Chapter 27 Summary on page 745 in your textbook.
- ❏ Key Concepts

Review and Activities

Complete the following questions and exercises on pages 746–747 in your textbook.
- ❏ Using Key Terms
- ❏ Understanding Accounting Concepts and Procedures
- ❏ Case Study
- ❏ Conducting an Audit with Alex
- ❏ Internet Connection
- ❏ Workplace Skills

Computerized Accounting

Read the Computerized Accounting information on page 748 in your textbook.
- ❏ *Making the Transition from a Manual to a Computerized System*
- ❏ *Setting Up a New Company in Peachtree*

Problems

Complete the following end-of-chapter problems for Chapter 27 in your textbook.
- ❏ Problem 27-4   *Dividing Partnership Earnings*
- ❏ Problem 27-5   *Calculating the Percentage of Each Partner's Capital Investment*
- ❏ Problem 27-6   *Recording Investments of Partners*
- ❏ Problem 27-7   *Sharing Losses Based on Capital Balances*
- ❏ Problem 27-8   *Partners' Withdrawals*
- ❏ Problem 27-9   *Preparing Closing Entries for a Partnership*

Challenge Problem

- ❏ Problem 27-10   *Evaluating Methods of Dividing Partnership Earnings*

Chapter Reviews and Working Papers

Complete the following exercises for Chapter 27 in your Chapter Reviews and Working Papers.
- ❏ Chapter Review
- ❏ Self-Test

CHAPTER 27 REVIEW   Introduction to Partnerships

Part 1   Accounting Vocabulary (2 points)

Total Points	35
Student's Score	

Directions: *Using terms from the following list, complete the sentences below. Write the letter of the term you have chosen in the space provided.*

> **A.** mutual agency   **B.** partnership agreement

_____   **1.** Within a partnership, the relationship that allows any partner to enter into agreements that are binding on all other partners is called _____.

_____   **2.** A _____ is a written document that sets out the terms under which a partnership will operate.

Part 2   Partnerships (10 points)

Directions: *Read each of the following statements to determine whether the statement is true or false. Write your answer in the space provided.*

True   **0.** Withdrawals within a partnership are recorded the same way as are the withdrawals of the owner of a sole proprietorship.

_____   **1.** When a partner invests assets in the partnership, he or she retains personal rights of ownership.

_____   **2.** A partnership may be formed through an oral agreement between two or more individuals.

_____   **3.** The amount of the withdrawals by each of the partners must always be equal.

_____   **4.** The partnership agreement should include the procedures for sharing profits and losses but should not include the investment of each partner.

_____   **5.** Noncash assets that are invested in the business are recorded at their current market value.

_____   **6.** Separate capital accounts are set up for each partner, but only one withdrawal account is used by all of the partners.

_____   **7.** Partners are not required to pay federal, state, or personal income taxes on their share of the business's net income.

_____   **8.** If a specific method for dividing net income or net loss among the partners is not set out in the partnership agreement, then the law provides that the division shall be equal among the partners.

_____   **9.** The division of partnership profits and losses is usually based on the partners' contributions of services and capital.

_____   **10.** When using the fractional-share basis, a partnership's net income or loss is divided equally among the partners.

Part 3 Analyzing Transactions for a Partnership (15 points)

Directions: *Teresa Hardee and Gail Taylor, two high school seniors, have formed a partnership to operate a baby-sitting service. Analyze the transactions below to determine the accounts to be debited and credited. Use the following account names.*

A. Cash in Bank	**D.** Teresa Hardee, Capital	**G.** Gail Taylor, Withdrawals
B. Baby Care Supplies	**E.** Gail Taylor, Capital	**H.** Income Summary
C. Play Equipment	**F.** Teresa Hardee, Withdrawals	

Debit	**Credit**	
A,B,C	D	**0.** Teresa Hardee invested cash, baby care supplies, and play equipment in the business.
_____	_____	**1.** Gail Taylor invested a slide in the business.
_____	_____	**2.** Recorded an additional cash investment by each partner.
_____	_____	**3.** Recorded a cash withdrawal by Teresa Hardee.
_____	_____	**4.** Gail Taylor withdrew baby care supplies.
_____	_____	**5.** Recorded the closing entry for the equal distribution of net income for each partner.
_____	_____	**6.** Recorded the closing entry for closing the withdrawals account for Gail Taylor.
_____	_____	**7.** Recorded the closing entry for closing the withdrawals account for Teresa Hardee.

Part 4 Advantages and Disadvantages of Partnerships (8 points)

Directions: *In the space provided below, indicate whether the statement expresses an advantage or a disadvantage of a partnership. Place an "A" in the space for an advantage or a "D" for a disadvantage.*

__A__	**0.** Opportunity to bring together abilities, experiences, and resources
_____	**1.** Ease of formation
_____	**2.** Limited life of the partnership
_____	**3.** Decision making without formal meetings
_____	**4.** Shared responsibility for the decision of one of the partners
_____	**5.** Inability to transfer one partner's interest in the partnership without the consent of the other partners
_____	**6.** No levying of federal and state income taxes against the partnership
_____	**7.** Personal liability for the debts of the partnership
_____	**8.** Few legal restrictions

Working Papers *for Section Problems*

Problem 27-1   Recording Partners' Investments

GENERAL JOURNAL   PAGE _____

	DATE	DESCRIPTION	POST. REF.	DEBIT	CREDIT	
1						1
2						2
3						3
4						4
5						5
6						6
7						7
8						8
9						9
10						10
11						11
12						12
13						13
14						14
15						15
16						16
17						17

Problem 27-2   Determining Partners' Fractional Shares

Ratio	Fractions
1. 3:1	_____
2. 5:3:1	_____
3. 3:2:2:1	_____
4. 2:1:1	_____
5. 2:1	_____

Problem 27-3   Analyzing a Source Document

GENERAL JOURNAL   PAGE ___*42*___

	DATE	DESCRIPTION	POST. REF.	DEBIT	CREDIT	
1						1
2						2
3						3
4						4
5						5

Computerized Accounting Using Peachtree

Software Objectives

When you have completed this chapter, you will be able to use Peachtree to:

1. Record investments in a partnership.
2. Record partner withdrawals.
3. Allocate profits and losses to the partners by different methods.

Problem 27-6   Recording Investments of Partners

INSTRUCTIONS

Beginning a Session

Step 1   Select the problem set: JR Landscaping (Prob. 27-6).
Step 2   Rename the company, change the accounting period to May 1, 2008 to May 31, 2008, and set the system date to May 31, 2008.

Completing the Accounting Problem

Step 3   Review the information provided in your textbook.
Step 4   Enter the transactions to record the investment by each partner.

TIP: Use the **General Journal Entry** option to record the entry for the partners' investment.

Step 5   Print a General Journal report and proof your work.
Step 6   Answer the Analyze question.

Ending the Session

Step 7   Click the **Close Problem** button in the Glencoe Smart Guide window to save your work.

> ### DO YOU HAVE A QUESTION?
>
> **Q.** *Does Peachtree support partnerships?*
>
> **A.** Yes, Peachtree does support the partnership form of business. When you create a new company Peachtree provides five choices for the form of business: corporation, S corporation, partnership, sole proprietorship, and limited liability company. Peachtree creates unique equity accounts depending on the type of business. For example, Peachtree includes an equity account called **Partners' Contributions** if you choose to let the program create the chart of accounts for you.

Problem 27-7   Sharing Losses Based on Capital Balances

INSTRUCTIONS

Beginning a Session

Step 1   Select the problem set: In Shape Fitness (Prob. 27-7).
Step 2   Rename the company, and set the system date to December 31, 2008.

Completing the Accounting Problem

Step 3   Review the information provided in your textbook.
Step 4   Record the entry to divide the loss between the two partners using the **General Journal Entry** option.
Step 5   Print a General Journal report and proof your work.
Step 6   Answer the Analyze question.

Checking Your Work and Ending the Session

Step 7   Click the **Close Problem** button in the Glencoe Smart Guide window.
Step 8   If your teacher has asked you to check your solution, select *Check my answer to this problem.*
Step 9   Click the **Close Problem** button and select a save option.

> ### Notes
>
> *Peachtree does not automatically divide net profit (loss) between partners when you close the fiscal year. You must manually perform this step.*

Problem 27-8   Partners' Withdrawals

INSTRUCTIONS

Beginning a Session

Step 1   Select the problem set: Travel Essentials (Prob. 27-8).

Step 2   Rename the company, and set the system date to December 31, 2008.

Completing the Accounting Problem

Step 3   Review the information provided in your textbook.

Step 4   Enter the transaction to record the partners' withdrawals using the **General Journal Entry** option.

Step 5   Print a General Journal report and proof your work.

 TIP: You can use the report options to print only the entries you recorded. Set the date range from 12/31/08 to 12/31/08.

Step 6   Answer the Analyze question.

Ending the Session

Step 7   Click the **Close Problem** button in the Glencoe Smart Guide window to save your work.

Problem 27-9   Preparing Closing Entries for a Partnership

INSTRUCTIONS

Beginning a Session

Step 1   Select the problem set: Travel Essentials (Prob. 27-9).

Step 2   Rename the company, and set the system date to December 31, 2008.

Completing the Accounting Problem

Step 3   Review the information provided in your textbook.

Step 4   Using the **General Journal Entry** form, manually record the closing entries to divide the net loss between the partners and to close the withdrawal accounts.

Step 5   Print a General Journal report and proof your work.

Step 6   Answer the Analyze question.

Ending the Session

Step 7   Click the **Close Problem** button in the Glencoe Smart Guide window to save your work.

Computerized Accounting Using Spreadsheets

Problem 27-5   Calculating the Percentage of Each Partner's Capital Investment

Completing the Spreadsheet

Step 1   Read the instructions for Problem 27-5 in your textbook. This problem involves calculating a partner's percentage ownership in a partnership.

Step 2   Open the Glencoe Accounting: Electronic Learning Center software.

Step 3   From the Program Menu, click on the **Peachtree Complete®️ Accounting Software and Spreadsheet Applications** icon.

Step 4   Log onto the Management System by typing your user name and password.

Step 5   Under the **Problems & Tutorials** tab, select template 27-5 from the Chapter 27 drop-down menu. The template should look like the one shown below.

```
PROBLEM 27-5
CALCULATING THE PERCENTAGE OF
EACH PARTNER'S CAPITAL INVESTMENT

(name)
(date)

                  INDIVIDUAL              TOTAL                  PARTNER'S
              PARTNER'S INVESTMENT   PARTNERSHIP INVESTMENT   PERCENTAGE OWNERSHIP
       1                                                             0.00%
       2                                                             0.00%
       3                                                             0.00%
       4                                                             0.00%
```

Step 6   Key your name and today's date in the cells containing the *(name)* and *(date)* placeholders.

Step 7   Enter each partner's individual investment and total partnership investment in the appropriate cells of the spreadsheet template. The spreadsheet template will automatically calculate the partner's percentage ownership.

Step 8   Save the spreadsheet using the **Save** option from the *File* menu. You should accept the default location for the save as this is handled by the management system.

Step 9   Print the completed spreadsheet.

Step 10   Exit the spreadsheet program.

Step 11   In the Close Options box, select the location where you would like to save your work.

Step 12   Answer the Analyze question from your textbook for this problem.

Spreadsheet Guide

What-If Analysis

If Partner 1's individual investment were $50,000 and the total partnership investment were $200,000, what would the partner's percentage of the total ownership be?

_____

_____

Working Papers *for End-of-Chapter Problems*

Problem 27-4   Dividing Partnership Earnings

Net Income	Share of Net Income		
	Partner 1	Partner 2	Partner 3
1. $45,000			
2. $89,700			
3. $22,000			
4. $32,000			
5. $92,700			

Analyze: _____

_____

_____

_____

Problem 27-5   Calculating the Percentage of Each Partner's Capital Investment

1. _____

2. _____

3. _____

4. _____

Analyze: _____

_____

_____

_____

Problem 27-6   Recording Investments of Partners

GENERAL JOURNAL   PAGE _____

	DATE	DESCRIPTION	POST. REF.	DEBIT	CREDIT	
1						1
2						2
3						3
4						4
5						5
6						6
7						7
8						8
9						9
10						10
11						11
12						12

Analyze: _____

_____

Problem 27-7   Sharing Losses Based on Capital Balances

Share percentages:
M. DeJesus = 35,000/80,000 = 0.4375
N. Faircloth = 45,000/80,000 = 0.5625

GENERAL JOURNAL   PAGE ___*14*___

	DATE	DESCRIPTION	POST. REF.	DEBIT	CREDIT	
1						1
2						2
3						3
4						4
5						5
6						6
7						7
8						8
9						9
10						10
11						11
12						12

Analyze: _____

_____

Problem 27-8 Partners' Withdrawals

GENERAL JOURNAL   PAGE ___42___

	DATE	DESCRIPTION	POST. REF.	DEBIT	CREDIT	
1						1
2						2
3						3
4						4
5						5
6						6
7						7
8						8
9						9
10						10
11						11
12						12

Analyze: _____

_____

_____

_____

Problem 27-9 Preparing Closing Entries for a Partnership

GENERAL JOURNAL   PAGE _____

	DATE	DESCRIPTION	POST. REF.	DEBIT	CREDIT	
1						1
2						2
3						3
4						4
5						5
6						6
7						7
8						8
9						9
10						10
11						11
12						12

Problem 27-9   (concluded)

GENERAL LEDGER (PARTIAL)

ACCOUNT __Barbara Scott, Capital__   ACCOUNT NO. ___301___

DATE		DESCRIPTION	POST. REF.	DEBIT	CREDIT	BALANCE DEBIT	BALANCE CREDIT
20--							
Dec.	31	Balance	✓				6731200

ACCOUNT __Barbara Scott, Withdrawals__   ACCOUNT NO. ___305___

DATE		DESCRIPTION	POST. REF.	DEBIT	CREDIT	BALANCE DEBIT	BALANCE CREDIT
20--							
Dec.	31	Balance	✓			660000	

ACCOUNT __Martin Towers, Capital__   ACCOUNT NO. ___303___

DATE		DESCRIPTION	POST. REF.	DEBIT	CREDIT	BALANCE DEBIT	BALANCE CREDIT
20--							
Dec.	31	Balance	✓				4960100

ACCOUNT __Martin Towers, Withdrawals__   ACCOUNT NO. ___307___

DATE		DESCRIPTION	POST. REF.	DEBIT	CREDIT	BALANCE DEBIT	BALANCE CREDIT
20--							
Dec.	31	Balance	✓			540000	

ACCOUNT __Income Summary__   ACCOUNT NO. ___310___

DATE	DESCRIPTION	POST. REF.	DEBIT	CREDIT	BALANCE DEBIT	BALANCE CREDIT

Analyze: _____

_____

_____

Problem 27-10   Evaluating Methods of Dividing Partnership Earnings

1. Garrity: _____

O'Riley: _____

White: _____

2. Garrity: _____

O'Riley: _____

White: _____

3. Garrity: _____

O'Riley: _____

White: _____

Analyze: _____

_____

_____

CHAPTER  Introduction to Partnerships

Self-Test

Part A Fill in the Missing Term

Directions: *Using the terms from the following list, complete the sentences below. Write the letter of the term you select in the space provided.*

> **A.** mutual agency   **B.** partnership agreement

Answer

_____ **1.** Within the partnership, the relationship that allows any partner to act on behalf of other partners is called _____.

_____ **2.** A _____ is a written document that sets out the terms under which the partnership will operate.

Part B True or False

Directions: *For each of the statements that follow, indicate in the space provided whether you agree or disagree with the statement by writing* True *if you agree or* False *if you disagree.*

Answer

_____ **1.** In a partnership, each partner can act as an agent and enter into contracts for the partnership.

_____ **2.** Regardless of how partnership net income is shared, net loss is always shared equally by the partners.

_____ **3.** A partnership has to pay federal income taxes.

_____ **4.** In a partnership, each individual partner is liable to creditors for debts of the partnership.

_____ **5.** A partnership must have a written agreement as to how the partners will share income and losses.

_____ **6.** A partnership agreement must be in writing to be valid.

_____ **7.** An association of two or more persons to carry on a business for a profit as co-owners is called a partnership.

_____ **8.** One of the major advantages of a partnership is the unlimited liability of the partners.

_____ **9.** One of the major advantages of a partnership is its ease of formation.

_____ **10.** The amount that a partner withdraws affects the division of net income.

Part C Analyze the Transactions

Directions: _Alice James and Ruth Simpson formed a partnership called Best Deal School Supplies. Analyze the transactions that follow to determine the accounts to be debited and credited. Use the account numbers listed below._

101	Cash in Bank	**302**	Alice James, Withdrawals
103	Merchandise Inventory	**303**	Ruth Simpson, Capital
104	Store Equipment	**304**	Ruth Simpson, Withdrawals
301	Alice James, Capital	**310**	Income Summary

Debit   **Credit**

_____ _____   **1.** Ruth Simpson invested cash in the partnership.

_____ _____   **2.** The partnership purchased store equipment for cash.

_____ _____   **3.** Alice James withdrew cash from the business.

_____ _____   **4.** Ruth Simpson withdrew merchandise from the business.

_____ _____   **5.** Record the closing entry for sharing the profits equally.

_____ _____   **6.** Record the closing entry for the withdrawals accounts.

_____ _____   **7.** Record the closing entry for sharing a loss.

CHAPTER 28 Financial Statements and Liquidation of a Partnership

Study Plan

Check Your Understanding

Section 1   *Read Section 1 on pages 754–755 and complete the following exercises on page 756.*
- ❑ Thinking Critically
- ❑ Communicating Accounting
- ❑ Problem 28-1   *Preparing the Income Statement and Balance Sheet for a Partnership*
- ❑ Problem 28-2   *Analyzing a Source Document*

Section 2   *Read Section 2 on pages 757–761 and complete the following exercises on page 762.*
- ❑ Thinking Critically
- ❑ Computing in the Business World
- ❑ Problem 28-3   *Recording a Loss and Gain on the Sale of Noncash Assets by a Partnership*

Summary   *Review the Chapter 28 Summary on page 763 in your textbook.*
- ❑ Key Concepts

Review and Activities   *Complete the following questions and exercises on pages 764–765 in your textbook.*
- ❑ Using Key Terms
- ❑ Understanding Accounting Concepts and Procedures
- ❑ Case Study
- ❑ Conducting an Audit with Alex
- ❑ Internet Connection
- ❑ Workplace Skills

Computerized Accounting   *Read the Computerized Accounting information on page 766 in your textbook.*
- ❑ *Making the Transition from a Manual to a Computerized System*
- ❑ *Setting Up General Ledger in Peachtree*

Problems   *Complete the following end-of-chapter problems for Chapter 28 in your textbook.*
- ❑ Problem 28-4   *Preparing an Income Statement and Balance Sheet for a Partnership*
- ❑ Problem 28-5   *Liquidating the Partnership with Losses on the Sale of Noncash Assets*
- ❑ Problem 28-6   *Recording a Gain or Loss on the Sale of Noncash Assets by a Partnership*
- ❑ Problem 28-7   *Preparing a Statement of Changes in Partners' Equity*
- ❑ Problem 28-8   *Liquidating the Partnership*

Challenge Problem   ❑ Problem 28-9   *Completing End-of-Period Activities for a Partnership*

Chapter Reviews and Working Papers   *Complete the following exercises for Chapter 28 in your Chapter Reviews and Working Papers.*
- ❑ Chapter Review
- ❑ Self-Test

CHAPTER 28 REVIEW

Financial Statements and Liquidation of a Partnership

Part 1   Understanding Financial Statements of a Partnership
(5 points)

Total Points	25
Student's Score	

Directions: *Read each of the following statements to determine whether the statement is true or false. Write your answer in the space provided.*

True   **0.** Partnerships involve more than one owner.

_____   **1.** The income statement for a partnership shows each partner's share of income or loss.

_____   **2.** Withdrawals by partners are shown on the income statement.

_____   **3.** The statement of changes in partners' equity is similar to the statement of stockholders' equity.

_____   **4.** The beginning capital balance for each partner in a partnership is reported in the Partners' Equity section of the balance sheet.

_____   **5.** Withdrawals and division of profits or loss are reported on the statement of changes in partners' equity.

Part 2   Analyzing Partnership Liquidation Transactions (10 points)

Directions: *Read each of the following statements to determine whether the statement is true or false. Write your answer in the space provided.*

True   **0.** Partners must agree to liquidate the business.

_____   **1.** In liquidation the debts of a partnership must be paid before any cash is distributed to the partners.

_____   **2.** In liquidation of a partnership the cash is distributed to partners based on the profit and loss sharing agreement.

_____   **3.** When noncash assets are sold at a gain the partners' capital accounts are increased.

_____   **4.** When noncash assets are sold at a loss the cash on hand is reduced.

_____   **5.** Each partner receives the same amount of cash as the final step in the partnership liquidation process.

_____   **6.** When noncash assets are sold for cash the partnership may suffer a loss or a gain.

_____   **7.** When a partnership stops operations and ends the partnership the partnership is liquidated.

_____   **8.** Losses from partnership liquidation are added to the partners' withdrawal accounts.

_____   **9.** The cash remaining in the partnership after the debts are paid is distributed to each partner based on the final balance in their capital accounts.

_____   **10.** If a noncash asset is sold at a gain, the assets of the partnership increase.

Part 3 Recording Liquidation Transactions (10 points)

Directions: _Using the following list of account names, determine the accounts to be debited and credited for the liquidating entries below._

A. Cash in Bank	**C.** Equipment	**E.** Roger Vogel, Capital
B. Inventory	**D.** Notes Payable	**F.** Lydia Parry, Capital

Debit	Credit	
_A_	_B_	**0.** Sold some of the inventory at carrying value.
____	____	**1.** Record a loss from the sale of equipment.
____	____	**2.** Record a gain from the sale of inventory.
____	____	**3.** Paid the note payable.
____	____	**4.** Paid cash to Roger Vogel.
____	____	**5.** Paid cash to Lydia Parry.

Working Papers *for Section Problems*

Problem 28-1   Preparing the Income Statement and
Balance Sheet for a Partnership

Problem 28-2   Analyzing a Source Document

Gain = $26,400
Shares of gain:

     Larry Bass = _____

     John Buie = _____

     Teri Anderson = _____

     Robert Norman = _____

     Paula Dunham = _____

     John Ruppe = _____

Problem 28-3 Recording a Loss and Gain on the Sale of Noncash Assets by a Partnership

GENERAL JOURNAL   PAGE _____

	DATE	DESCRIPTION	POST. REF.	DEBIT	CREDIT	
1						1
2						2
3						3
4						4
5						5
6						6
7						7
8						8
9						9
10						10
11						11
12						12
13						13
14						14
15						15
16						16
17						17
18						18
19						19
20						20

Computerized Accounting Using Peachtree

Software Objectives

When you have completed this chapter, you will be able to use Peachtree to:

1. Record transactions to account for partnership liquidation losses.
2. Record transactions to account for partnership liquidation gains.

Problem 28-5   Liquidating the Partnership with Losses on the Sale of Noncash Assets

INSTRUCTIONS

Beginning a Session

Step 1 Select the problem set: Pasta Mia Restaurant Supply (Prob. 28-5).
Step 2 Rename the company, and set the system date to September 21, 2008.

Completing the Accounting Problem

Step 3 Review the information provided in your textbook.
Step 4 Enter the transactions to liquidate the partnership.

TIP: Use the **General Journal Entry** option to record the entries to liquidate the partnership.

DO YOU HAVE A QUESTION

Q. *Can you print a Statement of Changes in Partners' Equity report using the Peachtree software?*

A. No, Peachtree does not include an option to print a Statement of Changes in Partners' Equity report. You have to manually prepare this report using the information found on the Balance Sheet for a partnership.

Step 5 Print a General Journal report and proof your work.
Step 6 Answer the Analyze question.

Checking Your Work and Ending the Session

Step 7 Click the **Close Problem** button in the Glencoe Smart Guide window.
Step 8 If your teacher has asked you to check your solution, select *Check my answer to this problem.* Review, print, and close the report.
Step 9 Click the **Close Problem** button and select a save option.

Problem 28-6 Recording a Gain or Loss on the Sale of Noncash Assets by a Partnership

INSTRUCTIONS

Beginning a Session

Step 1 Select the problem set: Industrial Tool & Machine (Prob. 28-6).
Step 2 Rename the company, and set the system date to June 4, 2008.

Completing the Accounting Problem

Step 3 Review the information provided in your textbook. Be sure to enter the transactions in the proper accounting period. Use the **System** option from the *Tasks* menu to change the accounting period before entering each transaction.
Step 4 Enter the transactions to liquidate the partnership.
Step 5 Print a General Journal report and proof your work.

 TIP: Set the General Journal report date range to include the entries from May 1 to June 30.

Step 6 Anwer the Analyze question.

Ending the Session

Step 7 Verify that the accounting period displayed on your screen is Period 6—6/1/08 to 6/30/08. Click the **Close Problem** button in the Glencoe Smart Guide window to save your work.

Problem 28-8 Liquidating the Partnership

INSTRUCTIONS

Beginning a Session

Step 1 Select the problem set: Alpine Gifts & Flowers (Prob. 28-8).
Step 2 Rename the company, and set the system date to October 15, 2008.

Completing the Accounting Problem

Step 3 Review the information provided in your textbook.
Step 4 Enter the transactions to liquidate the partnership.
Step 5 Print a General Journal report and proof your work.
Step 6 Anwer the Analyze question.

Ending the Session

Step 7 Click the **Close Problem** button in the Glencoe Smart Guide window to save your work.

Computerized Accounting Using Spreadsheets

Problem 28-4   Preparing an Income Statement and Balance Sheet for a Partnership

Completing the Spreadsheet

Step 1   Read the instructions for Problem 28-4 in your textbook. This problem involves preparing an income statement and a balance sheet for a partnership.

Step 2   Open the Glencoe Accounting: Electronic Learning Center software.

Step 3   From the Program Menu, click on the **Peachtree Complete® Accounting Software and Spreadsheet Applications** icon.

Step 4   Log onto the Management System by typing your user name and password.

Step 5   Under the **Problems & Tutorials** tab, select template 28-4 from the Chapter 28 drop-down menu. The template should look like the one shown below.

```
PROBLEM 28-4
PREPARING AN INCOME STATEMENT AND
BALANCE SHEET FOR A PARTNERSHIP

(name)
(date)


Joy Webster %                              AMOUNT
Diana Ruiz %                               AMOUNT


Net Income                                                      AMOUNT
Division of Net Income:
  Webster                                    0.00
  Ruiz                                       0.00


        Partners' Equity        Webster           Ruiz         Total Equity
Beginning Capital, January 1      0.00             0.00            0.00
Add: Net Income                   0.00             0.00            0.00
    Investments                 6,000.00         5,500.00        11,500.00
Subtotal                        6,000.00         5,500.00        11,500.00
Less: Withdrawals               1,800.00         1,200.00         3,000.00
Ending Capital, December 31     4,200.00         4,300.00         8,500.00
```

Step 6   Key your name and today's date in the cells containing the *(name)* and *(date)* placeholders.

Step 7   Joy Webster and Diana Ruiz share in the partnership equally. Therefore, each has a 50% share of the profits. Enter Webster's and Ruiz's partnership percentage in cells B10 and B11: **50**.

TIP: Cells B10 and B11 are formatted for percentages. Therefore, it is not necessary to enter a percent sign after the number, nor is it necessary to enter the number as a decimal.

Step 8 Now enter net income in cell C13: **5780**. Remember, it is not necessary to enter a comma or the decimal point and ending zeroes. The division of net income for Webster and Ruiz will be automatically calculated.

Step 9 Now scroll down below the division of net income and look at the balance sheet. The partners' equity section has been completed.

Step 10 Save the spreadsheet using the **Save** option from the *File* menu. You should accept the default location for the save as this is handled by the management system.

Step 11 Print the completed spreadsheet.

Step 12 Exit the spreadsheet program.

Step 13 In the Close Options box, select the location where you would like to save your work.

Step 14 Answer the Analyze question from your textbook for this problem.

What-If Analysis

If Webster's partnership percentage were 60% and Ruiz's partnership percentage were 40%, what would the division of net income be? How would this affect each partner's ending capital?

_____

_____

Problem 28-9   Completing End-of-Period Activities for a Partnership

Completing the Spreadsheet

Step 1   Read the instructions for Problem 28-9 in your textbook. This problem involves preparing the end-of-period financial statements for a partnership.

Step 2   Open the Glencoe Accounting: Electronic Learning Center software.

Step 3   From the Program Menu, click on the **Peachtree Complete® Accounting Software and Spreadsheet Applications** icon.

Step 4   Log onto the Management System by typing your user name and password.

Step 5   Under the **Problems & Tutorials** tab, select template 28-9 from the Chapter 28 drop-down menu. The template should look like the one shown below.

```
PROBLEM 28-9
COMPLETING END-OF-PERIOD ACTIVITIES FOR A PARTNERSHIP

(name)
(date)


Smooth %
Overhill %                                            AMOUNT
                                                      AMOUNT

R & C ROOFING
INCOME STATEMENT
FOR THE YEAR ENDED DECEMBER 31, 20--

Revenue:
  Consulting Fees                         15,900.00
  Roofing Fees                            62,750.00
Total Revenue                                         78,650.00
Expenses:
  Advertising Expense                      2,400.00
  Depreciation Expense - Office Equipment    185.00
  Depreciation Expense - Truck             3,900.00
  Depreciation Expense - Building          1,200.00
  Insurance Expense                        1,200.00
  Office Supplies Expense                    335.00
  Roofing Supplies Expense                11,470.00
  Salaries Expense                        28,109.00
  Truck Expense                            1,400.00
  Utilities Expense                        2,095.00
Total Expense                                         52,294.00
Net Income                                            26,356.00
Division of Net Income:
  Richard Smooth                              0.00
  Carrie Overhill                             0.00
Net Income                                                0.00
```

Spreadsheet Guide

Step 6   Key your name and today's date in the cells containing the *(name)* and *(date)* placeholders.

Step 7   Richard Smooth and Carrie Overhill agree to divide R & C Roofing's net income or loss on the following basis: Smooth, ¾; Overhill, ¼. Therefore, Smooth has a 75% partnership percentage, and Overhill has a 25% partnership percentage. Enter Smooth's and Overhill's partnership percentages in cells B9 and B10.

TIP: Cells B9 and B10 are formatted for percentages. Therefore, it is not necessary to enter a percent sign after the number, nor is it necessary to enter the number as a decimal.

Step 8   Now scroll down below the partnership percentages and look at the income statement, statement of changes in partners' equity, and balance sheet for R & C Roofing. Notice the financial statements are already completed, using information from the work sheet in your working papers and from the partnership percentages you entered in cells B9 and B10.

Step 9   Save the spreadsheet using the **Save** option from the *File* menu. You should accept the default location for the save as this is handled by the management system.

Step 10   Print the completed spreadsheet.

TIP: When printing a long spreadsheet with multiple parts, you may want to insert page breaks between the sections so that each one begins printing at the top of a new page. Page breaks have already been entered into this spreadsheet template. Check your program's Help file for instructions on how to enter page breaks.

Step 11   Exit the spreadsheet program.

Step 12   In the Close Options box, select the location where you would like to save your work.

Step 13   Answer the Analyze question from your textbook for this problem.

What-If Analysis

If Smooth and Overhill shared in the partnership equally, what would the division of net income be? How would this affect each partner's ending capital?

_____

_____

Working Papers *for End-of-Chapter Problems*

Problem 28-4   Preparing an Income Statement and Balance Sheet for a Partnership

Webster and Ruiz
Income Statement (partial)
For the Period Ending December 31, 20--

Webster and Ruiz
Balance Sheet (partial)
December 31, 20--

Analyze: _____

_____

_____

Problem 28-5 Liquidating the Partnership with Losses on the Sale of Noncash Assets

GENERAL JOURNAL   PAGE _____

	DATE	DESCRIPTION	POST. REF.	DEBIT	CREDIT	
1						1
2						2
3						3
4						4
5						5
6						6
7						7
8						8
9						9
10						10
11						11
12						12
13						13
14						14
15						15
16						16
17						17
18						18
19						19
20						20
21						21
22						22
23						23
24						24
25						25
26						26
27						27
28						28
29						29
30						30

Analyze: _____

_____

_____

Problem 28-6 Recording a Gain or Loss on the Sale of Noncash Assets by a Partnership

GENERAL JOURNAL

PAGE _____

	DATE	DESCRIPTION	POST. REF.	DEBIT	CREDIT	
1						1
2						2
3						3
4						4
5						5
6						6
7						7
8						8
9						9
10						10
11						11
12						12
13						13
14						14
15						15
16						16
17						17
18						18
19						19
20						20
21						21
22						22
23						23
24						24
25						25
26						26
27						27
28						28
29						29
30						30

Analyze: _____

_____

_____

Problem 28-7 Preparing a Statement of Changes in Partners' Equity

Analyze: _____

Problem 28-8   Liquidating the Partnership

GENERAL JOURNAL   PAGE ___78___

	DATE	DESCRIPTION	POST. REF.	DEBIT	CREDIT	
1						1
2						2
3						3
4						4
5						5
6						6
7						7
8						8
9						9
10						10
11						11
12						12
13						13
14						14
15						15
16						16
17						17
18						18
19						19
20						20
21						21
22						22
23						23
24						24
25						25
26						26
27						27
28						28
29						29
30						30

Analyze: _____

_____

_____

Problem 28-9   Completing End-of-Period Activities for a Partnership

R&C

Work

For the Year Ended

	ACCT. NO.	ACCOUNT NAME	TRIAL BALANCE DEBIT	TRIAL BALANCE CREDIT	ADJUSTMENTS DEBIT	ADJUSTMENTS CREDIT
1	101	Cash in Bank	17 928 00			
2	105	Accounts Receivable	4 310 00			
3	110	Office Supplies	495 00			(a) 335 00
4	115	Roofing Supplies	15 610 00			(b) 11 470 00
5	120	Prepaid Insurance	2 400 00			(c) 1 200 00
6	150	Office Equipment	2 650 00			
7	155	Accum. Depr.—Office Equip.		1 016 00		(d) 185 00
8	160	Truck	19 890 00			
9	165	Accum. Depr.—Truck		3 100 00		(e) 3 900 00
10	170	Building	30 000 00			
11	175	Accum. Depr.—Building		3 600 00		(f) 1 200 00
12	180	Land	10 000 00			
13	201	Accounts Payable		7 945 00		
14	301	R. Smooth, Capital		42 238 00		
15	305	R. Smooth, Withdrawals	8 700 00			
16	310	C. Overhill, Capital		17 538 00		
17	315	C. Overhill, Withdrawals	8 100 00			
18	320	Income Summary	—	—		
19	401	Consulting Fees		15 900 00		
20	405	Roofing Fees		62 750 00		
21	501	Advertising Expense	2 400 00			
22	505	Depr. Exp.—Office Equip.	—		(d) 185 00	
23	510	Depr. Expense—Truck	—		(e) 3 900 00	
24	515	Depr. Expense—Building	—		(f) 1 200 00	
25	520	Insurance Expense	—		(c) 1 200 00	
26	525	Office Supplies Expense	—		(a) 335 00	
27	530	Roofing Supplies Expense	—		(b) 11 470 00	
28	535	Salaries Expense	28 109 00			
29	540	Truck Expense	1 400 00			
30	545	Utilities Expense	2 095 00			
31			154 087 00	154 087 00	18 290 00	18 290 00
32		Net Income				
33						
34						

Problem 28-9   (continued)

Roofing

Sheet

December 31, 20--

	ADJUSTED TRIAL BALANCE		INCOME STATEMENT		BALANCE SHEET		
	DEBIT	CREDIT	DEBIT	CREDIT	DEBIT	CREDIT	
1	17928 00				17928 00		
2	4310 00				4310 00		
3	160 00				160 00		
4	4140 00				4140 00		
5	1200 00				1200 00		
6	2650 00				2650 00		
7		1201 00				1201 00	
8	19890 00				19890 00		
9		7000 00				7000 00	
10	30000 00				30000 00		
11		4800 00				4800 00	
12	10000 00				10000 00		
13		7945 00				7945 00	
14		42238 00				42238 00	
15	8700 00				8700 00		
16		17538 00				17538 00	
17	8100 00				8100 00		
18	—	—	—	—	—		
19		15900 00		15900 00			
20		62750 00		62750 00			
21	2400 00		2400 00				
22	185 00		185 00				
23	3900 00		3900 00				
24	1200 00		1200 00				
25	1200 00		1200 00				
26	335 00		335 00				
27	11470 00		11470 00				
28	28109 00		28109 00				
29	1400 00		1400 00				
30	2095 00		2095 00				
31	159372 00	159372 00	52294 00	78650 00	107078 00	80722 00	
32			26356 00			26356 00	
33			78650 00	78650 00	107078 00	107078 00	
34							

Problem 28-9 (continued)

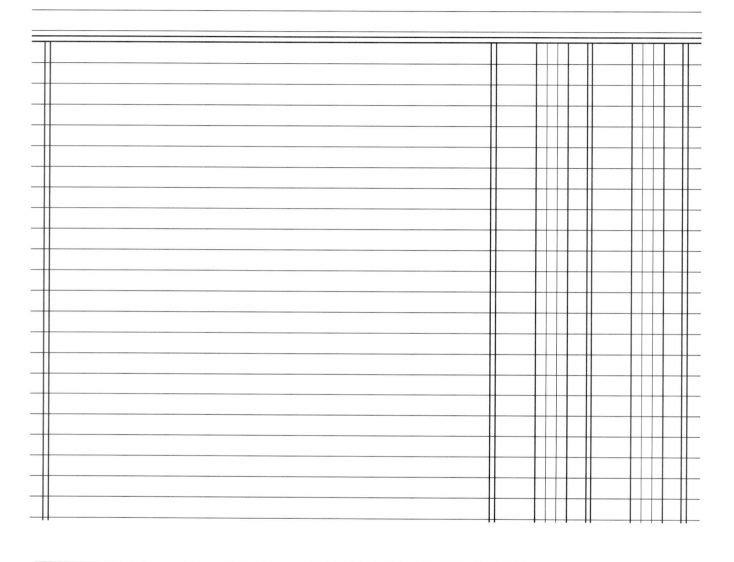

Problem 28-9   (continued)

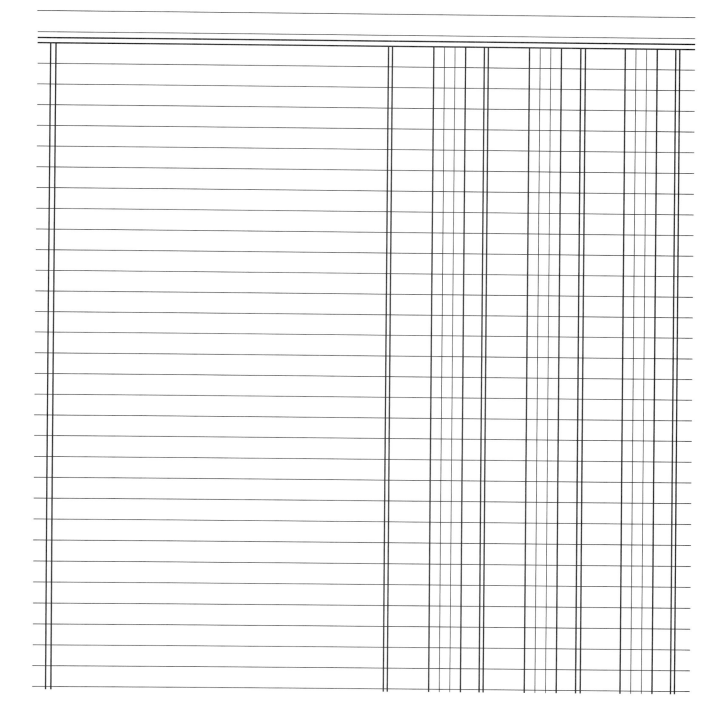

Problem 28-9 (concluded)

GENERAL JOURNAL   PAGE _____

	DATE	DESCRIPTION	POST. REF.	DEBIT	CREDIT	
1						1
2						2
3						3
4						4
5						5
6						6
7						7
8						8
9						9
10						10
11						11
12						12
13						13
14						14
15						15
16						16
17						17
18						18
19						19
20						20
21						21
22						22
23						23
24						24
25						25
26						26
27						27
28						28
29						29
30						30
31						31
32						32
33						33
34						34
35						35
36						36
37						37

Analyze: _____

CHAPTER 28   Financial Statements and Liquidation of a Partnership

Self-Test

Part A   Steps in Liquidating a Partnership

Directions: *Shown below are the four steps in liquidating a partnership. Rearrange these activities in the order in which they would be completed during the liquidation process.*

Answer

_____   **1.** All cash remaining after the creditors are paid is distributed to the partners based on the final balance in the partners' accounts.

_____   **2.** All gains and losses from the sale of noncash assets are added to or deducted from the capital accounts of the partners based on the partnership agreement.

_____   **3.** All noncash assets are sold for cash.

_____   **4.** All partnership creditors are paid.

Part B   True or False

Directions: *Read each of the following statements to determine whether the statement is true or false. Write your answer in the space provided.*

Answer

_____   **1.** Partnerships involve more than one owner.

_____   **2.** The balance sheet for a partnership shows each partner's share of income or loss.

_____   **3.** Withdrawals by partners are shown on the balance sheet.

_____   **4.** The statement of changes in partners' equity is similar to the statement of owner's equity.

_____   **5.** The ending capital balance for each partner in a partnership is reported in the Partners' Equity section of the balance sheet.

_____   **6.** Withdrawals and division of profits or loss are reported on the statement of changes in partners' equity.

_____   **7.** Partners must agree to liquidate the business.

_____   **8.** In liquidation the debts of a partnership are paid after cash is distributed to the partners.

_____   **9.** In liquidation of a partnership the cash is distributed to partners based upon the profit and loss sharing agreement.

_____ **10.** When noncash assets are sold at a loss the partners' capital accounts are decreased.

_____ **11.** When noncash assets are sold at a gain the cash on hand is increased.

_____ **12.** Each partner receives the same amount of cash as the final step in the partnership liquidation process.

_____ **13.** When noncash assets are sold for cash the partnership may suffer a loss or a gain.

_____ **14.** When a partnership stops operations and ends the partnership the partnership is liquidated.

_____ **15.** Gains from partnership liquidation are added to the partners' withdrawal accounts.

_____ **16.** The cash remaining in the partnership after the debts are paid is distributed to each partner based on the final balance in their capital accounts.

_____ **17.** If a noncash asset is sold at a loss, the assets of the partnership decrease.

MINI PRACTICE SET 6

Fine Finishes

CHART OF ACCOUNTS

ASSETS
101   Cash in Bank
105   Accounts Receivable—Mountain View City School District
120   Computer Equipment
130   Office Supplies
135   Office Equipment
140   Painting Supplies
145   Painting Equipment

LIABILITIES
205   Accounts Payable—Custom Color
210   Accounts Payable—J & J Hardware and Lumber
215   Accounts Payable—Paint Palace

PARTNERS' EQUITY
301   Laura Andersen, Capital
302   Laura Andersen, Withdrawals
303   David Ingram, Capital
304   David Ingram, Withdrawals
305   Sean Woo, Capital
306   Sean Woo, Withdrawals
310   Income Summary

REVENUE
401   Painting Fees
405   Consultation Fees

EXPENSES
505   Advertising Expense
510   Miscellaneous Expense
515   Rent Expense
520   Utilities Expense

Mini Practice Set 6   Source Documents

Instructions: *Use the following source documents to record the transactions for this practice set.*

---

fine finishes
755 Brewton Street
Forest Hills, AL 36105

MEMORANDUM 1

TO: *Accounting Clerk*
FROM: *Senior Accountant*
DATE: *February 1, 20--*
SUBJECT: *Partner Investment*

Record partners' investments with following amounts:

	Andersen	Ingram	Woo
Cash	$1,500.00	$1,000.00	$1,200.00
Computer Equip.	–	2,800.00	–
Office Equip.	100.00	–	–
Painting Supplies	150.00	–	225.00
Painting Equip.	1,375.00	–	1,675.00
Total	**$3,125.00**	**$3,800.00**	**$3,100.00**

---

fine finishes
755 Brewton Street
Forest Hills, AL 36105

1101
71-821
3321

DATE *February 1* 20--

PAY TO THE ORDER OF *Taft Leasing Co.* $ *1,500.00*

One thousand five hundred and ⁰⁰/₁₀₀ ——————— DOLLARS

✿ *Barclays Bank*

MEMO *Rent*   *Laura Andersen*

⑂332171821⑂ 4516 2133‖ 1101

---

fine finishes
755 Brewton Street
Forest Hills, AL 36105

1102
71-821
3321

DATE *February 1* 20--

PAY TO THE ORDER OF *Call an Expert* $ *25.00*

Twenty-five and ⁰⁰/₁₀₀ ——————— DOLLARS

✿ *Barclays Bank*

MEMO *Newspaper ad*   *Laura Andersen*

⑂332171821⑂ 4516 2133‖ 1102

---

fine finishes
755 Brewton Street
Forest Hills, AL 36105

1103
71-821
3321

DATE *February 1* 20--

PAY TO THE ORDER OF *City of Mountain View* $ *55.00*

Fifty-five and ⁰⁰/₁₀₀ ——————— DOLLARS

✿ *Barclays Bank*

MEMO *Business license*   *Laura Andersen*

⑂332171821⑂ 4516 2133‖ 1103

---

fine finishes
755 Brewton Street
Forest Hills, AL 36105

1104
71-821
3321

DATE *February 1* 20--

PAY TO THE ORDER OF *Western Utilities* $ *100.00*

One hundred and ⁰⁰/₁₀₀ ——————— DOLLARS

✿ *Barclays Bank*

MEMO *Utilities deposit*   *Laura Andersen*

⑂332171821⑂ 4516 2133‖ 1104

---

fine finishes
755 Brewton Street
Forest Hills, AL 36105

1105
71-821
3321

DATE *February 1* 20--

PAY TO THE ORDER OF *GTE* $ *175.00*

One hundred seventy-five and ⁰⁰/₁₀₀ ——————— DOLLARS

✿ *Barclays Bank*

MEMO *Telephone svc*   *Laura Andersen*

⑂332171821⑂ 4516 2133‖ 1105

---

fine finishes
755 Brewton Street
Forest Hills, AL 36105

RECEIPT
No. 01

February 2 20--

RECEIVED FROM *McGuires* $ *250.00*

Two hundred fifty and ⁰⁰/₁₀₀ ——————— DOLLARS

FOR *$250 deposit for McGuires contract*

RECEIVED BY *Laura Andersen*

Custom Color
3167 Turner Place, #1A
Wildwood, AL 36120

INVOICE NO. 742

DATE: *Feb. 2, 20--*
ORDER NO.:
SHIPPED BY:
TERMS:

TO *Fine Finishes*
    755 Brewton Street
    Forest Hills, AL 36105

QTY.	ITEM	UNIT PRICE	TOTAL
5	*Paint & Border Stencils*	$40.00	$200.00

Mini Practice Set 6   (continued)

fine finishes   **1106**
755 Brewton Street
Forest Hills, AL 36105   71-821 / 3321

DATE *February 4*   20– –

PAY TO THE ORDER OF *Office Max* $ *115.00*

One hundred fifteen and 00/100 — DOLLARS

✻ *Barclays Bank*

MEMO *Office supplies*   *Laura Andersen*

⑆332171821⑆ 4516 2133⑈ 1106

---

fine finishes   **RECEIPT**
755 Brewton Street
Forest Hills, AL 36105   No. 02

February 5   20 – –

RECEIVED FROM *McGuires* $ *450.00*

Four hundred fifty and 00/100 — DOLLARS

FOR *Balance on McGuires contract*

RECEIVED BY *Laura Andersen*

---

fine finishes   **1107**
755 Brewton Street
Forest Hills, AL 36105   71-821 / 3321

DATE *February 6*   20– –

PAY TO THE ORDER OF *Mountain View Chamber of Commerce* $ *45.00*

Forty-five and 00/100 — DOLLARS

✻ *Barclays Bank*

MEMO *Permit*   *Laura Andersen*

⑆332171821⑆ 4516 2133⑈ 1107

Paint Palace   **INVOICE NO. 1162**
612 James Avenue
Montgomery, AL 36105

DATE: Feb. 8, 20– –
ORDER NO.:
SHIPPED BY:
TERMS:

TO   *Fine Finishes*
755 Brewton Street
Forest Hills, AL 36105

QTY.	ITEM	UNIT PRICE	TOTAL
2	Painting equipment	$187.50	$375.00

---

fine finishes   **RECEIPT**
755 Brewton Street
Forest Hills, AL 36105   No. 03

February 10   20 – –

RECEIVED FROM *Prospective Client* $ *60.00*

Sixty and 00/100 — DOLLARS

FOR *Color and painting consultation*

RECEIVED BY *Laura Andersen*

---

fine finishes   **INVOICE NO. 101**
755 Brewton Street
Forest Hills, AL 36105

DATE: Feb. 12, 20– –
ORDER NO.:
TERMS:

TO   *Mountain View*
City School District

DATE	SERVICE	AMOUNT
	Cafeteria Painting at elementary school	$835.00

---

fine finishes   **1108**
755 Brewton Street
Forest Hills, AL 36105   71-821 / 3321

DATE *February 14*   20– –

PAY TO THE ORDER OF *Custom Color* $ *200.00*

Two hundred and 00/100 — DOLLARS

✻ *Barclays Bank*

MEMO *On account*   *Laura Andersen*

⑆332171821⑆ 4516 2133⑈ 1108

Mini Practice Set 6   (continued)

fine finishes
755 Brewton Street
Forest Hills, AL 36105

1109
71-821
3321

DATE *February 15*   20 — —

PAY TO THE ORDER OF *Laura Andersen* $ *650.00*

Six hundred fifty and ⁰⁰/₁₀₀ ———————— DOLLARS

⚙ *Barclays Bank*

MEMO *Personal withdrawal*   *Laura Andersen*

⑈332171821⑈ 4516 2133⑊ 1109

---

fine finishes
755 Brewton Street
Forest Hills, AL 36105

1110
71-821
3321

DATE *February 15*   20 — —

PAY TO THE ORDER OF *David Ingram* $ *650.00*

Six hundred fifty and ⁰⁰/₁₀₀ ———————— DOLLARS

⚙ *Barclays Bank*

MEMO *Personal withdrawal*   *Laura Andersen*

⑈332171821⑈ 4516 2133⑊ 1110

---

fine finishes
755 Brewton Street
Forest Hills, AL 36105

1111
71-821
3321

DATE *February 15*   20 — —

PAY TO THE ORDER OF *Sean Woo* $ *650.00*

Six hundred fifty and ⁰⁰/₁₀₀ ———————— DOLLARS

⚙ *Barclays Bank*

MEMO *Personal withdrawal*   *Laura Andersen*

⑈332171821⑈ 4516 2133⑊ 1111

---

fine finishes
755 Brewton Street
Forest Hills, AL 36105

RECEIPT
No. 04

February 15   20 — —

RECEIVED FROM *Wicker & Hartel Law Office* $ *1,000.00*

One thousand and ⁰⁰/₁₀₀ ——————— DOLLARS

FOR *Deposit on contract*

RECEIVED BY *Laura Andersen*

---

fine finishes
755 Brewton Street
Forest Hills, AL 36105

1112
71-821
3321

DATE *February 16*   20 — —

PAY TO THE ORDER OF *Odds & Ends (Painting Supplies)* $ *135.00*

One hundred thirty-five and ⁰⁰/₁₀₀ ———————— DOLLARS

⚙ *Barclays Bank*

MEMO *Painting supplies*   *Laura Andersen*

⑈332171821⑈ 4516 2133⑊ 1113

Custom Color
3167 Turner Place, #1A
Wildwood, AL 36120

INVOICE NO. 750

DATE: *Feb. 16, 20 — —*
ORDER NO.:
SHIPPED BY:
TERMS:

TO   *Fine Finishes*
755 Brewton Street
Forest Hills, AL 36105

QTY.	ITEM	UNIT PRICE	TOTAL
5	*Paint gallons*	$79.00	$395.00

---

fine finishes
755 Brewton Street
Forest Hills, AL 36105

1113
71-821
3321

DATE *February 16*   20 — —

PAY TO THE ORDER OF *Mountain View Realtors* $ *77.00*

Seventy-seven and ⁰⁰/₁₀₀ ———————— DOLLARS

⚙ *Barclays Bank*

MEMO *Advertisement*   *Laura Andersen*

⑈332171821⑈ 4516 2133⑊ 1113

---

fine finishes
755 Brewton Street
Forest Hills, AL 36105

RECEIPT
No. 05

February 18   20 — —

RECEIVED FROM *Prospective Client* $ *125.00*

One hundred twenty-five and ⁰⁰/₁₀₀ ——————— DOLLARS

FOR *Painting consultation*

RECEIVED BY *Laura Andersen*

Mini Practice Set 6   (continued)

fine finishes   **1114**
755 Brewton Street   71-821
Forest Hills, AL 36105   3321
                          DATE *February 19*   20 --
PAY TO THE
ORDER OF *A-1 Repair Services* $ *85.00*

Eighty-five and ⁰⁰/₁₀₀ ———————————— DOLLARS

  Barclays Bank
MEMO *Computer repair*   *Laura Andersen*
⑆332171821⑆ 4516 2133⑈ 1114

---

fine finishes   **RECEIPT**
755 Brewton Street
Forest Hills, AL 36105   No. 07

                                    February 25 20 --
RECEIVED FROM *Maintenance Service* $ *575.00*

Five hundred seventy-five and ⁰⁰/₁₀₀ ——————— DOLLARS

FOR *Minor repairs to garage*

                    RECEIVED BY *Laura Andersen*

---

fine finishes   **1115**
755 Brewton Street   71-821
Forest Hills, AL 36105   3321
                          DATE *February 21*   20 --
PAY TO THE
ORDER OF *Paint Palace* $ *375.00*

Three hundred seventy-five and ⁰⁰/₁₀₀ ————— DOLLARS

  Barclays Bank
MEMO *On account*   *Laura Andersen*
⑆332171821⑆ 4516 2133⑈ 1115

---

fine finishes   **1116**
755 Brewton Street   71-821
Forest Hills, AL 36105   3321
                          DATE *February 28*   20 --
PAY TO THE
ORDER OF *Laura Andersen* $ *650.00*

Six hundred fifty and ⁰⁰/₁₀₀ ——————————— DOLLARS

  Barclays Bank
MEMO *Personal withdrawal*   *David Ingram*
⑆332171821⑆ 4516 2133⑈ 1116

---

fine finishes   **RECEIPT**
755 Brewton Street
Forest Hills, AL 36105   No. 06

                                    February 22 20 --
RECEIVED FROM *Wicker & Hartel Law Office* $ *2,000.00*

Two thousand and ⁰⁰/₁₀₀ ————————————— DOLLARS

FOR *Final payment*

                    RECEIVED BY *Laura Andersen*

---

fine finishes   **1117**
755 Brewton Street   71-821
Forest Hills, AL 36105   3321
                          DATE *February 28*   20 --
PAY TO THE
ORDER OF *David Ingram* $ *650.00*

Six hundred fifty and ⁰⁰/₁₀₀ ——————————— DOLLARS

  Barclays Bank
MEMO *Personal withdrawal*   *Laura Andersen*
⑆332171821⑆ 4516 2133⑈ 1117

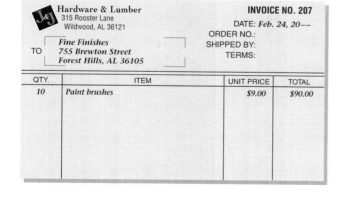

J&J Hardware & Lumber   **INVOICE NO. 207**
      315 Rooster Lane
      Wildwood, AL 36121   DATE: *Feb. 24, 20 --*
                                      ORDER NO.:
    Fine Finishes   SHIPPED BY:
TO *755 Brewton Street*   TERMS:
    Forest Hills, AL 36105

QTY.	ITEM	UNIT PRICE	TOTAL
10	*Paint brushes*	*$9.00*	*$90.00*

---

fine finishes   **1118**
755 Brewton Street   71-821
Forest Hills, AL 36105   3321
                          DATE *February 28*   20 --
PAY TO THE
ORDER OF *Sean Woo* $ *650.00*

Six hundred fifty and ⁰⁰/₁₀₀ ——————————— DOLLARS

  Barclays Bank
MEMO *Personal withdrawal*   *David Ingram*
⑆332171821⑆ 4516 2133⑈ 1118

Computerized Accounting Using Peachtree

Mini Practice Set 6

INSTRUCTIONS

Beginning a Session

Step 1   Select the problem set: Fine Finishes (MP-6).

Step 2   Rename the company, and set the system date to February 29, 2008.

Completing the Accounting Problem

Step 3   Review the transactions provided in your textbook.

Step 4   Record all of the business transactions using the **General Journal Entry** option.

TIP: As a shortcut, you can type just the day of the month when you enter a general journal entry.

TIP: You can use the [Row], [Add] and [Row], [Remove] buttons to edit a multi-line general journal entry.

Step 5   Print a General Journal report and proof your work.

TIP: Double-click a report title in the Select a Report window to go directly to that report.

Step 6   Print a General Ledger report.

TIP: Set the General Ledger report format option to "Summary by Transaction" to save paper when you print this report.

Step 7   Print a General Ledger Trial Balance.

Step 8   Print an Income Statement and a Balance Sheet.

TIP: Choose the custom Income Statement report, not the <Standard> Income Statement.

Step 9 Click the **Save Pre-closing Balances** button in the Glencoe Smart Guide window. It is important that you save your work before the closing process.

Closing the Period

Step 10 Close the current period. Use the **General Journal Entry** form to enter the closing entries manually.

IMPORTANT: Fine Finishes operates on a monthly fiscal period. Therefore, you must manually close the current period by entering the closing entries. Peachtree can close only a fiscal year.

Step 11 Print a General Journal report and proof your work.
Step 12 Print a Post-Closing Trial Balance.

TIP: Using the report options you can change the title of a report.

Step 13 Answer the Analyze questions.

Ending the Session

Step 14 Click the **Close Problem** button in the Glencoe Smart Guide window.

Mini Practice Set 6   (continued)

(1), (4), (9)

GENERAL LEDGER

ACCOUNT _____ ACCOUNT NO. _____

DATE	DESCRIPTION	POST. REF.	DEBIT	CREDIT	BALANCE	
					DEBIT	CREDIT

ACCOUNT _____ ACCOUNT NO. _____

DATE	DESCRIPTION	POST. REF.	DEBIT	CREDIT	BALANCE	
					DEBIT	CREDIT

Mini Practice Set 6 (continued)

ACCOUNT _____ ACCOUNT NO. _____

DATE	DESCRIPTION	POST. REF.	DEBIT	CREDIT	BALANCE	
					DEBIT	CREDIT

ACCOUNT _____ ACCOUNT NO. _____

DATE	DESCRIPTION	POST. REF.	DEBIT	CREDIT	BALANCE	
					DEBIT	CREDIT

ACCOUNT _____ ACCOUNT NO. _____

DATE	DESCRIPTION	POST. REF.	DEBIT	CREDIT	BALANCE	
					DEBIT	CREDIT

ACCOUNT _____ ACCOUNT NO. _____

DATE	DESCRIPTION	POST. REF.	DEBIT	CREDIT	BALANCE	
					DEBIT	CREDIT

ACCOUNT _____ ACCOUNT NO. _____

DATE	DESCRIPTION	POST. REF.	DEBIT	CREDIT	BALANCE	
					DEBIT	CREDIT

Mini Practice Set 6 (continued)

ACCOUNT _____ ACCOUNT NO. _____

DATE	DESCRIPTION	POST. REF.	DEBIT	CREDIT	BALANCE	
					DEBIT	CREDIT

ACCOUNT _____ ACCOUNT NO. _____

DATE	DESCRIPTION	POST. REF.	DEBIT	CREDIT	BALANCE	
					DEBIT	CREDIT

ACCOUNT _____ ACCOUNT NO. _____

DATE	DESCRIPTION	POST. REF.	DEBIT	CREDIT	BALANCE	
					DEBIT	CREDIT

ACCOUNT _____ ACCOUNT NO. _____

DATE	DESCRIPTION	POST. REF.	DEBIT	CREDIT	BALANCE	
					DEBIT	CREDIT

ACCOUNT _____ ACCOUNT NO. _____

DATE	DESCRIPTION	POST. REF.	DEBIT	CREDIT	BALANCE	
					DEBIT	CREDIT

Mini Practice Set 6   (continued)

ACCOUNT _____   ACCOUNT NO. _____

DATE	DESCRIPTION	POST. REF.	DEBIT	CREDIT	BALANCE	
					DEBIT	CREDIT

ACCOUNT _____   ACCOUNT NO. _____

DATE	DESCRIPTION	POST. REF.	DEBIT	CREDIT	BALANCE	
					DEBIT	CREDIT

ACCOUNT _____   ACCOUNT NO. _____

DATE	DESCRIPTION	POST. REF.	DEBIT	CREDIT	BALANCE	
					DEBIT	CREDIT

ACCOUNT _____   ACCOUNT NO. _____

DATE	DESCRIPTION	POST. REF.	DEBIT	CREDIT	BALANCE	
					DEBIT	CREDIT

Mini Practice Set 6 (continued)

ACCOUNT _____ ACCOUNT NO. _____

DATE	DESCRIPTION	POST. REF.	DEBIT	CREDIT	BALANCE	
					DEBIT	CREDIT

ACCOUNT _____ ACCOUNT NO. _____

DATE	DESCRIPTION	POST. REF.	DEBIT	CREDIT	BALANCE	
					DEBIT	CREDIT

ACCOUNT _____ ACCOUNT NO. _____

DATE	DESCRIPTION	POST. REF.	DEBIT	CREDIT	BALANCE	
					DEBIT	CREDIT

Mini Practice Set 6 (continued)

ACCOUNT _____ ACCOUNT NO. _____

DATE	DESCRIPTION	POST. REF.	DEBIT	CREDIT	BALANCE	
					DEBIT	CREDIT

ACCOUNT _____ ACCOUNT NO. _____

DATE	DESCRIPTION	POST. REF.	DEBIT	CREDIT	BALANCE	
					DEBIT	CREDIT

ACCOUNT _____ ACCOUNT NO. _____

DATE	DESCRIPTION	POST. REF.	DEBIT	CREDIT	BALANCE	
					DEBIT	CREDIT

ACCOUNT _____ ACCOUNT NO. _____

DATE	DESCRIPTION	POST. REF.	DEBIT	CREDIT	BALANCE	
					DEBIT	CREDIT

Mini Practice Set 6 (continued)

(2), (3), (9)

GENERAL JOURNAL

	DATE	DESCRIPTION	POST. REF.	DEBIT	CREDIT	
1						1
2						2
3						3
4						4
5						5
6						6
7						7
8						8
9						9
10						10
11						11
12						12
13						13
14						14
15						15
16						16
17						17
18						18
19						19
20						20
21						21
22						22
23						23
24						24
25						25
26						26
27						27
28						28
29						29
30						30
31						31
32						32
33						33
34						34
35						35
36						36
37						37

Mini Practice Set 6 (continued)

GENERAL JOURNAL

DATE		DESCRIPTION	POST. REF.	DEBIT	CREDIT	
						1
						2
						3
						4
						5
						6
						7
						8
						9
						10
						11
						12
						13
						14
						15
						16
						17
						18
						19
						20
						21
						22
						23
						24
						25
						26
						27
						28
						29
						30
						31
						32
						33
						34
						35
						36
						37
						38

Mini Practice Set 6   (continued)

GENERAL JOURNAL

PAGE _____

	DATE		DESCRIPTION	POST. REF.	DEBIT	CREDIT	
1							1
2							2
3							3
4							4
5							5
6							6
7							7
8							8
9							9
10							10
11							11
12							12
13							13
14							14
15							15
16							16
17							17
18							18
19							19
20							20
21							21
22							22
23							23
24							24
25							25
26							26
27							27
28							28
29							29
30							30
31							31
32							32
33							33
34							34
35							35
36							36
37							37
38							38

Mini Practice Set 6   (continued)

GENERAL JOURNAL   PAGE _____

	DATE	DESCRIPTION	POST. REF.	DEBIT	CREDIT	
1						1
2						2
3						3
4						4
5						5
6						6
7						7
8						8
9						9
10						10
11						11
12						12
13						13
14						14
15						15
16						16
17						17
18						18
19						19
20						20
21						21
22						22
23						23
24						24
25						25
26						26
27						27
28						28
29						29
30						30
31						31
32						32
33						33
34						34
35						35
36						36
37						37
38						38

Name Date Class

Mini Practice Set 6 (continued) **(5)**

	ACCT. NO.	ACCOUNT NAME	TRIAL BALANCE		INCOME STATEMENT		BALANCE SHEET		
			DEBIT	CREDIT	DEBIT	CREDIT	DEBIT	CREDIT	
1									1
2									2
3									3
4									4
5									5
6									6
7									7
8									8
9									9
10									10
11									11
12									12
13									13
14									14
15									15
16									16
17									17
18									18
19									19
20									20
21									21
22									22
23									23
24									24
25									25
26									26

Mini Practice Set 6   (continued)

(6)

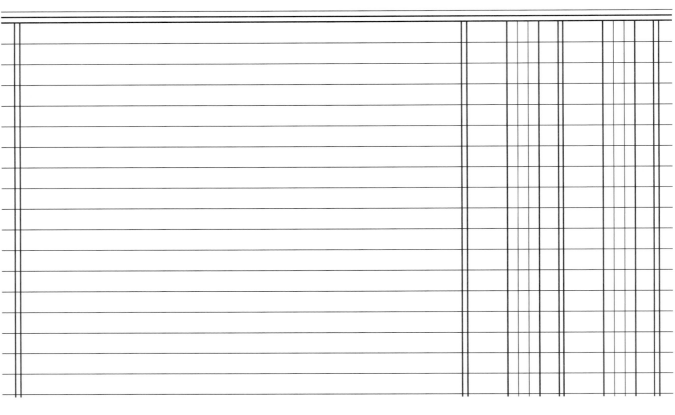

(7)

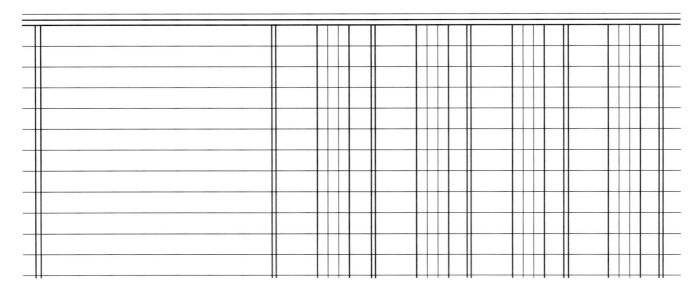

Mini Practice Set 6   (continued)
(8)

Mini Practice Set 6   (concluded)
(10)

_____
_____
_____

Analyze: **1.** _____

2. _____

3. _____

4. _____

MINI PRACTICE SET 6

Fine Finishes

Audit Test

Directions: *Use your completed solutions to answer the following questions.
Write the answer in the space to the left of each question.*

Answer

_____   **1.** How many checks were issued by Fine
                               Finishes in the month of February?

_____   **2.** What is the total ending balance of the
                               Accounts Payable accounts at February 28?

_____   **3.** What is the ending balance of the Accounts
                               Receivable account at February 28?

_____   **4.** What total amount was credited to the
                               Painting Fees revenue account for the
                               month?

_____   **5.** What were the total expenses for the
                               month?

_____   **6.** What amount was debited to the Income
                               Summary account to close the expense
                               accounts for the period?

_____   **7.** How many accounts were listed on the trial
                               balance dated February 28, 20--?

_____   **8.** What was the net income for the period?

_____   **9.** When net income was divided between
                               the partners, how much was allocated to
                               Laura Andersen?

_____ **10.** What total withdrawals were made by all
                               three partners for the period?

_____ **11.** What was the ending balance for the David Ingram, Capital account?

_____ **12.** What was the amount of total assets for the business at February 28?

_____ **13.** What was the amount of total liabilities for the business at February 28?

_____ **14.** How many accounts were listed on the post-closing trial balance?

_____ **15.** At month end, what debts remain unpaid by Fine Finishes?

_____ **16.** What was the total amount of debits to the Cash in Bank account for the period?

CHAPTER 29   Ethics in Accounting

Study Plan

Check Your Understanding

Section 1	*Read Section 1 on pages 776-779 and complete the following exercises on page 780.*

❏ Thinking Critically
❏ Computing in the Business World
❏ Problem 29-1   *Reporting Ethics Violations*
❏ Problem 29-2   *Exploring the Difference Between Ethics and Law*

Section 2	*Read Section 2 on pages 781-783 and complete the following exercises on page 784.*

❏ Thinking Critically
❏ Communicating Accounting
o Problem 29-3   *Promoting Principles of Conduct*

Summary

Review the Chapter 29 Summary on page 785 in your textbook
❏ Key Concepts

Review and Activities

Complete the following questions and exercises on pages 786-787 in your textbook.
❏ Using Key Terms
❏ Understanding Accounting Concepts and Procedures
❏ Case Study
❏ Conducting an Audit with Alex
❏ Internet Connection
❏ Workplace Skills

Problems

Complete the following end-of-chapter problems for Chapter 29 in your textbook.
❏ Problem 29-4   *Researching Ethics in the News*
❏ Problem 29-5   *Creating a Business Ethics Program*
❏ Problem 29-6   *Making Ethical Decisions*
❏ Problem 29-7   *Making Ethical Decisions*
❏ Problem 29-8   *Examining the Impact of Unethical Decisions*
❏ Problem 29-9   *Finding Out What Ethical Principles Mean to Your Classmates*
❏ Problem 29-10 *Finding Out What Ethical Principles Mean to Adults*
❏ Problem 29-11 *Applying a Code of Ethics to Personal Behavior*
❏ Problem 20-12 *Analyzing the AICPA Code of Ethics Preamble*

Chapter Reviews and Working Papers

Complete the following exercises for Chapter 29 in your Chapter Reviews and Working Papers.
❏ Chapter Review
❏ Self-Test

CHAPTER **29** REVIEW   Ethics in Accounting

Part 1   Accounting Vocabulary (9 points)

Total Points	44
Student's Score	

Directions: *Using terms from the following list, complete the sentences below. Write the letter of the term you have chosen in the space provided. The first sentence has been completed for you.*

A. business ethics	**D.** confidentiality	**G.** independence
B. code of ethics	**E.** ethics	**H.** integrity
C. competence	**F.** ethics officer	**I.** objectivity

_____*H*_____ **0.** Making false or misleading entries in a client's books violates the principle of _____.

_____ **1.** _____ refers to the knowledge, skills, and experience needed to complete a task.

_____ **2.** A formal policy of rules and standards that describes the ethical behaviors that a company expects from its management and employees is called a(n) _____.

_____ **3.** The study of _____ deals with our notions of right and wrong.

_____ **4.** A(n) _____ is the employee directly responsible for enforcing standards of conduct.

_____ **5.** An accountant who is impartial, honest, and free of conflicts of interest is adhering to the principle of _____.

_____ **6.** A CPA who serves as a director, officer, or employee of a company that he or she audits is violating the principle of _____.

_____ **7.** Information acquired in the course of an accountant's work is protected by the principle of _____.

_____ **8.** The policies and practices that reflect a company's core values such as honesty, trust, respect, and fairness are called _____.

_____ **9.** The principle of _____ requires that accountants choose what is right and just over what is wrong.

Part 2 The Nature of Ethics (10 points)

Directions: *Read each of the following statements to determine whether the statement is true or false. Write your answer in the space provided. The first statement is completed for you.*

False   **0.** The values found in a system of personal ethics are vastly different than those found in a system of business ethics.

_____   **1.** The way a business treats its employees and customers can illustrate the core values of the business.

_____   **2.** Our laws describe a minimum acceptable level of correct behavior for members of society.

_____   **3.** The topic of conflicts of interest is often addressed within a company's code of ethics.

_____   **4.** Effective communication of a company's code of ethics to its managers and employees is a key component to a successful business ethics program.

_____   **5.** Good business ethics can lead to improved company reputation and increased employee loyalty.

_____   **6.** When faced with an ethical dilemma, the accountant should always give preference to increased financial returns for stockholders of the company.

_____   **7.** The behaviors and actions of individuals do not impact the overall nature of society.

_____   **8.** Accountants who demonstrate ethical behaviors help secure public trust in the profession and in the business world.

_____   **9.** A company's code of ethics provides guidelines for suggested behavior within the organization, but violations are not punished.

_____   **10.** Ethics committees within a business can help settle disputed issues and resolve ethical problems.

Part 3 Key Principles (15 points)

Directions: *Read the statements below. Identify the key principle of conduct that has been violated. Write the letter of the term in the space provided. The first statement has been completed for you.*

A. competence	**D.** integrity
B. confidentiality	**E.** objectivity
C. independence	

_____B_____ **0**. A CPA reveals the details of a client's new marketing campaign at a company holiday party.

_____ **1**. A CPA owns stock in a company and also performs its annual audit.

_____ **2**. An accountant contracts with a client to perform services for which he holds inadequate training or experience.

_____ **3**. An accountant makes a misleading entry in the financial records of a client.

_____ **4**. A CPA makes a professional judgment in the course of an independent audit that reflects her personal interests.

_____ **5**. An accountant fails to maintain his or her continuing education in order to understand current accounting standards and procedures.

_____ **6**. An auditor fails to report false statements in a client's financial records.

_____ **7**. A CPA reveals an audit client's salary structure to a family member.

_____ **8**. An accountant destroys documents in order to hide questionable accounting practices of his or her client.

_____ **9**. An auditor signs off on financial statements known to contain overstatements of income.

_____ **10**. An accountant serves as an officer of a company for which he or she performs annual audits.

_____ **11**. An accountant renders professional services in an unorganized and unqualified manner.

_____ **12**. An auditor accepts a large gift from a client.

_____ **13**. A senior auditor assigns audit tasks to a clerical assistant who has not been trained in audit procedures.

_____ **14**. An auditor discusses a client's unpublished financial results with a golfing partner.

Part 4 Professional Organizations (10 points)

Directions: *Read each of the following statements to determine whether the statement is true or false. Write your answer in the space provided. The first statement is completed for you.*

False   **0.** Acceptance of the AICPA standards of professional behavior is mandatory for all certified public accountants.

_____   **1.** A member of the AICPA assumes an obligation of self-discipline above and beyond the requirements of laws and regulations.

_____   **2.** The AICPA is the national association of professional management accountants.

_____   **3.** Internal auditors work within a business to make sure the company operates according to its agreements with suppliers, government, and customers.

_____   **4.** Adherence to the Institute of Internal Auditors' code of ethics is necessary because of the public trust placed in internal auditors.

_____   **5.** The AICPA Code of Professional Conduct describes the responsibilities that the public owes to the CPA.

_____   **6.** Management accountants usually hold a job in a single company.

_____   **7.** Certified public accountants are expected to sacrifice personal advantage and give preference to honorable behavior at all times.

_____   **8.** The codes of ethics established by professional organizations cover every possible situation that might occur, providing detailed instructions for members.

_____   **9.** The Institute of Management Accountants is a professional organization devoted exclusively to management accountants and financial managers.

_____   **10.** Professional associations often create codes of ethics or codes of conduct for members.

Working Papers *for Section Problems*

Problem 29-1 Reporting Ethics Violations

_____

_____

_____

_____

_____

_____

_____

Problem 29-2 Exploring the Difference Between Ethics and Law

_____

_____

_____

_____

_____

_____

_____

_____

Problem 29-3 Promoting Principles of Conduct

_____

_____

_____

_____

_____

_____

_____

_____

Working Papers *for End-of-Chapter Problems*

Problem 29-4 Researching Ethics in the News

_____

_____

_____

_____

_____

Problem 29-5 Creating a Business Ethics Program

_____

_____

_____

_____

Problem 29-6 Making Ethical Decisions

_____

_____

_____

_____

_____

_____

Problem 29-7 Making Ethical Decisions

_____

_____

_____

_____

_____

_____

Working Papers *for End-of-Chapter Problems*

Problem 29-8 Examining the Impact of Unethical Decisions

Problem 29-9 Finding Out What Ethical Principles Mean to Your Classmates

Problem 29-10 Finding Out What Ethical Principles Mean to Adults

Problem 29-11 Applying a Code of Ethics to Personal Behavior

Problem 29-12 Analyzing the AICPA Code of Ethics Preamble

CHAPTER 29   Ethics in Accounting

Self-Test

Part A True or False

Directions: *Circle the letter T in the Answer column if the statement is true; circle the letter F if the statement is false.*

Answer

T   F   **1.** The ethical behavior of the accounting professional is important to securing public trust in the profession.

T   F   **2.** The law contains more subtle implications and is more complex than ethical concepts.

T   F   **3.** In order to be successful and profitable, businesses must favor the profit motive over ethical behavior.

T   F   **4.** Employees, managers, investors, and the general public rely on ethical practices in financial reporting.

T   F   **5.** The ethically trained accountant should adhere to the universal standards of what is right.

T   F   **6.** Ethics is the study of our notions of right and wrong.

T   F   **7.** The formal policy of rules and standards that describes the ethical behaviors that a company expects from its managers and employees is called an integrity policy.

T   F   **8.** Individuals who act ethically enjoy increased self-esteem and respect from their peers.

T   F   **9.** Competence is the principle that requires an accountant to be impartial, honest, and free of conflicts of interest.

Part B Fill in the Missing Term

Directions: *Using terms from the following list, complete the sentences below. Write the letter of the term you have chosen in the space provided.*

A. AICPA	**E.** enforcement	**I.** objectivity
B. business ethics	**F.** IMA	**J.** public
C. code of ethics	**G.** independence	
D. competence	**H.** integrity	

_____ **1.** After an ethics committee helps settle a disputed ethical problem, a(n) _____ phase may follow.

_____ **2.** A(n) _____ is a formal policy of rules and standards that describes the ethical behaviors that a company expects from its managers and employees.

_____ **3.** The ethically trained accountant should focus on optimizing the interests of the _____.

_____ **4.** The principle of _____ requires that accountants remain uninfluenced by personal interests when forming professional judgments.

_____ **5.** An accountant who accepts a personal loan from an audit client is in violation of the principle of _____.

_____ **6.** If an accountant renders services in a timely, careful, thorough, and technically correct manner, the principle of _____ has been observed.

_____ **7.** The _____ provides a code of professional conduct for its members, certified publicaccountants.

_____ **8.** Management accountants and financial managers often belong to the_____ , an organization devoted exclusively to this type accounting professional.

_____ **9.** The way a business attends to shareholder value and community service often reflects its commitment to _____.

_____ **10.** The key principles that provide accountants with the framework for rules of conduct are objectivity, independence, competence, confidentiality, and _____.

APPENDIX A

Adjustments for a Service Business Using a Ten-Column Work Sheet

Problem A-1   Analyzing Adjustments

1. Amount of Adjustment _____

Account Debited _____

Account Credited _____

2. Amount of Adjustment _____

Account Debited _____

Account Credited _____

Problem A-2   Adjusting Entries

1) _____

2) _____

3) _____

4) _____

_____

ACCT. NO.	ACCOUNT NAME	TRIAL BALANCE		ADJUSTMENTS		ADJUSTED TRIAL BALANCE	
		DEBIT	CREDIT	DEBIT	CREDIT	DEBIT	CREDIT
1							
2							
3							
4							
5							
6							
7							
8							
9							
10							
11							
12							
13							
14							
15							
16							
17							
18							
19							
20							
21							
22							
23							
24							
25							
26							

Problem A-4   Preparing a Ten-Column Worksheet

Ryan's

Work

For the Year Ended

	ACCT. NO.	ACCOUNT NAME	TRIAL BALANCE		ADJUSTMENTS	
			DEBIT	CREDIT	DEBIT	CREDIT
1	101	Cash in Bank				
2	105	Accts. Rec.—Nancy Burrows				
3	110	Accts. Rec.—Patrick Chang				
4	115	Supplies				
5	120	Prepaid Insurance				
6	125	Office Equipment				
7	130	Rental Equipment				
8	135	Canoes				
9	201	Accts. Pay.—South Canoe				
10	205	Accts. Pay.—H & T Adver.				
11	210	Accts. Pay.—Cassidy's Eqp.				
12	301	Ryan Gillespie, Capital				
13	305	Ryan Gillespie, With.				
14	310	Income Summary				
15	401	Rental Revenue				
16	402	Lesson Revenue				
17	501	Advertising Expense				
18	505	Insurance Expense				
19	510	Maintenance Expense				
20	515	Rent Expense				
21	520	Supplies Expense				
22	525	Utilities Expense				
23						
24						
25						
26						
27						
28						
29						
30						
31						
32						
33						

Canoe Rentals

Sheet

December 31, 20--

	ADJUSTED TRIAL BALANCE		INCOME STATEMENT		BALANCE SHEET		
	DEBIT	CREDIT	DEBIT	CREDIT	DEBIT	CREDIT	
							1
							2
							3
							4
							5
							6
							7
							8
							9
							10
							11
							12
							13
							14
							15
							16
							17
							18
							19
							20
							21
							22
							23
							24
							25
							26
							27
							28
							29
							30
							31
							32
							33

APPENDIX B   Using the Numeric Keypad

Using the 4-5-6 Keys

Exercises 1–6 are from your textbook. Exercises 7–24 are provided to give you additional practice.

Practice at a comfortable pace until you feel confident about each key's location.

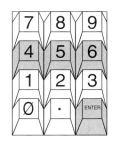

1.	**2.**	**3.**	**4.**	**5.**	**6.**
444	555	666	456	554	664
555	666	454	654	445	445
666	444	545	465	564	566
456	654	446	556	664	645
564	546	646	656	565	465
646	465	546	465	655	654
3,331	3,330	3,303	3,252	3,447	3,439

7.	**8.**	**9.**	**10.**	**11.**	**12.**
456	564	646	555	666	444
654	546	465	666	454	545
446	646	546	456	654	465
556	656	465	554	445	564
664	565	655	664	466	566
645	465	654	444	555	666
3,421	3,442	3,431	3,339	3,240	3,250

13.	**14.**	**15.**	**16.**	**17.**	**18.**
654	666	464	666	456	555
546	546	656	654	454	545
456	546	656	654	454	545
465	646	546	555	446	454
444	654	654	556	445	456
555	564	546	554	444	654
3,120	3,622	3,522	3,639	2,699	3,209

19.	**20.**	**21.**	**22.**	**23.**	**24.**
5,655	4,556	456	55	445	6,656
45	645	4,564	4	56	4,655
6	54	655	54	6,664	566
456	46	4,545	5,554	465	465
664	564	5,664	564	5,644	64
56	5	65	445	56	544
6,882	5,870	15,949	6,676	13,330	12,950

Using the 1, 7 and 0 Keys

Exercises 1–6 are from your textbook. Exercises 7–24 are provided to give you additional practice.

Practice at a comfortable pace until you feel confident about each key's location.

1.	2.	3.	4.	5.	6.
444	014	140	107	011	141
471	107	701	074	170	117
174	740	701	104	710	417
741	101	704	007	004	047
710	114	471	411	471	104
407	441	117	047	174	114
2,947	1,517	2,834	750	1,540	940

7.	8.	9.	10.	11.	12.
741	710	407	014	147	740
101	114	441	140	701	701
704	471	117	107	074	104
007	411	017	011	170	710
004	471	174	141	117	417
047	104	114	444	471	174
1,604	2,281	1,270	857	1,680	2,846

13.	14.	15.	16.	17.	18.
170	140	104	111	777	410
701	147	107	147	111	140
107	014	401	174	444	014
741	041	701	741	714	741
147	074	101	710	741	471
410	047	010	410	704	147
2,276	463	1,424	2,293	3,491	1,923

19.	20.	21.	22.	23.	24.
1,044	456	145	17	101	1,404
540	4,540	6,147	7,100	47	40
7,055	74	567	1,105	1,075	140
607	415	10	574	157	1,714
4,441	510	106	177	7,775	1,570
17	1,750	1,045	50	147	1,104
13,704	7,745	8,020	9,023	9,302	5,972

Using the 3 and 9 Keys

Exercises 1–6 are from your textbook. Exercises 7–12 are provided to give you additional practice.

Practice at a comfortable pace until you feel confident about each key's location.

1.	**2.**	**3.**	**4.**	**5.**	**6.**
666	669	339	966	939	699
999	663	363	393	363	936
333	936	336	966	393	939
963	396	936	633	639	336
639	936	636	393	369	696
399	363	996	993	369	939
3,999	3,963	3,606	4,344	3,072	4,545

7.	**8.**	**9.**	**10.**	**11.**	**12.**
963	639	399	669	663	936
396	936	363	339	363	336
936	636	993	966	393	966
633	393	993	939	363	393
639	369	369	699	936	939
336	696	939	666	999	333
3,903	3,669	4,056	4,278	3,717	3,903

Using the 2 and 8 Keys

Exercises 1–6 are from your textbook. Exercises 7–12 are provided to give you additional practice.

Practice at a comfortable pace until you feel confident about each key's location.

1.	**2.**	**3.**	**4.**	**5.**	**6.**
555	228	885	285	582	828
888	852	285	258	558	825
222	522	825	525	582	852
582	252	588	858	825	258
822	528	258	582	525	885
522	855	852	825	582	282
3,591	3,237	3,693	3,333	3,654	3,930

7.	**8.**	**9.**	**10.**	**11.**	**12.**
582	822	522	228	852	522
252	528	855	885	285	825
588	258	258	285	825	525
858	582	825	582	558	582
825	525	582	828	528	852
852	885	282	555	888	222
3,957	3,600	3,324	3,363	3,936	3,528

Using the Decimal Key

Exercises 1–18 are from your textbook. Exercises 19–24 are provided to give you additional practice.

Practice at a comfortable pace until you feel confident about each key's location.

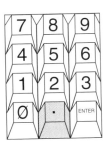

1.	**2.**	**3.**	**4.**	**5.**	**6.**
.777	.978	.998	8.78	7.88	8.79
.888	.987	.879	8.89	7.87	7.98
.999	.878	.787	8.87	8.97	9.89
.789	.987	.878	7.88	9.77	9.87
.897	.789	.797	9.87	7.97	7.89
.978	.797	.899	7.98	8.79	9.78
5.328	5.416	5.238	52.27	51.25	54.20

7.	**8.**	**9.**	**10.**	**11.**	**12.**
468.	48.2	.8	284.0	41.87	154.88
.489	2,537.	5,827.	100.	4,057.4	888.
214.2	852.	.024	8.45	89.45	.0082
7.12	3.978	18.73	56.0	2.25	200.08
6,394.4	257.0	85.00	23.00	20.0	632.48
.58	.2684	1.045	.89	36.248	64.1
7,084.789	3,698.4464	5,932.599	472.34	4,247.218	1,939.5482

13.	**14.**	**15.**	**16.**	**17.**	**18.**
267.50	425.21	1.25	467.54	65.27	9.78
4.19	414.50	0.18	95.14	102.38	5.94
87.64	1,684.84	585.56	6,926.95	8,216.58	652.25
654.84	49.95	7.50	35.00	1,852.84	3,782.70
1,750.67	720.65	11.60	7.13	4.60	39.25
141.82	77.61	23.55	154.95	79.15	36.87
2,906.66	3,372.76	629.64	7,686.71	10,320.82	4,526.79

19.	**20.**	**21.**	**22.**	**23.**	**24.**
61.93	446.00	20.67	78.30	519.02	8.68
1,296.13	4,466.73	216.37	14.39	3,113.56	62.36
200.00	61.45	445.39	14.60	102.55	43.12
157.43	258.16	5,650.00	352.49	39.95	611.18
14.97	900.62	1,426.15	617.32	1,874.05	21.67
869.42	25.15	109.15	803.70	3,130.78	1,972.32
2,599.88	6,158.11	7,867.73	1,880.80	8,779.91	2,719.33

APPENDIX C

Recording Transactions in the Combination Journal

Problem C-1 Analyzing Transactions for Combination Journal Entries

Date	Transaction	General Dr.	General Cr.	Accounts Receivable Dr.	Accounts Receivable Cr.	Sales Cr.	Sales Tax Payable Cr.	Accounts Payable Dr.	Accounts Payable Cr.	Purchases Dr.	Cash in Bank Dr.	Cash in Bank Cr.
Dec. 1	Debit	✓										
	Credit								✓			
2	Debit											
	Credit											
5	Debit											
	Credit											
10	Debit											
	Credit											
13	Debit											
	Credit											
15	Debit											
	Credit											
18	Debit											
	Credit											
22	Debit											
	Credit											
25	Debit											
	Credit											
25	Debit											
	Credit											
30	Debit											
	Credit											
31	Debit											
	Credit											

Problem C-2   Recording Transactions
Totaling, Proving, and Ruling the Combination Journal

COMBINATION

	DATE	ACCOUNT NAME	DOC. NO.	POST. REF.	GENERAL DEBIT	GENERAL CREDIT	ACCOUNTS RECEIVABLE DEBIT	ACCOUNTS RECEIVABLE CREDIT
1								
2								
3								
4								
5								
6								
7								
8								
9								
10								
11								
12								
13								
14								
15								
16								
17								
18								
19								
20								
21								
22								
23								
24								
25								
26								
27								
28								
29								
30								
31								
32								
33								

JOURNAL

PAGE

	MEDICAL FEES CREDIT	LABORATORY FEES CREDIT	ACCOUNTS PAYABLE DEBIT	ACCOUNTS PAYABLE CREDIT	MEDICAL SUPPLIES DEBIT	CASH IN BANK DEBIT	CASH IN BANK CREDIT	
1								1
2								2
3								3
4								4
5								5
6								6
7								7
8								8
9								9
10								10
11								11
12								12
13								13
14								14
15								15
16								16
17								17
18								18
19								19
20								20
21								21
22								22
23								23
24								24
25								25
26								26
27								27
28								28
29								29
30								30
31								31
32								32
33								33

Problem C-3 Recording Transactions
Totaling, Proving, and Ruling the Combination Journal

COMBINATION

	DATE	ACCOUNT NAME	DOC. NO.	POST. REF.	GENERAL DEBIT	GENERAL CREDIT	ACCOUNTS RECEIVABLE DEBIT	ACCOUNTS RECEIVABLE CREDIT
1								
2								
3								
4								
5								
6								
7								
8								
9								
10								
11								
12								
13								
14								
15								
16								
17								
18								
19								
20								
21								
22								
23								
24								
25								
26								
27								
28								
29								
30								
31								
32								
33								
34								
35								
36								

JOURNAL

SALES CREDIT	SALES TAX PAYABLE CREDIT	ACCOUNTS PAYABLE		PURCHASES DEBIT	CASH IN BANK		
		DEBIT	CREDIT		DEBIT	CREDIT	
							1
							2
							3
							4
							5
							6
							7
							8
							9
							10
							11
							12
							13
							14
							15
							16
							17
							18
							19
							20
							21
							22
							23
							24
							25
							26
							27
							28
							29
							30
							31
							32
							33
							34
							35
							36

Problem C-4   Recording Adjusting and Closing Entries in a Combination Journal

In Beat CD Shop

Work Sheet

For Period Ended December 31, 20--

#	ACCT. NO.	ACCOUNT NAME	ADJUSTMENTS DEBIT	ADJUSTMENTS CREDIT	INCOME STATEMENT DEBIT	INCOME STATEMENT CREDIT	BALANCE SHEET DEBIT	BALANCE SHEET CREDIT
1	120	Merchandise Inventory	(a) 200000				2600000	
2	125	Supplies		(b) 27000			9500	
3	130	Prepaid Insurance		(c) 70000			10000	
4								
5	137	Accum. Depr.—Store Equip.		(d) 180000				540000
6								
7								
8	205	Federal Income Tax Payable		(e) 60000				60000
9								
10								
11								
12	301	Capital Stock						2840000
13	305	Retained Earnings						840000
14	310	Income Summary		(a) 200000		200000		
15	401	Sales				4600000		
16	405	Sales Returns and Allowances			40500			
17	501	Purchases			1800000			
18	505	Transportation In			120000			
19	510	Purchases Discounts				22500		
20	515	Purchases Returns and Allow.				15000		
21	620	Depr. Expense—Store Equip.	(d) 180000		180000			
22	625	Federal Income Tax Expense	(e) 60000		140000			
23	630	Insurance Expense	(c) 70000		70000			
24	650	Miscellaneous Expense			50000			
25	655	Rent Expense			600000			
26	665	Supplies Expense	(b) 27000		27000			

Problem C-4 (concluded)
(1), (2)

COMBINATION JOURNAL (PARTIAL)

	DATE	ACCOUNT NAME	DOC. NO.	POST. REF.	GENERAL DEBIT	GENERAL CREDIT	ACCOUNTS RECEIVABLE DEBIT	ACCOUNTS RECEIVABLE CREDIT
1								
2								
3								
4								
5								
6								
7								
8								
9								
10								
11								
12								
13								
14								
15								
16								
17								
18								
19								
20								
21								
22								
23								
24								
25								
26								
27								
28								
29								
30								
31								
32								
33								

Notes

APPENDIX  The Accrual Basis of Accounting

Problem D-1   Identifying Accruals and Deferrals

Item	Prepaid Expense	Unearned Revenue	Accrued Expense	Accrued Revenue
1				
2				
3				
4				
5				
6				
7				
8				
9				
10				
11				

Problem D-2   Recording Adjusting Entries

GENERAL JOURNAL   PAGE _____

DATE		DESCRIPTION	POST. REF.	DEBIT	CREDIT	
						1
						2
						3
						4
						5
						6
						7
						8
						9
						10
						11
						12
						13
						14
						15
						16
						17
						18
						19
						20

Problem D-3   Recording Transactions for Notes Payable

GENERAL JOURNAL   PAGE _____

	DATE	DESCRIPTION	POST. REF.	DEBIT	CREDIT	
1						1
2						2
3						3
4						4
5						5
6						6
7						7
8						8
9						9
10						10
11						11
12						12
13						13
14						14
15						15
16						16
17						17
18						18
19						19
20						20

Problem D-4   Recording Accrued Expenses

GENERAL JOURNAL   PAGE _____

	DATE	DESCRIPTION	POST. REF.	DEBIT	CREDIT	
1						1
2						2
3						3
4						4
5						5
6						6
7						7
8						8
9						9
10						10
11						11
12						12
13						13
14						14
15						15
16						16
17						17
18						18
19						19
20						20

APPENDIX E   Federal Personal Income Tax

20-- Tax Table

Caution. Dependents, see the worksheet on page 20.

Example. Mr. Brown is single. His taxable income on line 6 of Form 1040EZ is $26,250. First, he finds the $26,250-26,300 income line. Next, he finds the "Single" column and reads down the column. The amount shown where the income line and filing status column meet → is $3,941. This is the tax amount he should enter on line 11 of Form 1040EZ.

At least	But less than	Single	Married filing jointlly
		Your tax is—	
26,200	26,250	3,934	3,934
26,250	26,300	(3,941)	3,941
26,300	26,350	3,949	3,949
26,350	26,400	3,956	3,956

If Form 1040EZ, line 6, is—		And you are—		If Form 1040EZ, line 6, is—		And you are—		If Form 1040EZ, line 6, is—		And you are—		If Form 1040EZ, line 6, is—		And you are—	
At least	But less than	Single	Married filing jointly	At least	But less than	Single	Married filing jointly	At least	But less than	Single	Married filing jointly	At least	But less than	Single	Married filing jointly
		Your tax is—				Your tax is—				Your tax is—				Your tax is—	
0	5	0	0	1,500	1,525	227	227	**3,000**				**6,000**			
5	15	2	2	1,525	1,550	231	231	3,000	3,050	454	454	6,000	6,050	904	904
15	25	3	3	1,550	1,575	234	234	3,050	3,100	461	461	6,050	6,100	911	911
25	50	6	6	1,575	1,600	238	238	3,100	3,150	469	469	6,100	6,150	919	919
50	75	9	9	1,600	1,625	242	242	3,150	3,200	476	476	6,150	6,200	926	926
75	100	13	13	1,625	1,650	246	246	3,200	3,250	484	484	6,200	6,250	934	934
100	125	17	17	1,650	1,675	249	249	3,250	3,300	491	491	6,250	6,300	941	941
125	150	21	21	1,675	1,700	253	253	3,300	3,350	499	499	6,300	6,350	949	949
150	175	24	24	1,700	1,725	257	257	3,350	3,400	506	506	6,350	6,400	956	956
175	200	28	28	1,725	1,750	261	261	3,400	3,450	514	514	6,400	6,450	964	964
200	225	32	32	1,750	1,775	264	264	3,450	3,500	521	521	6,450	6,500	971	971
225	250	36	36	1,775	1,800	268	268	3,500	3,550	529	529	6,500	6,550	979	979
250	275	39	39	1,800	1,825	272	272	3,550	3,600	536	536	6,550	6,600	986	986
275	300	43	43	1,825	1,850	276	276	3,600	3,650	544	544	6,600	6,650	994	994
300	325	47	47	1,850	1,875	279	279	3,650	3,700	551	551	6,650	6,700	1,001	1,001
325	350	51	51	1,875	1,900	283	283	3,700	3,750	559	559	6,700	6,750	1,009	1,009
350	375	54	54	1,900	1,925	287	287	3,750	3,800	566	566	6,750	6,800	1,016	1,016
375	400	58	58	1,925	1,950	291	291	3,800	3,850	574	574	6,800	6,850	1,024	1,024
400	425	62	62	1,950	1,975	294	294	3,850	3,900	581	581	6,850	6,900	1,031	1,031
425	450	66	66	1,975	2,000	298	298	3,900	3,950	589	589	6,900	6,950	1,039	1,039
450	475	69	69					3,950	4,000	596	596	6,950	7,000	1,046	1,046
475	500	73	73	**2,000**				**4,000**				**7,000**			
500	525	77	77	2,000	2,025	302	302	4,000	4,050	604	604	7,000	7,050	1,054	1,054
525	550	81	81	2,025	2,050	306	306	4,050	4,100	611	611	7,050	7,100	1,061	1,061
550	575	84	84	2,050	2,075	309	309	4,100	4,150	619	619	7,100	7,150	1,069	1,069
575	600	88	88	2,075	2,100	313	313	4,150	4,200	626	626	7,150	7,200	1,076	1,076
600	625	92	92	2,100	2,125	317	317	4,200	4,250	634	634	7,200	7,250	1,084	1,084
625	650	96	96	2,125	2,150	321	321	4,250	4,300	641	641	7,250	7,300	1,091	1,091
650	675	99	99	2,150	2,175	324	324	4,300	4,350	649	649	7,300	7,350	1,099	1,099
675	700	103	103	2,175	2,200	328	328	4,350	4,400	656	656	7,350	7,400	1,106	1,106
700	725	107	107	2,200	2,225	332	332	4,400	4,450	664	664	7,400	7,450	1,114	1,114
725	750	111	111	2,225	2,250	336	336	4,450	4,500	671	671	7,450	7,500	1,121	1,121
750	775	114	114	2,250	2,275	339	339	4,500	4,550	679	679	7,500	7,550	1,129	1,129
775	800	118	118	2,275	2,300	343	343	4,550	4,600	686	686	7,550	7,600	1,136	1,136
800	825	122	122	2,300	2,325	347	347	4,600	4,650	694	694	7,600	7,650	1,144	1,144
825	850	126	126	2,325	2,350	351	351	4,650	4,700	701	701	7,650	7,700	1,151	1,151
850	875	129	129	2,350	2,375	354	354	4,700	4,750	709	709	7,700	7,750	1,159	1,159
875	900	133	133	2,375	2,400	358	358	4,750	4,800	716	716	7,750	7,800	1,166	1,166
900	925	137	137	2,400	2,425	362	362	4,800	4,850	724	724	7,800	7,850	1,174	1,174
925	950	141	141	2,425	2,450	366	366	4,850	4,900	731	731	7,850	7,900	1,181	1,181
950	975	144	144	2,450	2,475	369	369	4,900	4,950	739	739	7,900	7,950	1,189	1,189
975	1,000	148	148	2,475	2,500	373	373	4,950	5,000	746	746	7,950	8,000	1,196	1,196
1,000				2,500	2,525	377	377	**5,000**				**8,000**			
1,000	1,025	152	152	2,525	2,550	381	381	5,000	5,050	754	754	8,000	8,050	1,204	1,204
1,025	1,050	156	156	2,550	2,575	384	384	5,050	5,100	761	761	8,050	8,100	1,211	1,211
1,050	1,075	159	159	2,575	2,600	388	388	5,100	5,150	769	769	8,100	8,150	1,219	1,219
1,075	1,100	163	163	2,600	2,625	392	392	5,150	5,200	776	776	8,150	8,200	1,226	1,226
1,100	1,125	167	167	2,625	2,650	396	396	5,200	5,250	784	784	8,200	8,250	1,234	1,234
1,125	1,150	171	171	2,650	2,675	399	399	5,250	5,300	791	791	8,250	8,300	1,241	1,241
1,150	1,175	174	174	2,675	2,700	403	403	5,300	5,350	799	799	8,300	8,350	1,249	1,249
1,175	1,200	178	178	2,700	2,725	407	407	5,350	5,400	806	806	8,350	8,400	1,256	1,256
1,200	1,225	182	182	2,725	2,750	411	411	5,400	5,450	814	814	8,400	8,450	1,264	1,264
1,225	1,250	186	186	2,750	2,775	414	414	5,450	5,500	821	821	8,450	8,500	1,271	1,271
1,250	1,275	189	189	2,775	2,800	418	418	5,500	5,550	829	829	8,500	8,550	1,279	1,279
1,275	1,300	193	193	2,800	2,825	422	422	5,550	5,600	836	836	8,550	8,600	1,286	1,286
1,300	1,325	197	197	2,825	2,850	426	426	5,600	5,650	844	844	8,600	8,650	1,294	1,294
1,325	1,350	201	201	2,850	2,875	429	429	5,650	5,700	851	851	8,650	8,700	1,301	1,301
1,350	1,375	204	204	2,875	2,900	433	433	5,700	5,750	859	859	8,700	8,750	1,309	1,309
1,375	1,400	208	208	2,900	2,925	437	437	5,750	5,800	866	866	8,750	8,800	1,316	1,316
1,400	1,425	212	212	2,925	2,950	441	441	5,800	5,850	874	874	8,800	8,850	1,324	1,324
1,425	1,450	216	216	2,950	2,975	444	444	5,850	5,900	881	881	8,850	8,900	1,331	1,331
1,450	1,475	219	219	2,975	3,000	448	448	5,900	5,950	889	889	8,900	8,950	1,339	1,339
1,475	1,500	223	223					5,950	6,000	896	896	8,950	9,000	1,346	1,346

Continued on page 25

- 24 -

Problem E-1 Preparing Form 1040EZ

Form 1040EZ

Department of the Treasury–Internal Revenue Service

Income Tax Return for Single and Joint Filers With No Dependents (P) **20--**

OMB No. 1545-0675

Label
(See Page 12.)

Use the IRS label.
Otherwise please print or type

L A B E L   H E R E

Your first name and initial   Last name

If a joint return, spouse's first name and initial   Last name

Home address (number and street). If you have a P.O. box, see page 7.   Apt. no.

City, town or post office, state, and ZIP code. If you have a foreign address, see page 7.

Your social security number

Spouse's social security number

▲ **Important!** ▲
you **must** enter your SSN(s) above.

Presidential Election Campaign
(See page 12.)

Note: *Checking "Yes" will not change your tax or reduce your refund.*
Do you, or spouse if a joint return, want $3 to go to this fund? ►

	You		Spouse	
	☐ Yes	☐ No	☐ Yes	☐ No

Income

Attach Form(s) W-2 here. Enclose, but do not attach, any payment.

1 Total wages, salaries, and tips. This should be shown in box 1 of your W-2 form(s). Attach your W-2 form(s).   **1**

2 Taxable interest. If the total is over $400, you cannot use Form 1040EZ   **2**

3 Unemployment compensation, qualified state tuition program earnings and Alaska Permanent Fund dividends (see page 14).   **3**

4 Add lines 1, 2, and 3. This is your **adjusted gross income.**   **4**

Note. You **must** check Yes or No

5 Can your parents (or someone else) claim you on their return?
☐ **Yes.** Enter amount from worksheet on back.   ☐ **No.** If **single,** enter 7,450.00. If **married,** enter 13,400.00 See back for explanation   **5**

6 Subtract line 5 from line 4. If line 5 is larger than line 4, enter 0. This is your **taxable income.** ►   **6**

Credits, payments, and tax

7 Rate reduction credit. See worksheet on page 14.   **7**

8 Enter your Federal income tax withheld from box 2 of your W-2 form(s).   **8**

9a **Earned income credit (EIC).** See page 15.   **9a**

b Nontaxable earned income.   **9b**

10 Add lines 7, 8, and 9a. These are your **total credits and payments.** ►   **10**

11 **Tax.** If you checked "Yes" on line 5, see page 20. Otherwise, use the amount on **line 6 above** to find your tax in the tax table on pages 24–28 of the booklet. Then, enter the tax from the table on this line.   **11**

Refund
Have it directly deposited! See Page 20 and fill in 12b, 12c, and 12d

12a If line 10 is larger than line 10, subtract line 10 from line 11. This is your **refund.** ► **12a**

► **b** Routing number ☐☐☐☐☐☐☐☐☐ ► **c** Type: ☐ Checking ☐ Savings

► **d** Account number ☐☐☐☐☐☐☐☐☐☐☐☐☐☐☐☐☐

Amount you owe

13 If line 11 is larger than line 10, subtract line 10 from line 11. This is **the amount you owe.** See page 21 for details on how to pay. ► **13**

Third party desgnee

Do you want to allow another person to discuss this return with the IRS (see page 22)? ☐ **Yes.** Complete the following. ☐ **No**

Designee's name ►   Phone no. ► ( )   Personal identification number(PIN) ► ☐☐☐☐☐

Sign here
Joint return? See page 11. Keep a copy for your records

Under penalties of perjury, I declare that I have examined this return, and to the best of my knowledge and belief, it is true, correct, and accurately lists all amounts and sources of income I received during the tax year. Declaration of preparer (other than the taxpayer) is based on all information of which the preparer has any knowledge.

Your signature   Date   Your occupation   Daytime phone number ( )

Spouse's signature if a joint return, **both** must sign.   Date   Spouse's occupation

Paid Preparer's use only

Preparer's signature ►   Date   Check if self-employed ☐   Preparer's SSN or PTIN

Firm's name (or yours if self-employed), address, and ZIP code ►   EIN   Phone no. ( )

For Disclosure, Privacy Act, and Paperwork Reduction Act Notice, see page 23   Cat. No. 12616G   Form **1040EZ** 20--

Problem E-1   (concluded)

Filling in your return For tips on how to avoid common mistakes, see page 30	If you received a scholarship or fellowship grant or tax–exempt interest income, such as on municipal bonds, see the booklet before filling in the form. Also, see the booklet if you received a Form 1099–INT showing Federal income tax withheld or if Federal income tax was withheld from your unemployment compensation or Alaska Permanent Fund dividends. **Remember,** you must report all wages, salaries, and tips even if you do not get a W–2 form from your employer. You must also report all your taxable interest, including interest from banks, savings and loans, credit unions, etc., even if you do not get a Form 1099–INT.

Worksheet for dependents who checked "Yes" on line 5
 	
(keep a copy for your records)
	Use this worksheet to figure the amount to enter on line 5 if someone can claim you (or your spouse if married) as a dependent, even if that person chooses not to do so. To find out if someone can claim you as a dependent, use TeleTax topic 354 (see page 6).
	
	A. Amount, if any, from line 1 on front . . . . + 250.00 Enter total ▶ **A.** _____
	B. Minimum standard deduction. . . . . **B.** ___750.00___
	C. Enter the **larger** of line A or line B here. . . . . **C.** _____
	D. Maximum standard deduction. If **single,** enter 4,550.00; if **married,** enter 7,600.00. . . . . **D.** _____
	E. Enter the **smaller** of line C or line D here. This is your standard deduction. . . . . **E.** _____
	F. Exemption amount.
	   • If single, enter 0
	   • If married and—
	     —both you and your spouse can be claimed as dependents, enter 0.
	     —only one of you can be claimed as a dependent, enter 2,900.00
	                          F. _____
	G. Add lines E and F. Enter the total here and on line 5 on the front. . . . . **G.** _____
	
	If you checked "No" on line 5 because no one can claim you (or your spouse if married) as a dependent, enter on line 5 the amount shown below that applies to you.
	• Single, enter 7,450.00. This is the total of your standard deduction (4,550.00) and your exemption (2,900.00).
	• Married, enter 13,400.00. This is the total of your standard deduction (7,600.00), your exemption (2,900.00), and your spouse's exemption (2,900.00).

Mailing return	Mail your return by **April 15, 20– –.** Use the envelope that came with your booklet. If you do not have that envelope, see the back cover for the address to use.

Form **1040EZ** (20– –)

Tax Computation Worksheet for Certain Dependents—Line 11

1. Figure the tax on the amount on Form 1040EZ, line 6.
   Use the Tax Table. . . . . **1.** _____

2. Is the amount on line 1 more than the amount shown below for your filing status?
   • Single — $900
   • Married filing jointly — $1,800
   ☐ **Yes.** If **single,** enter $300; if **married,** enter $600
   ☐ **No.** Divide the amount on line 1 by 3.0.   . . . . **2.** _____

3. Subtract line 2 from line 1. Enter the result here and
   on Form 1040EZ, line 11. . . . . **3.** _____

Problem E-2   Preparing Form 1040EZ

| Form **1040EZ** | Department of the Treasury–Internal Revenue Service **Income Tax Return for Single and Joint Filers With No Dependents** (P) **20--** | OMB No. 1545-0675 |

Label (See Page 12.)

Use the IRS label. Otherwise please print or type

L A B E L — Your first name and initial / Last name

If a joint return, spouse's first name and initial / Last name

H E R E — Home address (number and street). If you have a P.O. box, see page 7. Apt. no.

City, town or post office, state, and ZIP code. If you have a foreign address, see page 7.

Your social security number

Spouse's social security number

▲ **Important!** ▲ you **must** enter your SSN(s) above.

Presidential Election Campaign (See page 12.)

Note: *Checking "Yes" will not change your tax or reduce your refund.*
Do you, or spouse if a joint return, want $3 to go to this fund? ▶

| | You | Spouse |
| | ☐ Yes ☐ No | ☐ Yes ☐ No |

Income

Attach Form(s) W-2 here. Enclose, but do not attach, any payment.

Note. You **must** check Yes or No

1 Total wages, salaries, and tips. This should be shown in box 1 of your W-2 form(s). Attach your W-2 form(s). — 1

2 Taxable interest. If the total is over $400, you cannot use Form 1040EZ — 2

3 Unemployment compensation, qualified state tuition program earnings and Alaska Permanent Fund dividends (see page 14). — 3

4 Add lines 1, 2, and 3. This is your **adjusted gross income.** — 4

5 Can your parents (or someone else) claim you on their return?
Yes. ☐ Enter amount from worksheet on back.
No. ☐ If **single,** enter 7,450.00. If **married,** enter 13,400.00 See back for explanation — 5

6 Subtract line 5 from line 4. If line 5 is larger than line 4, enter 0. This is your **taxable income.** ▶ 6

Credits, payments, and tax

7 Rate reduction credit. See worksheet on page 14. — 7

8 Enter your Federal income tax withheld from box 2 of your W-2 form(s). — 8

9a **Earned income credit (EIC).** See page 15. — 9a

b Nontaxable earned income. — 9b

10 Add lines 7, 8, and 9a. These are your **total credits and payments.** ▶ 10

11 **Tax.** If you checked "Yes" on line 5, see page 20. Otherwise, use the amount on **line 6 above** to find your tax in the tax table on pages 24–28 of the booklet. Then, enter the tax from the table on this line. — 11

Refund Have it directly deposited! See Page 20 and fill in 12b, 12c, and 12d

12a If line 10 is larger than line 10, subtract line 10 from line 11. This is your **refund.** ▶ 12a

▶ b Routing number ☐☐☐☐☐☐☐☐☐ ▶ c Type: ☐ Checking ☐ Savings

▶ d Account number ☐☐☐☐☐☐☐☐☐☐☐

Amount you owe

13 If line 11 is larger than line 10, subtract line 10 from line 11. This is **the amount you owe.** See page 21 for details on how to pay. ▶ 13

Third party desgnee

Do you want to allow another person to discuss this return with the IRS (see page 22)? ☐ **Yes.** Complete the following. ☐ **No**
Designee's name ▶ | Phone no. ▶ ( ) | Personal identification number(PIN) ☐☐☐☐☐

Sign here
Joint return? See page 11. Keep a copy for your records

Under penalties of perjury, I declare that I have examined this return, and to the best of my knowledge and belief, it is true, correct, and accurately lists all amounts and sources of income I received during the tax year. Declaration of preparer (other than the taxpayer) is based on all information of which the preparer has any knowledge.

Your signature | Date | Your occupation | Daytime phone number ( )

Spouse's signature if a joint return, **both** must sign. | Date | Spouse's occupation

Paid Preparer's use only

Preparer's signature ▶ | Date | Check if self-employed ☐ | Preparer's SSN or PTIN

Firm's name (or yours if self-employed), address, and ZIP code ▶ | EIN | Phone no. ( )

For Disclosure, Privacy Act, and Paperwork Reduction Act Notice, see page 23 | Cat. No. 12616G | Form **1040EZ** 20--

Problem E-2 (concluded)

Filling in your return For tips on how to avoid common mistakes, see page 30	If you received a scholarship or fellowship grant or tax–exempt interest income, such as on municipal bonds, see the booklet before filling in the form. Also, see the booklet if you received a Form 1099–INT showing Federal income tax withheld or if Federal income tax was withheld from your unemployment compensation or Alaska Permanent Fund dividends. **Remember,** you must report all wages, salaries, and tips even if you do not get a W–2 form from your employer. You must also report all your taxable interest, including interest from banks, savings and loans, credit unions, etc., even if you do not get a Form 1099–INT.
Worksheet for dependents who checked "Yes" on line 5 (keep a copy for your records)	Use this worksheet to figure the amount to enter on line 5 if someone can claim you (or your spouse if married) as a dependent, even if that person chooses not to do so. To find out if someone can claim you as a dependent, use TeleTax topic 354 (see page 6). **A.** Amount, if any, from line 1 on front . . . . + 250.00 Enter total ▶ **A.**_____ **B.** Minimum standard deduction. . . . . **B.**___750.00___ **C.** Enter the **larger** of line A or line B here. . . . . **C.**_____ **D.** Maximum standard deduction. If **single,** enter 4,550.00; if **married,** enter 7,600.00. . . . . **D.**_____ **E.** Enter the **smaller** of line C or line D here. This is your standard deduction. . . . . **E.**_____ **F.** Exemption amount. 　•If single, enter 0 　•If married and— 　　—both you and your spouse can be claimed as dependents, enter 0. 　　—only one of you can be claimed as a dependent, enter 2,900.00　**F.**_____ **G.** Add lines E and F. Enter the total here and on line 5 on the front. . . . . **G.**_____ **If you checked "No" on line 5** because no one can claim you (or your spouse if married) as a dependent, enter on line 5 the amount shown below that applies to you. 　• Single, enter 7,450.00. This is the total of your standard deduction (4,550.00) and your exemption (2,900.00). 　• Married, enter 13,400.00. This is the total of your standard deduction (7,600.00), your exemption (2,900.00), and your spouse's exemption (2,900.00).
Mailing return	Mail your return by **April 15, 20– –.** Use the envelope that came with your booklet. If you do not have that envelope, see the back cover for the address to use.

Form **1040EZ** (20– –)

Tax Computation Worksheet for Certain Dependents—Line 11

1. Figure the tax on the amount on Form 1040EZ, line 6.
   Use the Tax Table. . . . . **1.**_____

2. Is the amount on line 1 more than the amount shown below for your filing status?
   • Single — $900
   • Married filing jointly — $1,800
   ☐ **Yes.** If **single,** enter $300; if **married,** enter $600 ⎫
   ☐ **No.** Divide the amount on line 1 by 3.0.　　　　　　 ⎬ . . . . **2.**_____
   　　　　　　　　　　　　　　　　　　　　　　　　　 ⎭

3. Subtract line 2 from line 1. Enter the result here and
   on Form 1040EZ, line 11. . . . . **3.**_____

APPENDIX F · Computerized Accounting Using Peachtree

Setting Up a New Company

Software Objectives

When you have completed this appendix, you will be able to use Peachtree to:

1. Create a new company.
2. Enter and edit a chart of accounts.
3. Record customer and vendor data.
4. Enter beginning balances.
5. Set the accounting options.

OVERVIEW

Peachtree Complete® Accounting guides you step-by-step through the process of creating a new company. Follow the steps below to create a new company, enter a chart of accounts, prepare customer/vendor records, and enter account balances.

Before you start the Peachtree software, gather all of the information you will need to set up your new company. Depending on the type of business you plan to create, you will need some or all of the following:

- Chart of Accounts
- Customer List
- Vendor List
- General Ledger, Customer Account Balances, and Vendor Account Balances
- Accounting Periods (first accounting period month and fiscal year range)

Note: You should be familiar with the Peachtree Complete Accounting software before attempting to create a new company. For example, you should be able to create a chart of accounts and enter customer and vendor records.

GETTING STARTED

Step 1   Start the Peachtree Complete Accounting software and choose either *Set up a new company* from the Peachtree Accounting window or select **New Company** from the *File* menu. The New Company Setup – Introduction window (as shown in Figure F1) will appear on your screen.

   If you are working with another company, the program will display the message *This will close the current company!* telling you that creating a new company will close the current open company. Make sure your have saved all your work in the company that will be closed. Press **OK** to continue.

Figure F1 *New Company Setup–Introduction Window*

Step 2   Read the information provided in the welcome screen.

Step 3   Click the **Next** button to move to the next screen and then review the information fields.

SETTING UP A NEW COMPANY: PART ONE

Step 4   Enter the company information, such as company name, address, telephone and fax number, business type (corporation, S corporation, partnership, sole proprietorship, or limited liability company), tax ID numbers, Web site, and E-mail (see Figure F2).

   The steps given in this appendix assume that you are setting up a corporation. Some of the steps may vary slightly from those shown here if you are creating a company of a different business type.

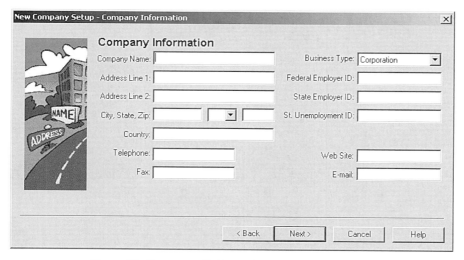

Figure F2 *Company Information Window*

Note: Don't worry if you can't complete the new company setup in one session. Once you enter the company name, chart of accounts, accounting method, posting method, accounting periods and defaults, you can quit the setup program at any time. The next time you choose to work on this company, the Setup Checklist will allow you to continue where you left off. To display the *Setup Checklist,* select **Setup Checklist** from the ***Maintain*** menu.

Step 5   Review the information you entered. Click the **Next** button when you are ready to move to the next screen.

Step 6   Review the Chart of Accounts information, and then choose the *Build your own company* option. (See Figure F3.)

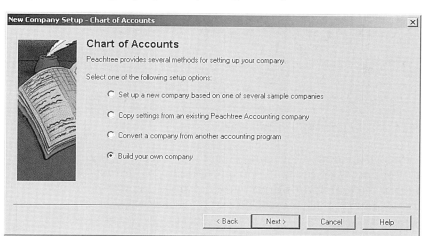

Figure F3   *Chart of Accounts Window*

You can choose from four different options to create a chart of accounts. If you were setting up a real business and did not have an existing chart of accounts, Peachtree provides many sample companies from which to choose a chart of accounts. Choose the first option, *Set up a new company based on one of several sample companies,* if you are setting up a new company and you don't have a chart of accounts.

Sometimes it's easier to copy a chart of accounts from an existing company and then update the chart of accounts. For example, you could copy a chart of accounts from one of the companies in the *Glencoe Accounting* text. In these instances, choose the *Copy settings from an existing Peachtree Accounting company* option. The setup program lets you copy the chart of accounts from any of the companies already installed on your computer.

If you were using a different accounting program, such as QuickBooks®, the setup program helps you convert those files to a Peachtree file format. These options can be real time savers if you are switching from another program to Peachtree.

Step 7   Click the **Next** button to move to the accounting method screen, and review the accrual and cash information.

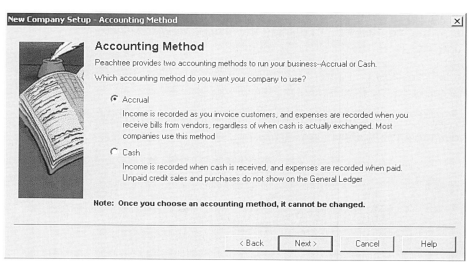

Figure F4 *Accounting Method Window*

Peachtree lets you set up a business using either the accrual or the cash accounting method. All of the concepts presented in your textbook are based on the accrual accounting method. Therefore, the steps presented here assume that you will be using the accrual method for your new business.

Step 8 Set the accounting method to *Accrual* (see Figure F4), and then click the **Next** button.

Step 9 Read the information that explains the two different processing methods—real time and batch. Choose the *Real Time* option to have Peachtree automatically update your accounting records each time you enter and post a transaction. (See Figure F5.)

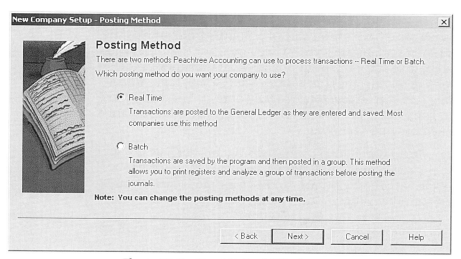

Figure F5 *Posting Method Window*

Step 10 Click **Next** to review the information that describes how Peachtree uses the accounting periods to process transactions and generate reports.

Step 11 Select the *12 monthly accounting periods* option. (See Figure F6.)

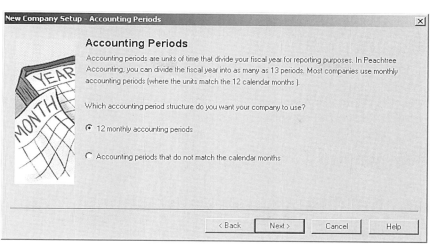

Figure F6 *Accounting Periods Window*

For most businesses, you should select the 12 monthly accounting periods—one period for each month.

Step 12 Click **Next** to continue setting up the accounting periods.

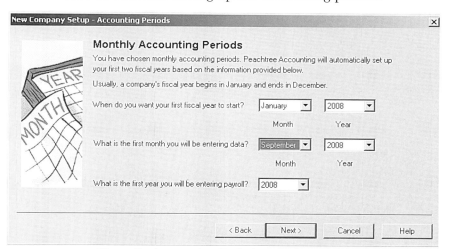

Figure F7 *Accounting Periods Window*

Step 13 Enter the start of your fiscal year.

Most businesses start the fiscal year in January, but the fiscal year can vary from business to business.

Step 14 Specify the first month in which data will be entered.

If you specified 12 monthly accounting periods and you will be entering data in September, enter September 2008 for the first period. (See Figure F7.)

Step 15 Enter the first year in which you will enter payroll.

Step 16 Click **Next** to continue, and select *Yes, I would like to follow the Setup Checklist.*

Step 17 Click the **Finish** button to create your new company.

     Appendix F ■ 905

The setup program will create the data files for your new company. Be patient as this process may take a few moments. The Creating Data Files dialog window will display status of completion.

Note: To turn off *Peachtree Today* click the **Close** button on the Setup Checklist. Then click on the *Preferences* tab, deselect "Display Peachtree Today each time this company is opened," click the **Save and Exit** button on the lower left corner of your screen, and then close the Peachtree Today window.

SETTING UP A NEW COMPANY: PART TWO

General Ledger

Step 18   Review the Setup Checklist that Peachtree displays after it creates the company data files. If you closed the Setup Checklist window in order to turn off *Peachtree Today,* select **Setup Checklist** from the **Maintain** menu to display the checklist. (See Figure F8.)

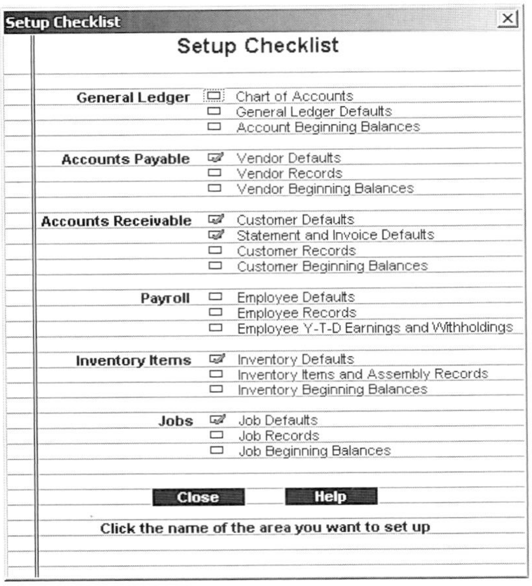

Figure F8 *Setup Checklist*

As you can see, there are quite a few items on the Setup Checklist. The items that you must complete depend on the company you are creating. For every new company, you **must** complete the Chart of Accounts and General Ledger Defaults on the General Ledger section of the Setup Checklist. If you plan on using subsidiary ledger accounts for customers and vendors, you must also complete the Accounts Payable and Accounts Receivable items. You will have to enter beginning balances too, unless you are beginning a new company from scratch.

Note: When the Setup Checklist displays you will notice that several tasks already have checkmarks next to the task name. Peachtree enters some default information automatically. You can go into these areas later and make any necessary changes. Refer to the software Help Index to learn more about setting up the payroll, inventory, or jobs modules. This information is not provided here.

Step 19   Choose the *Chart of Accounts* task on the Setup Checklist to begin entering the chart of accounts for your new company. Click on the words "Chart of Accounts." Do **not** click on the check box.

When you choose any of the checklist tasks, such as the option to enter a chart of accounts, the Peachtree program automatically opens the appropriate task window for you. You could also select the **Chart of Accounts** option from the *Maintain* menu.

Step 20   Enter the chart of accounts for your new company. (See the Maintain Chart of Accounts window in Figure F9.)

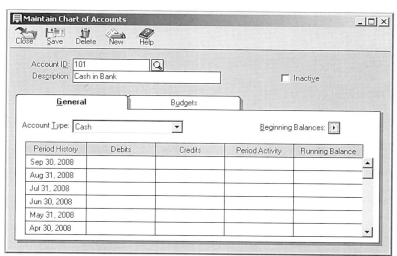

Figure F9 *Maintain Chart of Accounts Window*

Step 21   Click the **Close** button when you finish entering the chart of accounts.

Step 22   Click **Yes** whenever Peachtree asks if you want to mark a task as completed. Click the **No** button only if you did not complete the task. Peachtree updates the checklist to show you which items you have completed. **Note:** If you need to interrupt the second part of the setup process, click the **Close** button shown in the Setup Checklist window. When you begin the next session, select **Setup Checklist** from the *Maintain* menu to display the checklist.

Step 23   Choose the *General Ledger Defaults* task on the Setup Checklist.

Step 24   Enter the rounding account. Use the general ledger account you plan to assign to the *Retained Earnings* account. (Use the *Capital* account if you were setting up a sole proprietorship.) Click **OK**. **Note:** Click the **Help** button if you want to learn more about a particular option such as the rounding account.

Step 25   If your company is a new business, you do not have to enter beginning balances. Skip to Step 28 if you need to set up customer and vendor accounts. If you do not plan to have Accounts Receivable and Accounts Payable subsidiary ledger accounts, you are now finished with the setup process. Click the **Close** button in the Setup Checklist window and go to Step 46.

Step 26   Choose the *Account Beginning Balances* task on the Setup Checklist and follow these instructions to enter the beginning balances.

- Select the period to enter beginning balances.

    Choose the first period in which you will be entering transactions. For example, if you will be entering transactions starting in July, choose "From 7/1/08 through 7/31/08." Make sure that you choose a period for the current year. You will probably have to scroll down past several other periods to get to the current period. Select the period you want and click **OK**.

- Enter the beginning balances. (See Figure F10.) Click **OK** when you finish.

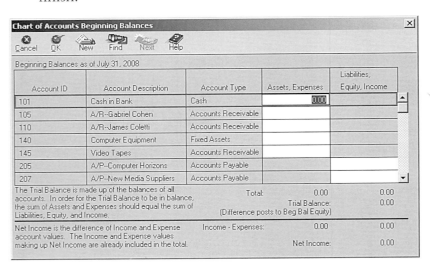

Figure F10 *Chart of Accounts Beginning Balances Window*

Step 27 Continue with Step 28 if you need to set up the Accounts Payable and Accounts Receivable modules. Otherwise, click the **Close** button on the Setup Checklist and then go to Step 46. You are now finished setting up the new company.

Accounts Payable

Step 28 Choose the *Vendor Defaults* task on the Setup Checklist. Review the information shown in the Vendor Defaults window shown in Figure F11. Peachtree automatically provides default information which you can change as needed.

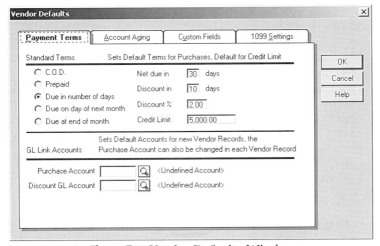

Figure F11 *Vendor Defaults Window*

Step 29 Enter the standard payment terms that are offered by most of the company's vendors.

For example, choose *Due in number of days* if most vendors offer a discount. You will also need to enter the number of days (Net due in and Discount in), discount percent (e.g., 2.00), and the credit limit. You can always change the terms offered by a particular vendor. The settings you enter here are the default settings.

Step 30 Enter the **Purchase Account** and **Discount GL Account** numbers. Peachtree needs these account numbers to integrate the Accounts Payable module with the General Ledger.

Step 31 You should not have to enter any account aging, custom fields, or 1099 settings. Click the **OK** button to accept the payment terms and choose to mark this item as completed.

Step 32 Choose the *Vendor Records* task on the Setup Checklist. Peachtree displays the Maintain Vendors window.

Step 33 Enter the vendor records. Complete as many of the fields as possible on the *General* tab. (See Figure F12.) Then, click the *Purchase Defaults* tab. Enter the default purchases account number and change the terms if this vendor offers terms different from the default terms you already set up.

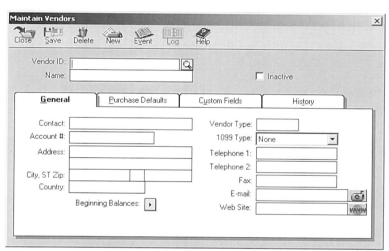

Figure F12 *Maintain Vendors Window*

Step 34 Click the **Close** button when you finish entering and saving the vendor records. Indicate that you want to mark this item as completed.

Step 35 Choose the *Vendor Beginning Balances* task on the Setup Checklist if you need to enter outstanding invoices. Otherwise, skip to Step 37.

Step 36 Select each vendor (shown in the Vendor Beginning Balances window) that has outstanding invoices. (See Figure F13.) Double click on the name to enter information. Enter the beginning balance information, then **Save** and close the window.

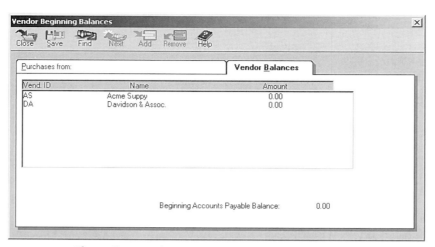

Figure F13 *Vendor Beginning Balances Window*

Accounts Receivable

Step 37 Choose the *Customer Defaults* task on the Setup Checklist.

Step 38 Enter the default payment terms, **GL Sales Account**, and **Discount GL Account**. Then, click the **OK** button to record the information.

> **Note:** If necessary, enter the account aging, custom fields, finance charges, and pay methods information.

Step 39 Choose the *Statement and Invoice Defaults* task on the Setup Checklist. Review the information in the Statement and Invoice Defaults window. You should not have to change any of these settings. Click **OK**.

Step 40 Choose the *Customer Records* task on the Setup Checklist. Peachtree displays the Maintain Customers/Prospects window.

Step 41 Enter the customer records. Complete as many of the fields as possible on the *General* tab. (See Figure F14.) Then, click the *Sales Defaults* tab. Verify the default sales account. Change the terms if the customer receives terms different from the default terms you already set up.

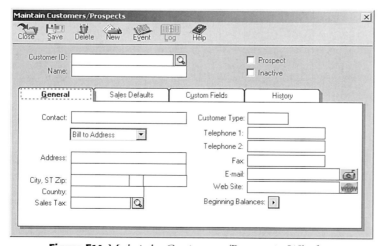

Figure F14 *Maintain Customers/Prospects Window*

Step 42   Click the **Close** button when you finish entering the customer records. Indicate that you want to mark this item as completed.

Step 43   Choose the *Customer Beginning Balances* task on the Setup Checklist if you need to enter outstanding invoices. Otherwise, skip to Step 45.

Step 44   Select each customer (shown in the Customer Beginning Balances window) that has outstanding invoices. Enter the beginning balance information. When you finish, save and close the window.

Step 45   You have completed all the required tasks on the checklist. (See Figure F15.) Peachtree automatically provides default information for *Inventory Defaults* and *Job Defaults,* so you are not required to zset up these tasks. Just a few more steps to go. Click **Close** to leave the Setup Checklist.

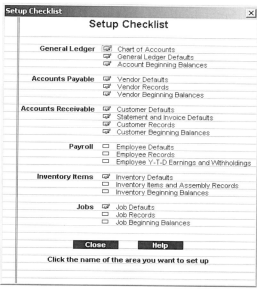

Figure F15 *Completed Setup Checklist*

Step 46   If you are creating a merchandising business, you need to set up the sales tax settings. If you are setting up a service business, you do not have to set up the sales tax setting. Go to Step 48.

Sales Tax Authorities

Step 47   Set up the sales tax settings.

- Click the ***Maintain*** menu, choose **Sales Taxes**, and then select **Sales Tax Authorities**.
- Enter a state sales tax authority and an exempt authority. You must enter an ID, description, tax rate, and **Sales Tax Payable GL Account** for both tax authorities. (See Figure F16 for an example tax authority.)
- Save, then close the Maintain Sales Tax Authorities window when you finish.

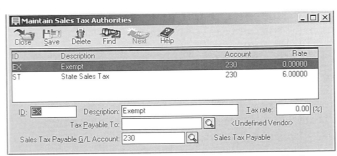

Figure F16 *Maintain Sales Tax Authorities Window*

Sales Tax Codes

- Click the *Maintain* menu, choose **Sales Taxes**, and then select **Sales Tax Codes**.
- Enter a sales tax code and an exempt code. You must enter a code and a description for each one. Then, complete the process by identifying the tax authority for each code. After entering the sale tax code and description, move your cursor to the blank field under the ID column and right-click the mouse to display the sales tax authorities that you already set up.

   Select the sales tax authority, click **OK**, then **Save** and close the window when you finish. (See Figure F17 for an example sales tax code.)

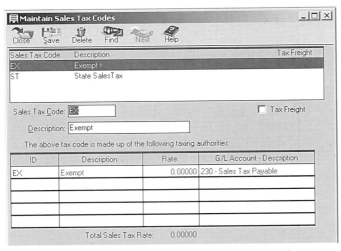

Figure F17 *Maintain Sales Tax Codes Window*

Step 48 **Important:** Back up your work to a floppy disk before you process any transactions. Once you enter any transactions, you cannot change the beginning balances.

Step 49 **Congratulations!** You can now begin working with and using the company data files you just set up.

APPENDIX  Additional Reinforcement Problems

Reinforcement Problem 3A   Determining the Effects of Business Transactions on the Accounting Equation

| Transaction | Assets | | | | | = | Liabilities | + | Owner's Equity |
	Cash in Bank	Accounts Receivable	Office Furniture	Computer Equipment	Office Equipment	=	Accounts Payable	+	Pamela Wong, Capital
1									
Balance						=		+	
2									
Balance						=		+	
3									
Balance						=		+	
4									
Balance						=		+	
5									
Balance						=		+	
6									
Balance						=		+	
7									
Balance						=		+	
8									
Balance						=		+	
9									
Balance						=		+	
10									
Balance						=		+	
11									
Balance						=		+	

Reinforcement Problem 3A   (continued)

Analyze: a. _____

         b. _____

         c. _____

         d. _____

Computerized Accounting Using Spreadsheets

Problem 3A   Determining the Effects of Business Transactions on the Accounting Equation

Completing the Spreadsheet

Step 1   Read the instructions for Additional Reinforcement Problem 3A in your textbook. This problem involves entering transactions in the accounting equation.

Step 2   Open the Glencoe Accounting: Electronic Learning Center software.

Step 3   From the Program Menu, click on the **Peachtree Complete®
Accounting Software and Spreadsheet Applications** icon.

Step 4   Log onto the Management System by typing your user name and password.

Step 5   Under the **Problems & Tutorials** tab, select template 3A from the drop-down menu. The template should look like the one shown below.

```
PROBLEM 3A
DETERMINING THE EFFECTS OF BUSINESS
TRANSACTIONS ON THE ACCOUNTING EQUATION

(name)
(date)
```

		ASSETS				LIABILITIES	OWNER'S EQUITY
Transaction	Cash in Bank	Accounts Receivable	Office Furniture	Computer Equipment	Office Equipment =	Accounts Payable +	Pamela Wong, Capital
1							
2							
3							
4							
5							
6							
7							
8							
9							
10							
11							
BALANCE	$0	$0	$0	$0	$0	$0	$0

```
TOTAL ASSETS                        $0

TOTAL LIABILITIES                   $0
TOTAL OWNER'S EQUITY                $0
TOTAL LIABILITIES + OWNER'S EQUITY  $0
```

Step 6   Key your name and today's date in the cells containing the *(name)* and *(date)* placeholders.

Step 7   In the first transaction, Ms. Wong, the owner of the business, opened a checking account for the business. Remember from Chapter 2 that two parts of the accounting equation are affected by this transaction: Cash in Bank and Pamela Wong, Capital. Cash in Bank is increasing, and Pamela Wong, Capital, is increasing. To record this transaction in the spreadsheet template, move the cell pointer to cell B12 and enter **48000**.

Appendix G ■ 915

TIP: Remember, to enter data into the cell, you must first key the data and then press **Enter**. Do *not* enter a dollar sign or a comma when you enter the data.

Step 8 Next, move the cell pointer to cell J12. Enter **48000** in cell J12 to record the increase in Pamela Wong, Capital. Do *not* include a dollar sign or a comma as part of the cell entry—the spreadsheet template will automatically format the data when it is entered. Move the cell pointer to cell J24. Notice that the spreadsheet automatically calculates the balance in each account as you enter the data.

Step 9 To check your work, look at rows 27 through 31 in column D. Total assets equal $48,000. Total liabilities plus owner's equity also equal $48,000. The accounting equation is in balance.

Step 10 Analyze the remaining transactions in Problem 3A and enter the appropriate data into the spreadsheet template.

TIP: To decrease an account balance, precede the amount entered by a minus sign. For example, to decrease Cash in Bank by $1,500, enter **–1500** in the Cash in Bank column.

Check the totals at the bottom of the spreadsheet after each transaction has been entered. Remember, total assets should always equal total liabilities plus owner's equity.

Step 11 Save the spreadsheet using the **Save** option from the *File* menu. You should accept the default location for the save as this is handled by the management system.

Step 12 Print the completed spreadsheet.

TIP: If your spreadsheet is too wide to fit on an 8.5-inch wide piece of paper, you can change your print settings to print the worksheet *landscape.* Landscape means that the worksheet will be printed broadside on the page.

Step 13 Exit the spreadsheet program.

Step 14 In the Close Options box, select the location where you would like to save your work.

Step 15 Answer the Analyze question from your textbook.

What-If Analysis

TIP: Always save your work before performing What-If Analysis. It is not necessary to save your work after performing What-If Analysis unless your teacher instructs you to do so.

If Ms. Wong paid $550 cash for a new printer for the business, what would the Cash in Bank balance be?

Reinforcement Problem 4A

(1), (2), (3)

Analyzing Transactions Affecting Assets, Liabilities, and Owner's Equity

(4) Sum of debit balances: _____

(5) Sum of credit balances: _____

Analyze: _____

Reinforcement Problem 5A   Analyzing Transactions Affecting Revenue, Expenses, and Withdrawals

Chart of Accounts (partial)

Cash in Bank

Accounts Receivable—
    Adams, Bell, and Cox, Inc.

Office Equipment

Computer Equipment

Accounts Payable—
    Computer Warehouse, Inc.

Pamela Wong, Capital

Pamela Wong, Withdrawals

Design Revenue

Maintenance Expense

Rent Expense

Utilities Expense

Reinforcement Problem 5A   (concluded)

(4)

Account Name	Debit Balances	Credit Balances

Analyze: _____

_____

Reinforcement Problem 6A   **Recording General Journal Transactions**

GENERAL JOURNAL   PAGE _____

	DATE	DESCRIPTION	POST. REF.	DEBIT	CREDIT	
1						1
2						2
3						3
4						4
5						5
6						6
7						7
8						8
9						9
10						10
11						11
12						12
13						13
14						14
15						15
16						16
17						17
18						18
19						19
20						20
21						21
22						22
23						23
24						24
25						25
26						26
27						27
28						28
29						29
30						30
31						31
32						32
33						33
34						34
35						35

Analyze: _____

Reinforcement Problem 7A Journalizing and Posting Transactions
(1), (3)

GENERAL LEDGER

ACCOUNT _____ ACCOUNT NO. _____

DATE	DESCRIPTION	POST. REF.	DEBIT	CREDIT	BALANCE	
					DEBIT	CREDIT

ACCOUNT _____ ACCOUNT NO. _____

DATE	DESCRIPTION	POST. REF.	DEBIT	CREDIT	BALANCE	
					DEBIT	CREDIT

ACCOUNT _____ ACCOUNT NO. _____

DATE	DESCRIPTION	POST. REF.	DEBIT	CREDIT	BALANCE	
					DEBIT	CREDIT

ACCOUNT _____ ACCOUNT NO. _____

DATE	DESCRIPTION	POST. REF.	DEBIT	CREDIT	BALANCE	
					DEBIT	CREDIT

ACCOUNT _____ ACCOUNT NO. _____

DATE	DESCRIPTION	POST. REF.	DEBIT	CREDIT	BALANCE	
					DEBIT	CREDIT

Reinforcement Problem 7A (continued)

ACCOUNT _____ ACCOUNT NO. _____

	DATE	DESCRIPTION	POST. REF.	DEBIT	CREDIT	BALANCE	
						DEBIT	CREDIT

ACCOUNT _____ ACCOUNT NO. _____

	DATE	DESCRIPTION	POST. REF.	DEBIT	CREDIT	BALANCE	
						DEBIT	CREDIT

ACCOUNT _____ ACCOUNT NO. _____

	DATE	DESCRIPTION	POST. REF.	DEBIT	CREDIT	BALANCE	
						DEBIT	CREDIT

ACCOUNT _____ ACCOUNT NO. _____

	DATE	DESCRIPTION	POST. REF.	DEBIT	CREDIT	BALANCE	
						DEBIT	CREDIT

ACCOUNT _____ ACCOUNT NO. _____

	DATE	DESCRIPTION	POST. REF.	DEBIT	CREDIT	BALANCE	
						DEBIT	CREDIT

ACCOUNT _____ ACCOUNT NO. _____

	DATE	DESCRIPTION	POST. REF.	DEBIT	CREDIT	BALANCE	
						DEBIT	CREDIT

Reinforcement Problem 7A (continued)

(2)

GENERAL JOURNAL   PAGE _____

	DATE	DESCRIPTION	POST. REF.	DEBIT	CREDIT	
1						1
2						2
3						3
4						4
5						5
6						6
7						7
8						8
9						9
10						10
11						11
12						12
13						13
14						14
15						15
16						16
17						17
18						18
19						19
20						20
21						21
22						22
23						23
24						24
25						25
26						26
27						27
28						28
29						29
30						30
31						31
32						32
33						33
34						34
35						35

Reinforcement Problem 7A   (concluded)

(4)

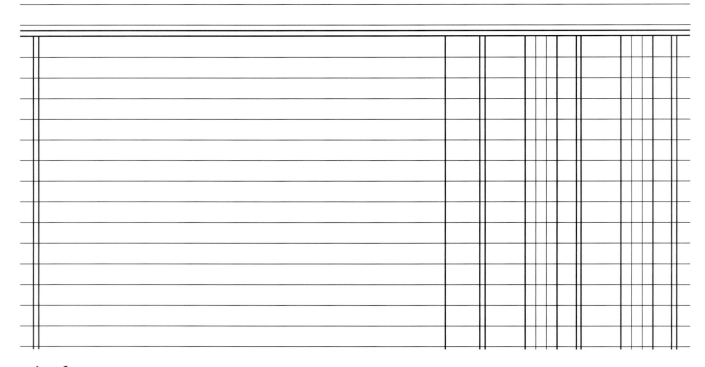

Analyze: _____

Reinforcement Problem 8A   Preparing a Six-Column Work Sheet

ACCT. NO.	ACCOUNT NAME	TRIAL BALANCE		INCOME STATEMENT		BALANCE SHEET	
		DEBIT	CREDIT	DEBIT	CREDIT	DEBIT	CREDIT

Reinforcement Problem 8A (concluded)

Analyze: _____

_____

_____

_____

Reinforcement Problem 9A Interpreting Financial Information

_____

_____

_____

_____

_____

_____

_____

_____

_____

_____

_____

_____

_____

_____

_____

_____

Reinforcement Problem 10A   Preparing Closing Entries

Thunder Graphics Desktop Publishing

Work Sheet

For the Year Ended December 31, 20--

ACCT. NO.	ACCOUNT NAME	TRIAL BALANCE DEBIT	TRIAL BALANCE CREDIT	INCOME STATEMENT DEBIT	INCOME STATEMENT CREDIT	BALANCE SHEET DEBIT	BALANCE SHEET CREDIT
101	Cash in Liberty State Bank	340000				340000	
105	Accts. Rec.—Adams, Bell, and Cox	140000				140000	
110	Accts. Rec.—Roger McFall	40000				40000	
113	Accts. Rec.—Designers Boutique	120000				120000	
115	Accts. Rec.—Pat Cooper	60000				60000	
120	Office Equipment	300000				300000	
125	Office Furniture	930000				930000	
130	Computer Equipment	2100000				2100000	
201	Accts. Pay.—Solutions Software		210000				210000
205	Accts. Pay.—Pro Internet Service		120000				120000
207	Accts. Pay.—Computer Warehouse		1700000				1700000
301	Pamela Wong, Capital		1430000				1430000
305	Pamela Wong, Withdrawals	800000				800000	
310	Income Summary						
401	Design Revenue		2400000		2400000		
405	Print Production Revenue		1500000		1500000		
501	Rent Expense	1200000		1200000			
515	Maintenance Expense	320000		320000			
520	Advertising Expense	200000		200000			
525	Utilities Expense	480000		480000			
540	Office Supplies Expense	180000		180000			
545	Miscellaneous Expense	150000		150000			
		7360000	7360000	2530000	3900000	4830000	3460000
	Net Income			1370000			1370000
				3900000	3900000	4830000	4830000

Reinforcement Problem 10A (concluded)

GENERAL JOURNAL

PAGE _____

	DATE	DESCRIPTION	POST. REF.	DEBIT	CREDIT	
1						1
2						2
3						3
4						4
5						5
6						6
7						7
8						8
9						9
10						10
11						11
12						12
13						13
14						14
15						15
16						16
17						17
18						18
19						19
20						20
21						21
22						22
23						23
24						24
25						25
26						26
27						27
28						28
29						29
30						30

Analyze: _____

_____

_____

Reinforcement Problem 11A   Recording Deposits in the Checkbook

(1)

	DOLLARS	CENTS
CASH		
CHECKS (List Singly)		
1		
2		
3		
4		
5		
6		
7		
8		
TOTAL		

Date _____ 20 _____

Checks and other items are received for deposit subject to the terms and conditions of this bank's collection agreement.

BE SURE EACH ITEM IS ENDORSED

P *First Pacific Bank*

⑆0650 0⑈334⑆ 27 ⑈0749⑈

(2), (3)

$ _____ No. 1068
Date _____ 20 ___
To _____
For _____

	Dollars	Cents
Balance brought forward		
Add deposits		
Total		
Less this check		
Balance carried forward		

THUNDER GRAPHICS
desktop publishing

1068

51-160
111

DATE _____ 20 ____

PAY TO THE
ORDER OF _____ $ _____

_____ DOLLARS

P *First Pacific Bank*

MEMO _____ _____

⑆0111 01602⑆ 749 2454⑈ 1068

(4)

$ _____ No. 1069
Date _____ 20 ___
To _____
For _____

	Dollars	Cents
Balance brought forward		
Add deposits		
Total		
Less this check		
Balance carried forward		

Reinforcement Problem 11A   (concluded)

(5)

GENERAL JOURNAL   PAGE ___*21*___

	DATE	DESCRIPTION	POST. REF.	DEBIT	CREDIT	
1						1
2						2
3						3
4						4
5						5
6						6
7						7
8						8
9						9
10						10

(6)

     a. _____

     b. _____

     c. _____

     d. _____

     e. _____

Analyze: _____

_____

Reinforcement Problem 12A   Preparing a Payroll Register

PAYROLL REGISTER

PAY PERIOD ENDING _____ 20 ____   DATE OF PAYMENT _____

EMPLOYEE NUMBER	NAME	MAR. STATUS	ALLOW.	TOTAL HOURS	RATE	EARNINGS			DEDUCTIONS							NET PAY	CK. NO.
						REGULAR	OVERTIME	TOTAL	SOC. SEC. TAX	MED. TAX	FED. INC. TAX	STATE INC. TAX	HOSP. INS.	OTHER	TOTAL		
1																	
2																	
3																	
4																	
25 TOTALS																	

Other Deductions: Write the appropriate code letter to the left of the amount: B—U.S. Savings Bonds; C—Credit Union; UD—Union Dues; UW—United Way.

Analyze: _____

Reinforcement Problem 13A   Recording Payroll Transactions

PAYROLL REGISTER

PAY PERIOD ENDING *December 17*   20 --   DATE OF PAYMENT *December 17, 20--*

EMPLOYEE NUMBER	NAME	MAR. STATUS	ALLOW.	TOTAL HOURS	RATE	EARNINGS REGULAR	OVERTIME	TOTAL	DEDUCTIONS SOC. SEC. TAX	MED. TAX	FED. INC. TAX	STATE INC. TAX	HOSP. INS.	OTHER	TOTAL	NET PAY	CK. NO.
173	Don Hoffman	M	1	22	6.95	152 90		152 90	9 48	2 22	24 00	2 29	7 85		45 84	107 06	
168	Manual Gongas	S	0	36	7.10	255 60		255 60	15 85	3 71	39 00	3 83	4 75	UD 3 25	70 39	185 21	
167	Riley Sullivan	M	2	40	7.40	296 00		296 00	18 35	4 29	43 00	4 44	7 85	UD 3 25	81 18	214 82	
175	Marcy Jackson	S	1	38	6.95	264 10		264 10	16 37	3 83	39 00	3 96	4 75	UD 3 25	71 16	192 94	
	TOTALS					968 60		968 60	60 05	14 05	145 00	14 52	25 20	9 75	268 57	700 03	

Other Deductions: Write the appropriate code letter to the left of the amount: B—U.S. Savings Bonds; C—Credit Union; UD—Union Dues; UW—United Way.

Reinforcement Problem 13A (concluded)

GENERAL JOURNAL   PAGE _____

	DATE	DESCRIPTION	POST. REF.	DEBIT	CREDIT	
1						1
2						2
3						3
4						4
5						5
6						6
7						7
8						8
9						9
10						10
11						11
12						12
13						13
14						14
15						15
16						16
17						17
18						18
19						19
20						20
21						21
22						22
23						23
24						24
25						25
26						26
27						27
28						28
29						29
30						30

Analyze: _____

_____

Reinforcement Problem 14A Recording and Posting Sales and Cash Receipt Transactions

(1), (4)

GENERAL LEDGER

ACCOUNT _____ ACCOUNT NO. _____

DATE	DESCRIPTION	POST. REF.	DEBIT	CREDIT	BALANCE DEBIT	BALANCE CREDIT

ACCOUNT _____ ACCOUNT NO. _____

DATE	DESCRIPTION	POST. REF.	DEBIT	CREDIT	BALANCE DEBIT	BALANCE CREDIT

ACCOUNT _____ ACCOUNT NO. _____

DATE	DESCRIPTION	POST. REF.	DEBIT	CREDIT	BALANCE DEBIT	BALANCE CREDIT

Reinforcement Problem 14A　(continued)

ACCOUNT _____　ACCOUNT NO. _____

DATE	DESCRIPTION	POST. REF.	DEBIT	CREDIT	BALANCE	
					DEBIT	CREDIT

ACCOUNT _____　ACCOUNT NO. _____

DATE	DESCRIPTION	POST. REF.	DEBIT	CREDIT	BALANCE	
					DEBIT	CREDIT

ACCOUNT _____　ACCOUNT NO. _____

DATE	DESCRIPTION	POST. REF.	DEBIT	CREDIT	BALANCE	
					DEBIT	CREDIT

Reinforcement Problem 14A (continued)

(2)

ACCOUNTS RECEIVABLE SUBSIDIARY LEDGER

Name _____

Address _____

DATE	DESCRIPTION	POST. REF.	DEBIT	CREDIT	BALANCE

Name _____

Address _____

DATE	DESCRIPTION	POST. REF.	DEBIT	CREDIT	BALANCE

Name _____

Address _____

DATE	DESCRIPTION	POST. REF.	DEBIT	CREDIT	BALANCE

Name _____

Address _____

DATE	DESCRIPTION	POST. REF.	DEBIT	CREDIT	BALANCE

Analyze: _____

Reinforcement Problem 14A   (concluded)

(3)

GENERAL JOURNAL   PAGE _____

	DATE	DESCRIPTION	POST. REF.	DEBIT	CREDIT	
1						1
2						2
3						3
4						4
5						5
6						6
7						7
8						8
9						9
10						10
11						11
12						12
13						13
14						14
15						15
16						16
17						17
18						18
19						19
20						20
21						21
22						22
23						23
24						24
25						25
26						26
27						27
28						28
29						29
30						30
31						31
32						32
33						33
34						34
35						35
36						36

Reinforcement Problem 15A   Recording Purchases and Cash Payment Transactions

General Ledger (partial)		Accounts Payable Subsidiary Ledger (partial)
Cash in Bank	Purchases	Carter Office Supply
Accounts Receivable	Purchases Discounts	Dancing Wind
Prepaid Insurance	Transportation In	Clothing Manufacturing
Supplies	Purchases Returns	Wilmington Shirt
Accounts Payable	and Allowances	Manufacturing
Sales Tax Payable		

GENERAL JOURNAL   PAGE _____

	DATE	DESCRIPTION	POST. REF.	DEBIT	CREDIT	
1						1
2						2
3						3
4						4
5						5
6						6
7						7
8						8
9						9
10						10
11						11
12						12
13						13
14						14
15						15
16						16
17						17
18						18
19						19
20						20
21						21
22						22
23						23
24						24
25						25
26						26
27						27
28						28

Reinforcement Problem 15A (concluded)

GENERAL JOURNAL

PAGE _____

	DATE	DESCRIPTION	POST. REF.	DEBIT	CREDIT	
1						1
2						2
3						3
4						4
5						5
6						6
7						7
8						8
9						9
10						10
11						11
12						12
13						13
14						14
15						15

Analyze: _____

_____

Reinforcement Problem 16A Recording and Posting Sales, Cash Receipts, and General Journal Transactions

(1)

GENERAL LEDGER

ACCOUNT _____ ACCOUNT NO. _____

DATE	DESCRIPTION	POST. REF.	DEBIT	CREDIT	BALANCE DEBIT	CREDIT

ACCOUNT _____ ACCOUNT NO. _____

DATE	DESCRIPTION	POST. REF.	DEBIT	CREDIT	BALANCE DEBIT	CREDIT

ACCOUNT _____ ACCOUNT NO. _____

DATE	DESCRIPTION	POST. REF.	DEBIT	CREDIT	BALANCE DEBIT	CREDIT

ACCOUNT _____ ACCOUNT NO. _____

DATE	DESCRIPTION	POST. REF.	DEBIT	CREDIT	BALANCE DEBIT	CREDIT

Reinforcement Problem 16A   (continued)

ACCOUNT _____   ACCOUNT NO. _____

DATE	DESCRIPTION	POST. REF.	DEBIT	CREDIT	BALANCE DEBIT	BALANCE CREDIT

ACCOUNT _____   ACCOUNT NO. _____

DATE	DESCRIPTION	POST. REF.	DEBIT	CREDIT	BALANCE DEBIT	BALANCE CREDIT

ACCOUNT _____   ACCOUNT NO. _____

DATE	DESCRIPTION	POST. REF.	DEBIT	CREDIT	BALANCE DEBIT	BALANCE CREDIT

(9)

Reinforcement Problem 16A (continued)

(2)

ACCOUNTS RECEIVABLE SUBSIDIARY LEDGER

Name _____

Address _____

DATE	DESCRIPTION	POST. REF.	DEBIT	CREDIT	BALANCE

Name _____

Address _____

DATE	DESCRIPTION	POST. REF.	DEBIT	CREDIT	BALANCE

Name _____

Address _____

DATE	DESCRIPTION	POST. REF.	DEBIT	CREDIT	BALANCE

Name _____

Address _____

DATE	DESCRIPTION	POST. REF.	DEBIT	CREDIT	BALANCE

Reinforcement Problem 16A   (continued)

Name _____

Address _____

	DATE	DESCRIPTION	POST. REF.	DEBIT	CREDIT	BALANCE

Name _____

Address _____

	DATE	DESCRIPTION	POST. REF.	DEBIT	CREDIT	BALANCE

Name _____

Address _____

	DATE	DESCRIPTION	POST. REF.	DEBIT	CREDIT	BALANCE

Reinforcement Problem 16A   (continued)

SALES JOURNAL

PAGE _____

	DATE	SALES SLIP NO.	CUSTOMER'S ACCOUNT DEBITED	POST. REF.	SALES CREDIT	SALES TAX PAYABLE CREDIT	ACCOUNTS RECEIVABLE DEBIT	
1								1
2								2
3								3
4								4
5								5
6								6
7								7
8								8
9								9
10								10
11								11
12								12
13								13
14								14
15								15
16								16
17								17
18								18
19								19
20								20
21								21
22								22
23								23
24								24
25								25
26								26
27								27
28								28
29								29
30								30
31								31
32								32
33								33
34								34

Reinforcement Problem 16A   (continued)

CASH RECEIPTS JOURNAL

DATE	DOC. NO.	ACCOUNT NAME	POST. REF.	GENERAL CREDIT	SALES CREDIT	SALES TAX PAYABLE CREDIT	ACCOUNTS RECEIVABLE CREDIT	SALES DISCOUNTS DEBIT	CASH IN BANK DEBIT	
------	----------	--------------	------------	----------------	--------------	--------------------------	----------------------------	-----------------------	--------------------	
										1
										2
										3
										4
										5
										6
										7
										8
										9
										10
										11
										12
										13
										14
										15
										16
										17
										18
										19
										20
										21
										22
										23
										24
										25

Reinforcement Problem 16A   (concluded)

GENERAL JOURNAL   PAGE _____

	DATE		DESCRIPTION	POST. REF.	DEBIT	CREDIT	
1							1
2							2
3							3
4							4
5							5
6							6
7							7
8							8
9							9
10							10
11							11
12							12
13							13
14							14
15							15
16							16
17							17
18							18
19							19
20							20
21							21
22							22
23							23
24							24
25							25
26							26
27							27
28							28
29							29
30							30
31							31
32							32
33							33

Analyze: _____

_____

Reinforcement Problem 17A Recording Special Journal and General Journal Transactions

SALES JOURNAL

PAGE _____

	DATE	SALES SLIP NO.	CUSTOMER'S ACCOUNT DEBITED	POST. REF.	SALES CREDIT	SALES TAX PAYABLE CREDIT	ACCOUNTS RECEIVABLE DEBIT	
1								1
2								2
3								3
4								4
5								5
6								6
7								7
8								8
9								9
10								10
11								11
12								12
13								13
14								14
15								15
16								16
17								17
18								18
19								19
20								20
21								21
22								22
23								23
24								24
25								25
26								26
27								27
28								28
29								29
30								30
31								31
32								32
33								33
34								34

Reinforcement Problem 17A (continued)

CASH RECEIPTS JOURNAL

DATE	DOC. NO.	ACCOUNT NAME	POST. REF.	GENERAL CREDIT	SALES CREDIT	SALES TAX PAYABLE CREDIT	ACCOUNTS RECEIVABLE CREDIT	SALES DISCOUNTS DEBIT	CASH IN BANK DEBIT

Reinforcement Problem 17A   (continued)

PAGE _____

PURCHASES JOURNAL

DATE	INVOICE NO.	CREDITOR'S ACCOUNT CREDITED	POST. REF.	ACCOUNTS PAYABLE CREDIT	PURCHASES DEBIT	GENERAL ACCOUNT DEBITED	POST. REF.	DEBIT
1								
2								
3								
4								
5								
6								
7								
8								
9								
10								
11								
12								
13								
14								
15								
16								
17								
18								
19								
20								
21								
22								
23								
24								
25								
26								

Reinforcement Problem 17A   (continued)

CASH PAYMENTS JOURNAL

PAGE _____

DATE	DOC. NO.	ACCOUNT NAME	POST. REF.	GENERAL DEBIT	GENERAL CREDIT	ACCOUNTS PAYABLE DEBIT	PURCHASES DISCOUNTS CREDIT	CASH IN BANK CREDIT	
20-- July 1	534	Sullivan Screen Printers				325 00		325 00	1
									2
									3
									4
									5
									6
									7
									8
									9
									10
									11
									12
									13
									14
									15
									16
									17
									18
									19
									20
									21
									22
									23
									24
									25
									26

Reinforcement Problem 17A (continued)

GENERAL JOURNAL

PAGE _____

	DATE	DESCRIPTION	POST. REF.	DEBIT	CREDIT	
1						1
2						2
3						3
4						4
5						5
6						6
7						7
8						8
9						9
10						10
11						11
12						12
13						13
14						14
15						15
16						16
17						17
18						18
19						19
20						20
21						21
22						22
23						23
24						24
25						25
26						26
27						27
28						28
29						29
30						30
31						31
32						32
33						33
34						34
35						35

Reinforcement Problem 17A   (concluded)

Analyze: _____

Notes

Reinforcement Problem 18A   Calculating Adjustments and Preparing the Ten-Column Work Sheet

T-Shirt

Work

For the Year Ended

	ACCT. NO.	ACCOUNT NAME	TRIAL BALANCE DEBIT	TRIAL BALANCE CREDIT	ADJUSTMENTS DEBIT	ADJUSTMENTS CREDIT
1	101	Cash in Bank	1472900			
2	115	Accounts Receivable	570200			
3	130	Merchandise Inventory	5121500			
4	135	Supplies	319700			
5	140	Prepaid Insurance	180000			
6	145	Office Equipment	783700			
7	150	Store Equipment	1850400			
8	155	Delivery Equipment	1175400			
9	201	Accounts Payable		1303900		
10	204	Fed. Corporate Income Tax Pay.				
11	210	Employees' Fed. Inc. Tax Pay.		63600		
12	211	Employees' State Inc. Tax Pay.		11700		
13	212	Social Security Tax Payable		47900		
14	213	Medicare Tax Payable		11300		
15	215	Sales Tax Payable		393100		
16	216	Fed. Unemployment Tax Pay.		7900		
17	217	State Unemployment Tax Pay.		31500		
18	301	Capital Stock		5000000		
19	305	Retained Earnings		2542500		
20	310	Income Summary				
21	401	Sales		13312300		
22	405	Sales Discounts	25800			
23	410	Sales Returns and Allow.	134200			
24	501	Purchases	7251000			
25	505	Transportation In	114100			
26	510	Purchases Discounts		129200		
27	515	Purchases Returns and Allow.		57100		
28	601	Advertising Expense	420500			
29		Carried Forward	19419400	22912000		
30						
31						
32						

Trends

Sheet

December 31, 20– –

ADJUSTED TRIAL BALANCE		INCOME STATEMENT		BALANCE SHEET		
DEBIT	CREDIT	DEBIT	CREDIT	DEBIT	CREDIT	
						1
						2
						3
						4
						5
						6
						7
						8
						9
						10
						11
						12
						13
						14
						15
						16
						17
						18
						19
						20
						21
						22
						23
						24
						25
						26
						27
						28
						29
						30
						31
						32

Reinforcement Problem 18A   (concluded)

T-Shirt

Work Sheet

For the Year Ended

	ACCT. NO.	ACCOUNT NAME	TRIAL BALANCE DEBIT	TRIAL BALANCE CREDIT	ADJUSTMENTS DEBIT	ADJUSTMENTS CREDIT
1		**Brought Forward**	19419400	22912000		
2						
3	605	**Bankcard Fees Expense**	61900			
4	630	**Fed. Corporate Income Tax Exp.**	248000			
5	640	**Insurance Expense**				
6	655	**Maintenance Expense**	132200			
7	660	**Miscellaneous Expense**	377200			
8	663	**Payroll Tax Expense**	125100			
9	665	**Rent Expense**	1090000			
10	670	**Salaries Expense**	1198900			
11	675	**Supplies Expense**				
12	685	**Utilities Expense**	259300			
13			22912000	22912000		
14						
15						
16						
17						
18						
19						
20						
21						
22						
23						
24						
25						
26						
27						
28						
29						
30						
31						
32						

Trends

(continued)

December 31, 20– –

	ADJUSTED TRIAL BALANCE		INCOME STATEMENT		BALANCE SHEET		
	DEBIT	CREDIT	DEBIT	CREDIT	DEBIT	CREDIT	
1							1
2							2
3							3
4							4
5							5
6							6
7							7
8							8
9							9
10							10
11							11
12							12
13							13
14							14
15							15
16							16
17							17
18							18
19							19
20							20
21							21
22							22
23							23
24							24
25							25
26							26
27							27
28							28
29							29
30							30
31							31
32							32

Analyze:

Reinforcement Problem 19A   Preparing Financial Statements

T-Shirt

Work

For the Year Ended

	ACCT. NO.	ACCOUNT NAME	TRIAL BALANCE DEBIT	TRIAL BALANCE CREDIT	ADJUSTMENTS DEBIT	ADJUSTMENTS CREDIT
1	101	Cash in Bank	19731 00			
2	115	Accounts Receivable	6462 00			
3	130	Merchandise Inventory	25192 00		(a) 1228 00	
4	135	Supplies	4669 00			(b) 2938 00
5	140	Prepaid Insurance	2400 00			(c) 625 00
6	145	Office Equipment	14895 00			
7	150	Store Equipment	25223 00			
8	155	Delivery Truck	12750 00			
9	201	Accounts Payable		15824 00		
10	210	Fed. Corporate Income Tax Pay.				(d) 142 00
11	211	Employees' Fed. Inc. Tax Pay.		534 00		
12	212	Employees' State Inc. Tax Pay.		151 00		
13	213	Social Security Tax Payable		451 00		
14	214	Medicare Tax Payable		180 00		
15	215	Sales Tax Payable		2413 00		
16	216	Fed. Unemployment Tax Pay.		54 00		
17	217	State Unemployment Tax Pay.		282 00		
18	301	Capital Stock		50000 00		
19	305	Retained Earnings		10811 00		
20	310	Income Summary				(a) 1228 00
21	401	Sales		131551 00		
22	405	Sales Discounts	196 00			
23	410	Sales Returns and Allow.	1668 00			
24	501	Purchases	65819 00			
25	505	Transportation In	1321 00			
26	510	Purchases Discounts		789 00		
27	515	Purchases Returns and Allow.		967 00		
28	601	Advertising Expense	2117 00			
29		Carried Forward	182443 00	214007 00	1228 00	4933 00
30						
31						
32						

Trends

Sheet

December 31, 20--

ADJUSTED TRIAL BALANCE		INCOME STATEMENT		BALANCE SHEET		
DEBIT	CREDIT	DEBIT	CREDIT	DEBIT	CREDIT	
1973100				1973100		1
646200				646200		2
2642000				2642000		3
173100				173100		4
177500				177500		5
1489500				1489500		6
2522300				2522300		7
1275000				1275000		8
	1582400				1582400	9
	14200				14200	10
	53400				53400	11
	15100				15100	12
	45100				45100	13
	18000				18000	14
	241300				241300	15
	5400				5400	16
	28200				28200	17
	5000000				5000000	18
	1081100				1081100	19
	122800		122800			20
	13155100		13155100			21
19600		19600				22
166800		166800				23
6581900		6581900				24
132100		132100				25
	78900		78900			26
	96700		96700			27
211700		211700				28
18010800	21537700	7112100	13453500	10898700	8084200	29
						30
						31
						32

Reinforcement Problem 19A   (continued)

T-Shirt

Work Sheet

For the Year Ended

	ACCT. NO.	ACCOUNT NAME	TRIAL BALANCE		ADJUSTMENTS	
			DEBIT	CREDIT	DEBIT	CREDIT
1		**Brought Forward**	18244300	21400700	122800	493300
2						
3	605	**Bankcard Fees Expense**	32800			
4	630	**Fed. Corporate Income Tax Exp.**	351000		(d) 14200	
5	640	**Insurance Expense**			(c) 62500	
6	655	**Maintenance Expense**	135000			
7	660	**Miscellaneous Expense**	93100			
8	663	**Payroll Tax Expense**	83400			
9	665	**Rent Expense**	1270000			
10	670	**Salaries Expense**	723400			
11	675	**Supplies Expense**			(b) 293800	
12	685	**Utilities Expense**	46770 0			
13			21400700	21400700	493300	493300
14		**Net Income**				
15						
16						
17						
18						
19						
20						
21						
22						
23						
24						
25						
26						
27						
28						
29						
30						
31						
32						

Name   **Date**   **Class**

Trends
(continued)
December 31, 20– –

| ADJUSTED TRIAL BALANCE | | INCOME STATEMENT | | BALANCE SHEET | | |
DEBIT	CREDIT	DEBIT	CREDIT	DEBIT	CREDIT	
180108 00	215377 00	71121 00	134535 00	108987 00	80842 00	1
						2
328 00		328 00				3
3652 00		3652 00				4
625 00		625 00				5
1350 00		1350 00				6
931 00		931 00				7
834 00		834 00				8
12700 00		12700 00				9
7234 00		7234 00				10
2938 00		2938 00				11
4677 00		4677 00				12
215377 00	215377 00	106390 00	134535 00	108987 00	80842 00	13
		28145 00			28145 00	14
		134535 00	134535 00	108987 00	108987 00	15
						16
						17
						18
						19
						20
						21
						22
						23
						24
						25
						26
						27
						28
						29
						30
						31
						32

Reinforcement Problem 19A   (continued)

(1)

Reinforcement Problem 19A (continued)

(2)

(3)

Reinforcement Problem 19A   (concluded)

Analyze: _____

_____

_____

_____

_____

Reinforcement Problem 20A   Journalizing Closing Entries

Account Names and Balances as of December 31, 20--:

Sales	$94,412.00
Purchases Discounts	750.00
Purchases Returns and Allowances	455.00
Sales Discounts	867.00
Purchases	35,000.00
Transportation In	1,700.00
Advertising Expense	900.00
Bankcard Fees Expense	647.00
Federal Income Tax Expense	5,343.00
Insurance Expense	1,200.00
Miscellaneous Expense	369.00
Rent Expense	18,000.00
Supplies Expense	2,612.00
Utilities Expense	4,200.00
Sales Returns and Allowances	1,735.00

Reinforcement Problem 20A   (concluded)

GENERAL JOURNAL   PAGE _____

	DATE	DESCRIPTION	POST. REF.	DEBIT	CREDIT	
1						1
2						2
3						3
4						4
5						5
6						6
7						7
8						8
9						9
10						10
11						11
12						12
13						13
14						14
15						15
16						16
17						17
18						18
19						19
20						20
21						21
22						22
23						23
24						24
25						25
26						26
27						27
28						28
29						29
30						30
31						31
32						32
33						33

Analyze: _____

_____

Reinforcement Problem 21A   Recording Stockholders' Equity Transactions

GENERAL JOURNAL   PAGE _____

	DATE	DESCRIPTION	POST. REF.	DEBIT	CREDIT	
1						1
2						2
3						3
4						4
5						5
6						6
7						7
8						8
9						9
10						10
11						11
12						12
13						13
14						14
15						15
16						16
17						17
18						18
19						19
20						20
21						21
22						22
23						23
24						24
25						25
26						26
27						27
28						28
29						29
30						30
31						31
32						32
33						33
34						34
35						35
36						36
37						37

Reinforcement Problem 21A   (concluded)

GENERAL JOURNAL   PAGE _____

	DATE	DESCRIPTION	POST. REF.	DEBIT	CREDIT	
1						1
2						2
3						3
4						4
5						5
6						6
7						7
8						8
9						9
10						10
11						11
12						12
13						13
14						14
15						15

Analyze: _____

_____

_____

Reinforcement Problem 22A   Maintaining a Petty Cash Register

PAGE

PETTY CASH REGISTER

DATE	VOU. NO.	DESCRIPTION	PAYMENTS	OFFICE SUPPLIES	DELIVERY EXPENSE	MISC. EXPENSE	GENERAL ACCOUNT NAME	AMOUNT	
									1
									2
									3
									4
									5
									6
									7
									8
									9
									10
									11
									12
									13
									14
									15
									16
									17
									18
									19
									20
									21
									22
									23
									24

DISTRIBUTION OF PAYMENTS

Analyze:

Reinforcement Problem 23A   Calculating and Recording Depreciation Expense

(1)

Date	Cost	Annual Depreciation	Accumulated Depreciation	Book Value

(2)

GENERAL JOURNAL   PAGE _____

	DATE	DESCRIPTION	POST. REF.	DEBIT	CREDIT	
1						1
2						2
3						3
4						4
5						5
6						6
7						7
8						8
9						9
10						10

Reinforcement Problem 23A   (concluded)

ACCOUNT ___*Accumulated Depreciation—Delivery Truck*___   ACCOUNT NO. ____*155*____

DATE	DESCRIPTION	POST. REF.	DEBIT	CREDIT	BALANCE	
					DEBIT	CREDIT

(3) a. _____

_____

_____

b. _____

c. _____

d. _____

Analyze: _____

_____

_____

_____

Computerized Accounting Using Spreadsheets

Problem 23A   Calculating and Recording Depreciation Expense

Completing the Spreadsheet

Step 1   Read the instructions for Problem 23A in your textbook. This problem involves preparing a depreciation schedule.

Step 2   Open the Glencoe Accounting: Electronic Learning Center software.

Step 3   From the Program Menu, click on the **Peachtree Complete®
Accounting Software and Spreadsheet Applications** icon.

Step 4   Log onto the Management System by typing your user name and password.

Step 5   Under the **Problems & Tutorials** tab, select template 23A from the drop-down menu. The template should look like the one shown below.

```
PROBLEM 23A
CALCULATING AND RECORDING
DEPRECIATION EXPENSE

(name)
(date)
```

Date	Cost	Annual Depreciation	Accumulated Depreciation	Book Value
Oct. 12, 2004	AMOUNT			
Dec. 31, 2004		$0	$0	AMOUNT
Dec. 31, 2005		$0	$0	$0
Dec. 31, 2006		$0	$0	$0
Dec. 31, 2007		$0	$0	$0
				AMOUNT

Step 6   Key your name and today's date in the cells containing the *(name)* and *(date)* placeholders.

Step 7   Enter the cost, original book value, and salvage value of the delivery truck in the cells containing the AMOUNT placeholders. The annual depreciation, accumulated depreciation, and book value for each year will be automatically computed.

Step 8   Save the spreadsheet using the **Save** option from the *File* menu. You should accept the default location for the save as this is handled by the management system.

Step 9   Print the completed spreadsheet.

Step 10   Exit the spreadsheet program.

Step 11   In the Close Options box, select the location where you would like to save your work.

Step 12   Answer the Analyze question from your textbook for this problem.

TIP: If your spreadsheet is too wide to fit on an 8.5-inch wide piece of paper, you can change your print settings to print the work sheet *landscape*. Landscape means the work sheet will be printed broadside on the page.

TIP: Always save your work before performing What-If Analysis. It is not necessary to save your work after performing What-If Analysis unless your teacher instructs you to do so.

What-If Analysis

If the salvage value were $1,500, what would the annual depreciation for each year be?

_____

_____

Reinforcement Problem 24A   Calculating and Recording Uncollectible Accounts Expense

(1)

Estimate of uncollectible accounts: _____

(2)

GENERAL JOURNAL   PAGE _____

	DATE	DESCRIPTION	POST. REF.	DEBIT	CREDIT	
1						1
2						2
3						3
4						4
5						5
6						6

(3)

GENERAL LEDGER (PARTIAL)

ACCOUNT _____ ACCOUNT NO. _____

DATE	DESCRIPTION	POST. REF.	DEBIT	CREDIT	BALANCE DEBIT	BALANCE CREDIT

ACCOUNT _____ ACCOUNT NO. _____

DATE	DESCRIPTION	POST. REF.	DEBIT	CREDIT	BALANCE DEBIT	BALANCE CREDIT

(4)

Book value of accounts receivable: _____

Analyze: _____

Reinforcement Problem 25A   Accounting for Inventories

Cost of Ending Inventory:

Specific Identification _____

_____

_____

_____

_____

FIFO _____

_____

_____

_____

LIFO _____

_____

_____

_____

Weighted Average Cost _____

_____

_____

_____

Analyze: _____

_____

_____

Reinforcement Problem 26A   Calculating Current and Future Interest

	Maturity Date	Interest	
		Current Year	Following Year
1.	_____	_____	_____
2.	_____	_____	_____
3.	_____	_____	_____
4.	_____	_____	_____
5.	_____	_____	_____
6.	_____	_____	_____
7.	_____	_____	_____
8.	_____	_____	_____
9.	_____	_____	_____
10.	_____	_____	_____

Analyze: _____

_____

_____

_____

Reinforcement Problem 26B Recording Noninterest-Bearing Notes Payable

GENERAL JOURNAL   PAGE _____

	DATE		DESCRIPTION	POST. REF.	DEBIT	CREDIT	
1							1
2							2
3							3
4							4
5							5
6							6
7							7
8							8
9							9
10							10
11							11
12							12
13							13
14							14
15							15
16							16
17							17
18							18
19							19
20							20
21							21
22							22
23							23
24							24
25							25
26							26

Analyze: _____

Computerized Accounting Using Spreadsheets

Problem 26A   Calculating Current and Future Interest

Completing the Spreadsheet

Step 1   Read the instructions for Problem 26A in your textbook. This problem involves determining the maturity date and interest expense to be paid in the current and following year for ten notes.

Step 2   Open the Glencoe Accounting: Electronic Learning Center software.

Step 3   From the Program Menu, click on the **Peachtree Complete®** **Accounting Software and Spreadsheet Applications** icon.

Step 4   Log onto the Management System by typing your user name and password.

Step 5   Under the **Problems & Tutorials** tab, select template 26A from the drop-down menu. The template should look like the one shown below.

```
PROBLEM 26A
CALCULATING CURRENT AND
FUTURE INTEREST

(name)
(date)
```

	Amount	Issue Date	Interest Rate	Term	Maturity Date	Interest Current Year	Following Year
1		10-Dec		30 days	9-Jan	$0.00	$0.00
2		21-Nov		60 days	20-Jan	$0.00	$0.00
3		10-Oct		90 days	8-Jan	$0.00	$0.00
4		5-Dec		60 days	3-Feb	$0.00	$0.00
5		10-Nov		120 days	10-Mar	$0.00	$0.00
6		8-Sep		180 days	7-Mar	$0.00	$0.00
7		17-Nov		70 days	26-Jan	$0.00	$0.00
8		1-Oct		6 months	1-Apr	$0.00	$0.00
9		1-Dec		3 months	1-Mar	$0.00	$0.00
10		1-Aug		9 months	1-May	$0.00	$0.00

Step 6   Key your name and today's date in the cells containing the *(name)* and *(date)* placeholders.

Step 7   Enter the amount and interest rate for the first note in the appropriate cells of the spreadsheet template. The spreadsheet template will automatically calculate the interest for the current year and following year for the first note.

TIP: When entering the interest rates, it is not necessary to enter a percent sign after the number, nor is it necessary to enter the number as a decimal. For example, enter 9% as **9** in cell D11. The spreadsheet will automatically format this as a percent.

Spreadsheet Guide

Step 8 Continue to enter the amount and interest rate for the remaining notes. The current year interest and following year interest will be automatically calculated for each note.

Step 9 Save the spreadsheet using the **Save** option from the *File* menu. You should accept the default location for the save as this is handled by the management system.

Step 10 Print the completed spreadsheet.

Step 11 Exit the spreadsheet program.

Step 12 In the Close Options box, select the location where you would like to save your work.

Step 13 Answer the Analyze question from your textbook for this problem.

 TIP: Always save your work before performing What-If Analysis. It is not necessary to save your work after performing What-If Analysis unless your teacher instructs you to do so.

What-If Analysis

Suppose the note amount for Note #10 were $3,333. What would the current year interest be? What would the following year interest be?

_____

_____

Reinforcement Problem 27A   Recording Partners' Investments

GENERAL JOURNAL

PAGE _____

	DATE	DESCRIPTION	POST. REF.	DEBIT	CREDIT	
1						1
2						2
3						3
4						4
5						5
6						6
7						7
8						8
9						9
10						10
11						11
12						12
13						13
14						14
15						15
16						16
17						17
18						18
19						19
20						20
21						21
22						22
23						23
24						24
25						25
26						26

Analyze: _____

_____

Reinforcement Problem 28A   Liquidation of a Partnership

GENERAL JOURNAL   PAGE _____

	DATE	DESCRIPTION	POST. REF.	DEBIT	CREDIT	
1						1
2						2
3						3
4						4
5						5
6						6
7						7
8						8
9						9
10						10
11						11
12						12
13						13
14						14
15						15
16						16
17						17
18						18
19						19
20						20
21						21
22						22
23						23
24						24
25						25
26						26
27						27
28						28
29						29
30						30
31						31
32						32

Analyze: _____

_____

Notes

Notes

Notes